Polymer Synthesis: Theory and Practice

Dietrich Braun • Harald Cherdron
Matthias Rehahn • Helmut Ritter
Brigitte Voit

Polymer Synthesis: Theory and Practice

Fundamentals, Methods, Experiments

Fifth Edition

Springer

Dietrich Braun
Technische Universität Darmstadt
Darmstadt
Germany

Matthias Rehahn
Technische Universität
Darmstadt
Ernst-Berl-Institut für
Technische und
Makromolekulare Chemie
FG der Polymeren
Darmstadt
Germany

Brigitte Voit
Leibniz-Institut für
Polymerforschung Dresden e.V.
Dresden
Germany

Harald Cherdron
Wiesbaden
Germany

Helmut Ritter
Heinrich-Heine-Universität
Düsseldorf
Institut für Organische Chemie
und Makromolekulare Chemie
Düsseldorf
Germany

ISBN 978-3-642-43505-8   ISBN 978-3-642-28980-4 (eBook)
DOI 10.1007/978-3-642-28980-4
Springer Heidelberg New York Dordrecht London

Preface to the Fifth Edition

The focus of the previous edition was on a broader description of the general methods and techniques for the synthesis, modifications and characterization of macromolecules. This was supplemented by selected and detailed experiments and by sufficient theoretical treatment so that no additional textbook be needed in order to understand the experiments. In addition to the preparative aspects, we had also tried to give the reader an impression of the relation of chemical structure and morphology of polymers to their properties. This concept is also maintained in this edition.

In recent years, the so-called functional polymers (which have special electrical, electronic, optical and biological properties), have attracted a lot of interest. We, therefore, felt, that a textbook should exist that would contain recipes which describe the synthesis of these materials. For this reason, we added a new chapter "Functional Polymers".

Together with new experiments in Chaps. 3, 4 and 5, the book now contains more than 120 recipes that describe a wide range of macromolecules.

The completion of the 5th Edition was aided by the contribution of a number of scientists. Our special thanks are due to Dr. M. Tabatabai for her arduous task of checking, assembling and formatting the manuscript. We wish to express our sincerest appreciation to Prof. Dr. M. Buchmeiser for contribution chapter 3.3.3, Dr. B. Müller, Dr. J. Zhou and Ms. Dr. Z. Rezaie for production of chapter 2.3.4 and 6.1; Dr. J. Pionteck, Dr. P. Pötschke, and Dr. F. Böhme for contributing to and reading of the chapter 1 and 5. We also thank the Chemistry Editorial and Production Department of Springer-Verlag for an excellent cooperation.

September 2012

D. Braun, Darmstadt
H. Cherdron, Wiesbaden
M. Rehahn, Darmstadt
H. Ritter, Düsseldorf
B. Voit, Dresden

Contents

Relevant SI Units and Conversions

Relevant SI units and conversions

Name	SI unit		Multiples	Units outside[a] SI	
Length	Meter	m	km, cm, mm	Ångstrom[a]	$Å=10^{-10}$ m
			μm, nm	Foot	ft=30.48 cm
				Inch	in=2.54 cm
				Micron	$μ=10^{-9}$ m
				Mil	mil=2.54 10^{-5} m
				Statute mile	mile=1609.3 m
Mass	Kilogram	kg	Mg, g, mg, μg	Ounce	oz=28.349 g
				Pound (US, 16 oz)	lb=453.592 g
				Ton (metric)	t=1,000 kg
				Long ton (UK)	ton=1.02 10^3 kg
				Short ton (US)	ton=9.1 10^2 kg
Time	Second	s	ks, ms, μs, ns	Minute[a]	min=60 s
				Hour[a]	h=3,600 s
				Day[a]	d=86,400 s
Temperature	Kelvin	K		Degree Celsius[a]	0°C=273.15 K
				Degree Fahrenheit	°F=9/5 °C+32
Density	kg/m^3	kg/m^3	mg/m^3, kg/dm^3, g/cm^3	t/m^3, kg/l[a]	
Volume	Cubic meter	m^3	dm^3, cm^3, mm^3	liter[a]	l, L=1 dm^3
Force	Newton	N	MN, kN, mN, μN	dyn	$dyn=10^{-5}$ N
				kg force	

(continued)

Name	SI unit		Multiples	Units outside[a] SI	
					kgf (Kp)= 9.81 N
Pressure	Pascal	Pa	GPa, MPa, KPa, mPa, μPa	Bar	bar=10^5 Pa
				Atmosphere	atm=$1.101 \cdot 10^5$ Pa
				mm Hg	torr=133.3 Pa
				lbs/sq.in.	psi=6.8910^3 Pa
Viscosity (dynamic)		Pa·s	mPa·s	Poise	P=0.1 Pa s
				Centipoise	cP=10^{-3} Pa s
Viscosity (kinematic)		m^2/s	mm^2/s	Stokes	St=10^{-4} m^2/s
				Centistokes	cSt=10^{-6} m^2/s
Electric charge	Coulomb	C=A·s			
Electric current	Ampere	A			
Electric potential	Volt	V=J·C^{-1}	MV, kV, mV		
Energy, work, heat	Joule	J	TJ, GJ, MJ, kJ, mJ	Electron volt	eV=1.60 10^{-19} J
Power (radiant flux)	Watt	W (=V·A= J/s)	GW, MW, kW, mW		

[a]Accepted for SI

List of General Abbreviations

$[\eta]$	Intrinsic viscosity, Staudinger index
AIBN	2,2´-azobis(isobutyronitrile)
Ar	Aryl
BDDA	Butandioldiacrylate
bp	Boiling point
BPO	Benzoyl peroxide
cat	Catalyst
CRU	Constitutional repeating unit
DMF	Dimethyl formamide
DMSO	Dimethyl sulfoxide
DP	Degree of polymerization
DSC	Differential scanning calorimetry
Et	Ethyl
g/l	Grams per liter
GPC	Gel permeation chromatography (alternatively: SEC)
h	Hour(s)
IR	Infrared
l	Liter(s)
LCST	Lower critical solution temperature
LED	Light emitting diode
Me	Methyl
min	Minute(s)
M_n	Number average molecular weight
mol	Mole(s)
mp	Melting point
MTB	Methyl-*tert*-butyl
MW	Molecular weight
M_w	Weight average molecular weight
NLO	Nonlinear optical
NMP	*N*-methylpyrrolidone
NMR	Nuclear magnetic resonance
OFET	Organic field-effect transistor
OLED	Organic light emitting diode

OPV	Organic photo voltage
OSC	Organic solar cell
OTFT	Organic thin-film transistor
Ph	Phenyl
POLED	Polymer based organic light emitting diode
ppm	Parts per million
s	Second(s)
SC	Solar cell
SEC	Size exclusion chromatography (alternatively: GPC)
SMP	Shape memory polymer
T_c	Crystallization temperature
T_d	Degradation temperature
T_g	Glass transition temperature
TGA	Thermogravimetric analysis
THF	Tetrahydrofuran
T_m	Melting temperature
UCST	Upper critical solution temperature
UV	Ultraviolet
η_{inh}	Inherent viscosity
η_{rel}	Relative viscosity
η_{sp}	Specific viscosity

Abbreviations for Technically Important Polymers

ABS	Acrylonitrile butadiene styrene copolymer blend
ACM, ANM	Acrylic ester rubber
APP	Atactic PP
ASA	Acrylonitrile styrene acrylic ester copolymer blend
BR	Polybutadiene rubbers (*cis-* or *trans-*1,4; 1,2)
CA	Cellulose acetate
CAB	Cellulose acetobutyrate
CAP	Cellulose acetopropionate
CF	Cresol–formaldehyde resin
CFC	Carbon fiber reinforced carbon
CFR	Carbon reinforced resin
CMC	Carboxymethylcellulose
CN	Cellulose nitrate
COC	Cycloolefin copolymers
CP	Cellulose propionate
CR	Chloroprene rubber
CSM	Chlorosulfonated polyethylene
CTA	Cellulose triacetate
EC	Ethyl cellulose
ECTFE	Ethylene chlorotrifluoroethylene copolymers
EEA	Ethylene ethyl acrylate copolymers
EP	Epoxide resin
EPDM	Ethylene propylene diene terpolymers
EPM	Ethylene propylene rubbers
EPS	Expandable polystyrene
ETFE	Ethylene tetrafluoroethylene copolymers
EVA, EVAC	Ethylene vinyl acetate copolymers
EVAL	Ethylene vinyl alcohol copolymers
FEP	Tetrafluoroethylene perfluoropropylene copolymers
FF	Furan resins
FRP	Glass fiber reinforced polymers
GR-I, IIR	Butyl rubber
HIPS	High impact polystyrene

HT-PDDT	Head-to-tail polydodecylthiophene
IPP	Isotactic PP
IR	Polyisoprene rubber (*cis*- or *trans*-1,4; -1,2; -3,4)
LCP	Liquid crystalline polymer
MBS	Methyl methacrylate butadiene styrene copolymer blend
MC	Methylcellulose
MF	Melamin–formaldehyde resin
MPF	Melamin-phenol-formaldehyde resin
MPPE	Poly(phenylene ether) polystyrene blend
NBR	Nitrile rubber
NC	Nitrocellulose
NR	Natural rubber
PA	Polyamides
PAEK	Poly(aryl ether ketone)
PAI	Poly(amide imide)
PAN	Polyacrylonitrile
PB	Polybutylene
PBI	Poly(benzimidazole)
PBT	Poly(butylene terephthalate)
PC	Polycarbonate
PCTFE	Poly(chlorotrifluoroethylene)
PE	Polyethylene
PEC	Chlorinated polyethylene
PEHD	High-density PE
PEI	Poly(ether imide)
PELD	Low-density PE
PELLD	Linear low density PE
PEMD	Medium-density PE
PEO, PEOX	Poly(ethylene oxide)
PES	Poly(ether sulfone)
PET	Poly(ethylene terephthalate)
PEVLD	Very low density PE
PF	Phenol–formaldehyde resin
PFA	cf. TFA
PFEP	Tetrafluoroethylene perfluoropropylene copolymers
PI	Polyimide
PIB	Poly(isobutylene)
PMI	Poly(methacrylimide)
PMMA	Poly(methyl methacrylate)
PMP, TPX	Poly(4-methylpent-1-ene)
PNiPAAM	Poly(*N*-isopropylacrylamide)
PO	Polyolefins
POM	Poly(oxymethylene), polyformaldehyde, polyacetal
PP	Polypropylene
PPC	Chlorinated PP

PPE, PPO	Poly(phenylene ether)
PPOX (PPO)	Poly(propylene oxide)
PPS	Poly(phenylene sulfide)
PPSU, PSU	Poly(phenylene sulfone)
PPV	Polyphenylenvinylen
PS	Polystyrene
PTFE	Poly(tetrafluoroethylene)
PTP	Poly(terephthalate)
PUR	Polyurethane
PVA, PVAC	Poly(vinyl acetate)
PVAL	Poly(vinyl alcohol)
PVB	Poly(vinylbutyral)
PVC	Poly(vinyl chloride)
PVCC	Chlorinated PVC
PVDC	Poly(vinylidene chloride)
PVDF	Poly(vinylidene fluoride)
PVF	Poly(vinyl fluoride)
PVFM	Poly(vinylformal)
PVP	Poly(vinylpyrrolidone)
RF	Resorcin–formaldehyde resin
SAN	Styrene acrylonitrile copolymer
SBR	Styrene butadiene rubber
SBS	Styrene butadiene styrene triblock copolymers (ABA)
SI	Silicone plastics
SIR	Silicone rubbers
SPP	Syndiotactic PP
TFA	Tetrafluoroethylene perfluoropropyl vinyl ether copolymers
TPE	Thermoplastic elastomers
TPU	Thermoplastic polyurethanes
UF	Urea–formaldehyde resins
UP	Unsaturated polyester
VCE	Vinyl chloride ethylene copolymers

Introduction

The origin of polymer science as a part of organic chemistry goes back to the end of the nineteenth century when chemists detected that the properties of many substances with colloidal properties are connected with their molecular size. As a result of these and preferably of his own studies Hermann Staudinger (1881–1965) concluded in the early 1920s that substances like natural rubber, cellulose, and proteins but also many synthetic resins obtained by so-called polyreactions consist of large molecules, for which Staudinger proposed the term "macromolecules". Nowadays, macromolecules and polymers are synonyms for substances with especially high molecular masses. However a sharp boundary cannot be drawn between low-molecular-weight and macromolecular substances; rather there is a gradual transition between them. One can say that macromolecules consist of a minimum of several hundred atoms. Accordingly, the lower limit for their molecular mass can be taken as around 10^3 g mol^{-1}.

Staudinger postulated that such macromolecules were built up from small so-called monomer units that were linked together by normal chemical binding forces and not by physical associations of small molecules. This hypothesis was by no means self-evident at that time and therefore became an object of many academic controversies between chemists and physicists until about the late 1930s. In the meantime Staudinger's basic ideas of macromolecular chemistry are generally accepted.

In 1953 Staudinger received the Nobel prize in chemistry "for his discoveries in the field of macromolecular chemistry". The scientific reputation of this new and rather young branch of chemistry is also confirmed by several other Nobel prizes, so to Paul J. Flory (in 1974) for his theoretical and experimental work in physical chemistry of macromolecules and to Karl Ziegler and Giulio Natta who in 1963 shared the Nobel prize for their discoveries in the field of chemistry and technology of polymers. More recently P.-G. de Gennes (1991, for discovering methods for order phenomena in liquid crystals and polymers); A.J. Heeger, A.G. MacDiarmid and H. Shirakawa (2000, for discovering and development of conducting polymers) and R.H. Grubbs, R.R. Schrock and Y. Chauvin (2005, catalyst development, e.g. for Ring Opening Metathesis Polymerization ROMP) were awarded.

D. Braun et al., *Polymer Synthesis: Theory and Practice*, DOI 10.1007/978-3-642-28980-4_1, © Springer-Verlag Berlin Heidelberg 2013

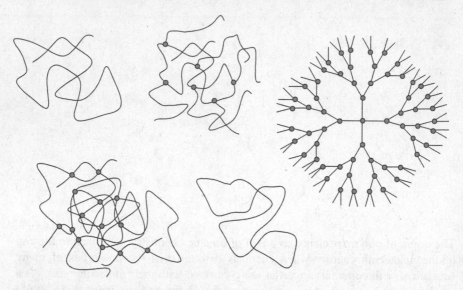

Fig. 1.1 Macromolecular architectures

Due to their high molecular masses, macromolecular substances (polymers) show particular properties not observed for any other class of materials. In many cases, the chemical nature, the size, and the structure of these giant molecules result in excellent mechanical and technical properties. They can display very long linear chains, but also cyclic, branched, crosslinked, hyperbranched, and dendritic architectures as well. The thermoplastic behaviour or the possibility of crosslinking of polymeric molecules allow for convenient processing into manifold commodity products as plastics, synthetic rubber, films, fibres, and paints (Fig. 1.1).

Man, however was by far not the first to recognize the tremendous potential of giant chain architectures: millions of years ago, nature developed macromolecules for many specific purposes. Cellulose, as example, is a substance which – due to its extraordinary stress-stability-guarantees the shape and stability of the thinnest blade of grass and the largest tree even in a gust or strong storm.

Moreover, transformation of small molecules into high-molecular-weight materials changes solubility dramatically. Nature takes advantage of this effect for storage of energy by converting sugar into starch or glycogen, for example. Also, thin polymeric fibres and films are widely used in nature: spiders apply them to catch insects, silkworms to build their cocoons, crustaceans form their outer shell of it, birds their feathers, and mammals their fur. Last but not least, nature uses macromolecules to store the key information of life – the genetic code – by means of a polymer called DNA.

These few examples are ample evidence that nature benefitted from the advantages of long chain molecules for variety of central applications long before man discovered the use of polymer materials for similar purposes: for the longest time in our history we were unable to produce tailor-made macromolecules for protection, clothes and housing. Instead, we applied the polymeric material as it was provided by nature as wool, leather, cotton, wood or straw.

The first macromolecular substances which found technical interest were based on chemically modified natural materials, for example cellulose nitrate (Celluloid) or crosslinked casein (Galalith). Only with the onset of industrialisation in the nineteenth century did these renewable raw materials become the limiting factor for further growth, and chemists began developing artificial macromolecules based on fossil carbon sources like coal, oil, and gas. Polymers like condensation products from phenol and formaldehyde (Bakelite) started the "plastics age" in 1910 and polymers of styrene or vinyl chloride were used since about 1930 and until nowadays as important plastics. Presently, worldwide more than 260 million tons polymers per year are produced and used as plastics, films, fibres, and synthetic rubber.

More recently, so-called functional polymers with special physical or chemical properties have replaced other materials in many electrical or optical applications for microelectronic applications due to their electronic properties or have been used for biochemical purposes.

To sum up, macromolecular science covers a fascinating field of research and technology, focused on the creation, the understanding, and the tailoring of materials formed out of very high-molecular-weight molecules.

1.1   Some Definitions

Prior to a profound discussion of the means of generating, characterizing, processing, and recycling macromolecules, some basic definitions and explanations should be provided.

1.1.1 Monomers

In a chemical reaction between two molecules, the constitution of the reaction product can be unequivocally deduced if the starting materials possess functional groups that react selectively under the chosen conditions. If an organic compound contains one reactive group that can give rise to one linkage in the intended reaction, it is called monofunctional; for two, three, or more groups it is called bi-, tri-, or oligo-functional, respectively. However, this statement concerning the functionality of a compound is only significant in relation to a specific reaction. For example, the primary amino group is monofunctional with respect to the formation of an acid amide, but up to trifunctional when reacted with alkyl halides. Monounsaturated compounds, epoxides, and cyclic esters are monofunctional in their addition reactions with monofunctional compounds, but bifunctional in chain growth polymerizations.

Molecules suitable for the formation of macromolecules must be at least bifunctional with respect to the desired polymerization; they are termed monomers. Linear macromolecules result from the coupling of bifunctional molecules with each other or with other bifunctional molecules; in contrast, branched or crosslinked polymers are formed when tri- or poly-functional compounds are involved.

1.1.2 Oligomers

Medium-size members of homologous polymeric series such as dimers, trimers, etc. are called oligomers. They can be linear or cyclic and are often found as byproducts of polymer syntheses, e.g., in cationic polymerizations of trioxane or in polycondensations of ε-aminocaproic acid (see Example 4.9). For the preparation of linear oligomers with two generally reactive end groups, the so-called telechelics, special methods, i.e., oligomerizations, were developed.

1.1.3 Polymers

As already shown, conventional macromolecules (or polymers) consist of a minimum of a several hundred covalently linked atoms and have molar masses clearly above 10^3 g/mol. The degree of polymerization, P, and the molecular weight, M, are the most important characteristics of macromolecular substances because nearly all properties in solution and in bulk depend on them. The *degree of polymerization* indicates how many monomer units are linked to form the polymer chain. The *molecular weight* of a homopolymer is given by Eq. 1.1.

$$M = P \cdot M_{ru} \tag{1.1}$$

where M_{ru} stands for the molar mass of the monomer repeating unit. While pure low-molecular-weight substances consist of molecules of identical structure and size, this is generally not the case for macromolecular substances. They, instead, consist of mixtures of macromolecules of similar structure but different degrees of polymerizations and molecular weights. Therefore, they are called polydisperse. As a result of this polydispersity, the values of P and M are only mean values, called P and M.

High molar masses and chain-like architectures result in properties quite different from those of low-molecular-weight substances. This may be demonstrated for the case of polyethylene:

$$- - -CH_2-CH_2-CH_2-CH_2-CH_2-CH_2- -$$

While chains having molecular weights of a few thousands only form brittle waxes, polyethylenes having molar masses of above hundred thousand show much better mechanical properties. They can be processed into films, pipes, and other performance products. When molar mass is further increased up to several millions, even higher impact strengths and abrasion resistances are achieved which enable these materials to be used in heavy-duty applications like skating floors and artificial hips.

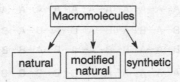

Fig. 1.2 Classification of macromolecules I

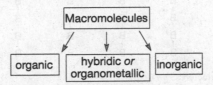

Fig. 1.3 Classification of macromolecules II

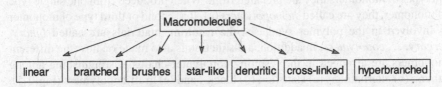

Fig. 1.4 Classification of macromolecules III

Macromolecules may be classified according to different criteria. One criterion is whether the material is *natural or synthetic* in origin. Cellulose, lignin, starch, silk, wool, chitin, natural rubber, polypeptides (proteins), polyesters (polyhydroxybutyrate), and nucleic acids (DNA, RNA) are examples of naturally occurring polymers while polyethylene, polystyrene, polyurethanes, or polyamides are representatives of their synthetic counterparts. When natural polymers are modified by chemical conversions (cellulose → cellulose acetate, for example), the products are called *modified natural polymers*.

Another criterion is the chemical composition of the macromolecules: when containing only carbon, hydrogen, oxygen, nitrogen, halogens, and phosphorus, they are called *organic*. If they additionally contain metal atoms, or if they have a carbon-free main chain but organic lateral substituents – such as polysiloxanes, polysilanes, and polyphosphazenes – they are called *organometallic* or *hybridic*. Finally, if they do not contain carbon atoms at all – such as polymeric sulfur – they are called *inorganic* (Figs. 1.2 and 1.3).

At the same time, the macromolecules might be classified according to whether their chains have only one kind of atoms – like carbon – in the backbone (*isochains*) or different elements (*heterochains*). Concerning their chain architecture, polymers are subdivided into *linear, branched, comb-like, crosslinked, dendritic*, or *star-like* systems (Fig. 1.4).

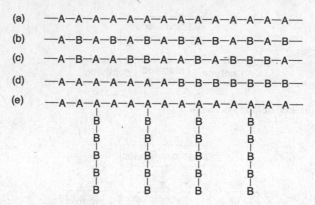

Fig. 1.5 Copolymer architectures

Moreover, polymers are quite often classified according to the number of different types of monomers they are prepared from. When produced from one single type of monomer, they are called *homopolymers* (a). If a second or third type of monomer is involved in the polymer synthesis, the resulting materials are called *binary*, *ternary*, ... *copolymers*. In addition, a distinction is also made on how the different monomers are arranged in the resulting copolymer chains, distinguishing among others: (b) alternating-, (c) statistic-, (d) block-, and (e) graft-copolymers (Fig. 1.5).

Finally, for practical reasons it is useful to classify polymeric materials according to where and how they are employed. A common subdivision is that into *structural polymers* and *functional polymers*. Structural polymers are characterized by – and are used because of – their good mechanical, thermal, and chemical properties. Hence, they are primarily used as construction materials in addition to or in place of metals, ceramics, or wood in applications like plastics, fibers, films, elastomers, foams, paints, and adhesives. Functional polymers, in contrast, have completely different property profiles, for example, special electrical, optical, or biological properties. They can assume specific chemical or physical functions in devices for microelectronic, biomedical applications, analytics, synthesis, cosmetics, or hygiene.

1.2 Chemical Structure and Nomenclature of Macromolecules

The Commission of Nomenclature of the Macromolecular Division of IUPAC (*International Union of Pure and Applied Chemistry*) formulated general rules for the nomenclature of polymers (relevant publications see section "Publications About Nomenclature"). Selected recommendations are explained in the following paragraph.

A *polymer* is defined as a substance consisting of molecules that are characterized by multiple repetitions of one or more species of atoms or groups of atoms. These repeating species of atoms or groups of atoms are designated *constitutional units*. A *regular polymer* can be described by a certain sequence of such constitutional units, whereas this is impossible with an *irregular polymer*. The smallest constitutional unit that leads through repetition to a regular polymer is the *constitutional repeating unit*. Accordingly, the following polymer chaincontains – among others – the following constitutional units:

$$---CH_2-CH\left[-CH_2-CH\right]_n CH_2-CH---$$
$$\qquad R \qquad R \qquad R$$

$$-CH-CH_2- \qquad -CH_2-CH- \qquad -CH- \qquad -CH_2- \quad etc.$$
$$\;\;R \qquad\qquad\qquad R \qquad\qquad R$$
$$\quad (a) \qquad\qquad\qquad (b)$$

However, only (a) and (b) are constitutional repeating units, describing the polymer's constitution precisely and completely. The polyamide prepared from hexamethylenediamine and adipic acid has the following *constitutional repeating* unit:

$$-NH-(CH_2)_6-NH-CO-(CH_2)_4-CO-$$

According to the above definition it has two constitutional units:and

$$-NH-(CH_2)_6-NH-$$

$$-CO-(CH_2)_4-CO-$$

In contrast to this, a *statistic copolymer* (often also called *random copolymer*; see Table 1.1), schematically described as follows:

$$-- \bigwedge\bigwedge\bigwedge\bigwedge\bigwedge\bigwedge --$$
$$\quad A \quad B \quad B \quad A \quad A \quad B \quad A$$

cannot be represented by one single constitutional *repeating* unit. Hence, it is an *irregular polymer*.

The systematic nomenclature of regular single-stranded polymers starts by naming the constitutional repeating unit as a group with two free valences, conforming as far as possible to the nomenclature rules of organic chemistry. The name of the polymer is then simply obtained by adding the prefix "poly". The direction and sequence of the constitutional repeating units according to which the polymer is named are also defined by rules: subunits are arranged in decreasing priority from left to right, for example:

Table 1.1 Nomenclature of copolymers

Name	Nomenclature
Binary copolymer (general)	Poly(A-*co*-B)
	e.g. Poly(styrene-*co*-butadiene)
Statistic copolymer	Poly(A-*stat*-B)
	e.g., Poly(styrene-*stat*-butadiene)
Random copolymer[a]	Poly(A-*ran*-B)
	e.g., Poly(styrene-*ran*-butadiene)
Alternating copolymer	Poly(A-*alt*-B)
	e.g., Poly(ethylene-*alt*-tetrafluoroethylene)
Diblock copolymer	Poly(A)-*block*-Poly(B)
	e.g., Poly(styrene)-*block*-Poly(butadiene)
Graft copolymer	Poly(A)-*graft*-Poly(B)
	e.g., Poly(styrene)-*graft*-Poly(butadiene)

[a]In many copolymerizations, growth is influenced by the terminal (active) monomer unit. This can be described by Markov trials: Zero order (or Bernoullian mechanism) means that the terminal unit of the growing chain does not influence the addition (rate, stereoregularity, etc.) of the next monomer molecule. Such copolymerizations often are called "random".

poly(1-phenylethylene) = polystyrene

poly(oxy-1,4-phenylene) = poly(phenylene oxide)

$$-\!\!\left[NH-(CH_2)_6-NH-CO-(CH_2)_4-CO\right]_n\!\!-$$

poly(hexamethylene adipamide) (polyamide 66, nylon 66)

The IUPAC names for polymers are often very complicated and lengthy. Therefore, parallel to the systematic names, some semi-systematic or trivial names are allowed. Here, in most cases, the name of the basic monomer is used in combination with the prefix "poly". Polystyrene may serve as an example. Brackets are used for the name of the monomer when it contains more than one word such as poly(vinyl chloride):

poly(vinyl chloride)

The part of a macromolecule corresponding to the smallest molecule or to a molecule from which the macromolecule is or could be built is designated as a *monomer unit*. In vinyl polymers such as poly(vinyl chloride), the monomer unit contains two chain atoms, but monomer units with one, three, or even more chain atoms are also known:

$$\text{---}CH_2\text{---}CH_2\boxed{\text{---}CH_2\text{---}}CH_2\text{---}CH_2\text{---}CH_2\text{---}$$

monomer unit with one chain atom (polymethylene)

$$\text{---}CH_2\boxed{\text{---}CH_2\text{---}CH_2\text{---}}CH_2\text{---}CH_2\text{---}CH_2\text{---}$$

monomer unit with two chain atoms (polyethylene)

$$\text{---}CH_2\text{---}CH_2\boxed{\text{---}O\text{---}CH_2\text{---}CH_2\text{---}}O\text{---}$$

monomer unit with three chain atoms [poly(ethylene oxide)].

Constitutional repeating unit and monomer unit can be identical as in the case of homopolymers of vinyl or acryl compounds:

$$\text{---}CH_2\text{---}\underset{R}{CH}\left[CH_2\text{---}\underset{R}{CH}\right]_n CH_2\text{---}\underset{R}{CH}\text{---}$$

monomer unit identical to the constitutional repeating unit

However, a constitutional repeating unit can also contain several monomer units. This is the case in alternating copolymers and in many macromolecules obtained via step-growth polymerization:

$$\left[NH-(CH_2)_6-NH-CO-(CH_2)_4-CO\right]_n$$

constitutional repeating unit consisting of two different monomer units.

Macromolecules having identical constitutional repeating units can nevertheless differ as a result of isomerism. For example, linear, branched, and crosslinked polymers of the same monomer are considered as structural isomers. Another type of structural isomerism occurs in the chain polymerization of vinyl or vinylidene monomers. Here, there are two possible orientations of the monomers when they add to the growing chain end. Therefore, two possible arrangements of the constitutional repeating units may occur:

H = Head
T = Tail

In general, the head-to-tail structure is the by far most predominant motif. The proportion of head-to-head structure is small and can only be determined experimentally in some specific cases. Further types of structural isomerism are found in polymeric conjugated dienes: addition of a monomer to the chain end can occur in 1,2- and in 1,4-position. Moreover, in the case of nonsymmetric dienes, 3,4-addition is a further possibility:

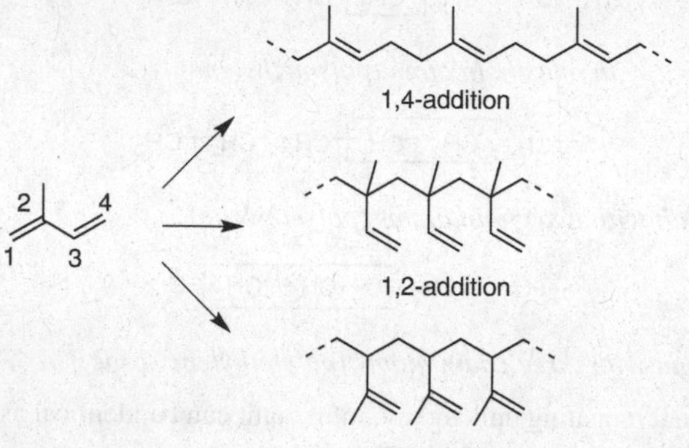

1,4-addition

1,2-addition

3,4-addition

When polymers have double bonds within their main chains – such as in poly (1,4-isoprene) – there arises a further kind of isomerism, i.e., *cis/trans (Z/E)* isomerism:

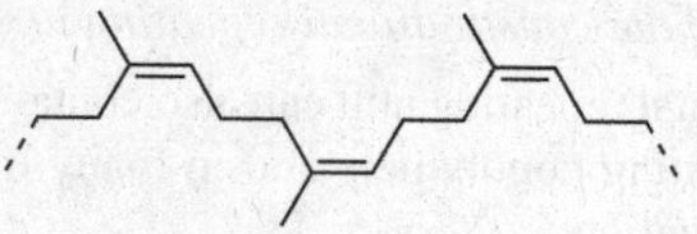

poly(cis-1,4-isoprene) (natural rubber)

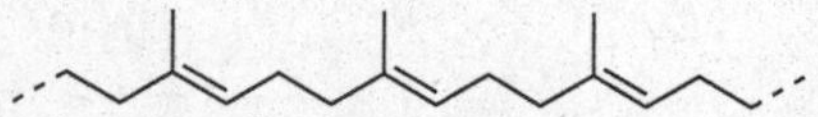

poly(trans-1,4-isoprene) (gutta percha, balata)

The structural uniformity of synthetic polymers is in general not as perfect as in the case of their natural counterparts. However, using special initiators and optimized polymerization conditions, it is possible to prepare quite homogeneous *cis*-1,4-polyisoprene ("synthetic natural rubber").

Linear macromolecules having a constitutional repeating unit such as -CH$_2$-CHX- (X $\neq$ H) show two further stereoisomerisms, i.e., optical isomerism and tacticity. The stereoisomerism named "tacticity" has its origin in the different spatial arrangements of the substituents X. When we arrange the carbon atoms of the polymer main chain in a planar zigzag conformation in the paper plane, X is either above or below that plane ("Natta projection"). If the substituents X of the vinyl polymer are either all above or all below that plane (case I), the polymers are

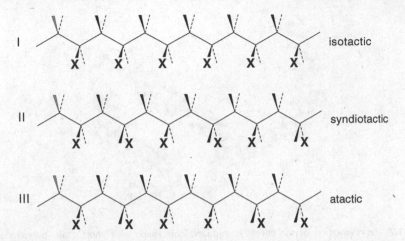

Fig. 1.6 Tacticity in macromolecules of substituted vinyl monomers (Natta projection)

called *isotactic*. Their chains consist of a regular sequence of constitutional repeating units containing carbon atoms with the same configuration. When the substituted carbon atoms have alternating configurations (case II), the polymer is called *syndiotactic*. Here, the substituents X are alternating above and below the paper plane. If there is a random spatial orientation (configuration) of the substituents X (case III), the polymers are called *atactic*.

Optical isomerism is possible whenever the substituents X contain centers of asymmetry: polymers obtained from pure enantiomeric monomers are optically active. However, the specific rotation of the polymers is in general clearly different from that of the monomers. Optical isomerism is also possible when asymmetrically substituted carbon atoms are placed in the main chain (see Example 3.25a) (Fig. 1.6).

The above considerations concerning structural isomerism and stereoisomerism are not restricted to homopolymers but can occur in copolymers as well. Here, moreover, structural isomerism can have its origin additionally in different distributions of two (or more) types of constitutional repeating units within the polymer chain.

For the nomenclature of *copolymers* it is common either to use the full name, for example "statistic copolymer of styrene and butadiene", or to use the abbreviations recommended by IUPAC. Some cases are listed in Table 1.1.

1.3 States of Order in Polymers

Aside from chemical composition and chain length the properties of macromolecular substances are substantially determined by the conformation and configuration of the individual macromolecules. Isolated macromolecules do not take up a

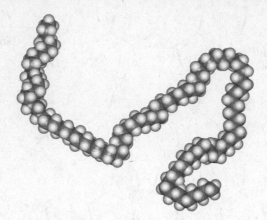

Fig. 1.7 Representation of the energy minimized coil shape of a short chain polyethylene ($P_n = 55$) as obtained via mean-field calculations

precisely defined three-dimensional shape; they rather assume a statistically most probable form which approximates to the state of maximum possible entropy. This is neither a compact sphere nor an extended rigid chain, but rather a more or less loose statistical coil (Fig. 1.7).

In solution the form of such a coil is subject to dynamic changes. In contrast, the shape of the coil in the solid state is largely fixed due to the immobility of the individual macromolecules. Real chain molecules are not able to assume the shape of an ideal statistical coil, as would be expected by random-flight statistics, since molecular parameters such as fixed bond angles and restricted rotation about the bonds affect the shape of the coil. Branching and incorporation of rigid chain components (aromatic rings, heterocyclic rings) also influence the coil form. Finally, the shape of the coil is determined by intermolecular or intramolecular interactions, such as H-bridges or electrostatic forces.

Polymers can exist as liquids, as elastomers or as solids but can be transferred into the gaseous state only under very special conditions as are realized in, for example, MALDI mass spectrometry. This is because their molecular weight is so high that thermal degradation sets in before they start to evaporate. Only a few polymers are technically applied in the liquid state (silicon oils, specialty rubbers) but most polymers are applied either as elastomers, or as rigid amorphous or semicrystalline solids.

1.3.1 Macromolecules in Solution

Polymer solutions are important because many polymer syntheses as well as most procedures for their molecular characterization are carried out in solution. Polymer solutions are furthermore essential in the processing of some polymers to fibers, preparation of polymer blends, coatings, and adhesives. Moreover, polymer solutions are applied because of their high viscosity (thickeners). Last but not least, also mixtures of polymers might be considered as solutions: polymer blends represent (homogeneous or heterogeneous) "solutions" of a high-molecular-weight solute in a high-molecular-weight "solvent". Because of this importance of polymer solutions, in the following some ideas are given on how polymers in solution can be understood.

In dilute solution, the behavior of macromolecules is quite different to that of common low-molecular-weight molecules. For example, the shape of a macromolecular coil is subject to permanent dynamic changes, and the coils are in a more or less swollen state when compared to their "unperturbed" (solid state) dimensions. As a result, polymer solutions tend to be viscous even at low concentrations. This again can have some influence on their reactivity (in macromolecular substitution reactions as well as during their chain growth), and change their physical properties to a certain extent. However, the most fundamental difference between low- and high-molecular-weight materials is that polymeric substances are not composed of structurally and molecularly uniform molecules. Thus, even if they have an identical analytical composition, the individual chain molecules differ in their structure, configuration, conformation, as well as in their molecular size. Hence, there is a mixture of molecules of different size, i.e., a molecular-weight distribution, and the compounds are called *polydisperse*. As a consequence, it is evident that the expression "identical" is not, in practice, applicable to the individual macromolecules of a polymer sample. It follows that most physical measurements on polymers only give average values. The aforementioned peculiarities cause that methods suitable for the characterization of low-molecular-weight compounds are frequently not applicable – or applicable only in a substantially modified form – to polymers. In many cases, moreover, completely new methods are required. Many of these methods require the polymers to be in the dissolved state.

As already mentioned, real chain molecules do not reach their ideal coil shape as would be expected by random-flight statistics. This is because molecular parameters such as fixed bond angles and restricted rotation about the bonds affect the coil conformation as well as branches and incorporated rigid chain segments (aromatic rings, heterocyclic rings). Also, the shape of the coil is determined by intermolecular or intramolecular interactions, such as hydrogen bridges or electrostatic forces.

In the case where macromolecules do not have strong interactions, neither with other macromolecules nor with solvent molecules, are build of simple linear nonbranched chains, and where the single bonds within the polymer main chain are sufficiently rotationally free, then the molecules assume – at least in highly diluted solution – a shape which resembles a coiled thread. In order to describe this statistic coil it is common to use quantities such as the number of bonds in the chain, bond lengths, valence angles, and rotational angles. The average conformation of the coil is then equivalent to the average overall shapes of the macromolecule which are randomly formed. The fully stretched chain or the collapsed coil rarely occur. Loosely coiled species, on the other hand, are quite common.

The simplest theoretical chain model of a macromolecule assumes the polymer to be a linear chain composed of $N + 1$ chain atoms, connected by N bonds of identical length b. The mathematically maximum possible chain length is $L = Nb$.

However, this length cannot be reached because of the valence angles at the chain atoms. IUPAC designates the physically possible maximum length of a chain as the *contour length*, r_{cont}. For an *all-trans* configured chain with bond angles τ it is:

$$r_{cont} = Nb \cdot \sin\frac{\tau}{2} = Nb_e \tag{1.2}$$

Whereby b_e is the effective bond length. However, the *all-trans* chain represents a highly improbable situation: dissolved polymer chains are much more probable in a randomly coiled chain conformation, and these conformations change permanently. Thus a polymer chain in solution is better characterized by its average end-to-end distance in the coil, or by its average radius of gyration. These data give valuable information about the behavior of chain molecules in the respective solution: the end-to-end distance is the spatial distance between the end groups of a linear chain in its (randomly) coiled state. The mean-square radius of gyration, $<s^2>$, of a dissolved macromolecule is defined as the mass-average of R_i^2 (R_i being the distance of an entity of the chain from the coil's center of gravity) of all entities:

$$\langle s^2 \rangle \equiv \frac{\left\langle \sum_i m_i R_i^2 \right\rangle}{\sum_i m_i} \tag{1.3}$$

Mean-square end-to-end distances, $<r^2>$, of single linear chains are not experimentally accessible but can be calculated by various models. Assuming no specific interaction between the chain segments (ϑconditions), the position of segments in the coil are distributed at random in space and time. Such coils are called *random coils* or *statistical coils*. The models developed for random coils with only short-range interactions of the chain segments differ in the restrictions applied to the valence angles and torsion angles. For very bulky substituents, the chains no longer form random coils but *worm-like chains* are obtained which can be characterized by the *Kuhn length*, L_K. Such chains are often called semiflexible.

End-to-end distances are parameters that are easy to calculate by theoretical approaches but cannot be measured directly. On the other hand, the radius of gyration is a quantity readily accessible by experiment. For $M \to \infty$ general relationships exist between the radius of gyration and the end-to-end distance for all linear chains with only short-range interactions, for example $<r^2> = 6 <s^2>$.

1.3.1.1 Solvents and Solubility

In some specific cases, dissolved macromolecules take up the shape predicted by the above theories of isolated chain molecules. In general, however, the interaction between solvent molecules and macromolecules has significant effects on the chain dimensions. In "poor" solvents, the interactions between polymer segments and solvent molecules are not that much different from those between different chain segments. Hence, the coil dimensions tend towards those of an unperturbed chain: if the dimension of the unperturbed coil is identical to that in solution, the solution

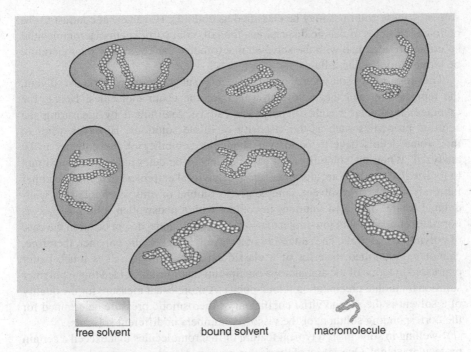

Fig. 1.8 Schematic representation of a dilute solution of macromolecules

conditions are called ϑ conditions (ϑ solvent, ϑ temperature). In "good" solvents, on the other hand, the interaction of chain segments and solvent molecules is preferred over the interaction between chain segments, and a good solvation of the macromolecules is given. Here, the coils contain a considerable amount of trapped solvent and thus are drastically expanded in their dimensions with respect to the unperturbed coil. The trapped solvent is in a state of continuous exchange (by diffusion) with the surrounding solvent, but it is nevertheless fixed to an extent that, in many situations, it may be regarded as moving with the coil as a whole. The macromolecular coils are thus comparable with small swollen gel particles that, like a fully soaked sponge, consist of a framework (the coiled macromolecules) and the embedded solvent. This concept is schematically illustrated in Fig. 1.8.

With increasing concentration of the polymer solution, the coils take up a greater proportion of the total volume until finally, at a "critical" concentration c*, there is mutual contact between the coils. At still higher concentrations the coils interpenetrate or, if this is not possible on account of incompatibility effects, the interaction may be confined to the boundary regions.

Depending on the molecular weight of the macromolecules and the quality of the solvent, the coil volume of macromolecules in solution may be 20–1,000 times larger than the chain volume itself. Thus, such a swollen gel particle may consist of more than 99% solvent. Since the diameter of such gel coils may be between ten and several hundred nanometers – again depending on the molecular weight and solvent – these solutions may be classified as colloids. However, in contrast to the colloidal particles of classic dispersions, the colloidal particles in macromolecular solutions are identical with the solvated macromolecular coils: they may therefore be termed "molecular colloids".

The properties of solutions of macromolecular substances depend on the solvent, the temperature, and the molecular weight of the chain molecules. Hence, the (average) molecular weight of polymers can be determined by measuring the solution properties such as the viscosity of dilute solutions. However, prior to this, some details have to be known about the solubility of the polymer to be analyzed. When the solubility of a polymer has to be determined, it is important to realize that macromolecules often show behavioral extremes; they may be either infinitely soluble in a solvent, completely insoluble, or only swellable to a well-defined extent. Saturated solutions in contact with a nonswollen solid phase, as is normally observed with low-molecular-weight compounds, do not occur in the case of polymeric materials. The suitability of a solvent for a specific polymer, therefore, cannot be quantified in terms of a classic saturated solution. It is much better expressed in terms of the amount of a precipitant that must be added to the polymer solution to initiate precipitation (cloud point). A more exact measure for the quality of a solvent is the second virial coefficient of the osmotic pressure determined for the corresponding solution, or the viscosity numbers in different solvents.

Swelling in solvents is a typical feature of macromolecules that exceed a certain molecular weight. One aspect of this is that macromolecular compounds can take up large amounts of solvent, forming a gel, with a marked increase of volume. If this process does not lead to a homogeneous solution at the end, it is called "limited swelling"; unlimited swelling, on the other hand, is synonymous with complete dissolution. The extent of swelling depends on the chemical nature of the polymer, the molecular weight, the swelling medium, and the temperature. For crosslinked polymers, which are of course insoluble, it is a measure of the degree of crosslinking.

Although many thermodynamic theories for the description of polymer solutions are known, there is still no full understanding of these systems and quite often, one needs application of empirical rules and conclusions by analogy. As a rough guide, some solvents and non-solvents are indicated in Table 2.7 (Sect. 2.3.1) for various polymers. However, not all combinations of solvent and non-solvent lead to efficient purification of a polymer via dissolution and reprecipitation, and trial experiments are required therefore.

Considering the rather complicated processes that take place during dissolution, it is not surprising that some systems show peculiar behavior. For example, while solubility generally increases with temperature, there are also polymers that exhibit a negative temperature coefficient of solubility in certain solvents. Thus, poly (ethylene oxide), poly(N-isopropylacrylamide), or poly(methyl vinyl ether) dissolve in water at room temperature but precipitate upon warming. This behavior is found for all polymer–solvent systems showing a lower critical solution temperature (LCST). It can be explained by the temperature-dependent change of the structure of water clusters. The LCST may be increased or lowered by adding special salts.

Surprising effects can also be observed when solvent mixtures are used to dissolve a polymer. There are examples where mixtures of two non-solvents act as a solvent; vice versa, a mixture of two solvents may behave like a non-solvent. For example, polyacrylonitrile is insoluble in both, nitromethane and water, but it dissolves in a mixture of the two solvents. Similar behavior can be observed for polystyrene/acetone/hexane and poly(vinyl chloride)/acetone/carbon disulfide. Examples of systems where the polymer dissolves in two pure solvents but not in their mixture are polyacrylonitrile/malonodinitrile/dimethylformamide and poly (vinyl acetate)/formamide/acetophenone. These peculiarities are especially to be taken into account if one wants to adjust certain solution properties (e.g., for fractionation) by adding one solvent to another.

Finally, we should mention the phenomenon of incompatibility of mixtures of polymer solutions. It applies to nearly all combinations of polymer solutions: when the homogeneous solutions of two different polymers in the same solvent are mixed, phase separation occurs. For example, 10% solutions of polystyrene and poly(vinyl acetate), each in benzene, form two separated phases upon mixing. One phase contains mainly the first polymer, the other phase mainly the second polymer, but in both phases there is a certain amount of the other polymer present. This limited compatibility of polymer mixtures can be explained thermodynamically and depends on various factors, such as the structure of the macromolecule, the molecular weight, the mixing ratio, the overall polymer concentration, and the temperature.

1.3.1.2 Polyelectrolytes

Electrolytes are compounds that break apart into positive and negative ions when dissolved in polar solvents like water. Sodium chloride, for example, splits into positive sodium ions and oppositely charged chloride ions when dissolved in water. Polyelectrolytes are polymers that do the same: they fall apart into charged polyions and many oppositely charged counterions. Then, all the charges attached to the polymer chain repel each other. Thus at low salt concentrations the polyelectrolyte's random coils expand tremendously. This is because the like charges on the polymer chain repel each other. Expansion allows these charges to be as far apart as possible. When the polyelectrolyte chain stretches out it takes up more space, and is more effective at resisting the flow of the solvent molecules around it. Therefore, the solution becomes thick and syrupy. When, on the other hand, the concentration of low-molecular-weight salt is increased by adding, for example, sodium chloride, the polymer chains will collapse back into random coils because the range of the intramolecular coulomb force decreases with increasing salt concentration. Consequently, the intrinsic viscosity of the solution decreases as well (see Example 3.5). When polyelectrolytes are studied, it is important to remember that they behave different from conventional polymers in many differently ways. For example, the polyions have high charges because a single polyelectrolyte molecule may have many thousands of ionizable groups. Moreover, the like ions on the chain can only separate to a certain extent because they are connected to each other by the polymer backbone chain. Effects produced by ionic charge interaction will not vanish as concentration decreases to infinite dilution as in the case of low-molecular-weight electrolytes: in low-molecular-weight electrolytes, if the solution is dilute enough, the charged groups will be too far apart to interact. But with polyelectrolytes, even if there is only one polyelectrolyte molecule in the solution, the charges on that molecule will interact with each other. In a highly dilute solution, the individual polyelectrolyte molecule is an area of increased charge density. Counterions are attracted to these pockets of charge density. Out in the bulk solvent, far away from the polyelectrolyte molecule, counterion concentration will be very low. For all of these reasons, it is not appropriate to speak of the "ionic strength" of a polyelectrolyte solution. The ionic strength may be different in some regions than in others. It will be high in and around the polyelectrolyte coils, but lower in the bulk phase far from the polyelectrolyte molecules.

1.3.2 Macromolecules in the Molten State

As explained earlier (Sect. 1.3.1), macromolecules in a low-molecular-weight solvent prefer a coiled chain conformation (random coil). Under special conditions (theta state) the macromolecule finds itself in a force-free state and its coil assumes the unpertubed dimensions. This is also exactly the case for polymers in an amorphous melt or in the glassy state: their segments cannot decide whether neighboring chain segments (which replace all the solvent molecules in the bulk phase) belong to its own chain or to another macromolecule (having an identical constitution, of course). Therefore, here too, it assumes the unperturbed (Θ) dimensions.

Another important aspect of polymer melts is that macromolecular substances do not change abruptly from the solid state into a low-viscous liquid; instead, the change proceeds over a finite temperature range. This observation can be attributed to the fact that chain molecules undergo different types of motion, i.e., micro- and macro-Brownian motions. While in the former case only some chain segments carry out fluctuations and motions (the length of the moving segments depends on the respective temperature), in the latter case the whole macromolecule moves relative to the other macromolecules (reptation).

Far below the glass transition temperature (T_g, see Sect. 2.3.5.3) the macro-Brownian motions are frozen in completely, and most of the micro-Brownian motions are frozen in as well ("glassy state"). Near T_g, the micro-Brownian motions set in and become stronger with increasing temperature. The material softens. Finally, upon further raise of temperature, the macro-Brownian motions set in as well, and the polymer can be deformed by applying an external force. Finally, it behaves like a liquid provided the chain length is not too long. Just around T_g, some physical properties change distinctively such as the specific volume, the expansion coefficient, the specific heat, the elastic modulus, and the dielectric constant. Determination of the temperature dependence of these quantities can thus be used to determine T_g.

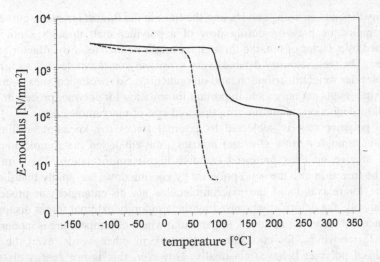

Fig. 1.9 Dependence of elastic modulus on temperature in amorphous (--) and in crystalline (—) poly(ethylene terephthalate)

Softening as a result of micro-Brownian motion occurs in amorphous and crystalline polymers, even if they are crosslinked. However, there are characteristic differences in the temperature dependence of mechanical properties like hardness, elastic modulus, or mechanic strength when different classes of polymers change into the molten state. In amorphous, non-crosslinked polymers, raise of temperature to values above T_g results in a decrease of viscosity until the material starts to flow. Parallel to this softening, the elastic modulus and the strength decrease (see Fig. 1.9).

In semicrystalline, non-crosslinked polymers (thermoplastics) where T_g is considerably below T_m (and below room temperature), rigidity and elastic modulus decrease as well with increasing temperature to some extent, but they dwindle abruptly as soon as the crystalline structure begins to collapse at temperatures near T_m. In crosslinked polymers, where the network segments can undergo micro-Brownian but not macro-Brownian motion, softening but neither flow nor melting is observed when the raising temperature crosses T_g. Therefore, only small changes in the abovementioned mechanical properties occur upon heating; a marked decline does not set in until the decomposition temperature is approached. In industry, the softening temperature (e.g., Vicat temperature) is often quoted instead of T_g since it is easier to measure; for amorphous polymers it lies close to T_g (see Sect. 2.3.5.4).

The flow behavior of molten macromolecular substances is generally quite different from that of low-molecular-weight compounds. This is obvious, for example, from the shape of the flow curves. Moreover, flow orientation can be observed. In ideal liquids (water, glycerol, sulfuric acid, etc.), the viscosity is a characteristic quantity which does not depend on the shear rate γ ("Newtonian flow"): if the shear force τ is plotted vs. γ, a straight line is obtained with slope η (Newtonian viscosity). Macromolecular melts behave differently since their melt viscosity depends on both, τ and γ, and the lines in the flow diagram are curved. If, for example, the pressure during flow of a polymer melt through a nozzle is increased by a factor of ten, the throughput might be increased by a factor of one hundred. The melt viscosity depends very sensitively on the molecular weight, on the molecular weight distribution, and on branching. So rheological measurements on polymer melts not only yield important information for polymer processing, but also allow deductions about the structure and size of macromolecules.

If a polymer melt is subjected to external forces (by kneading, rolling, or extrusion through narrow slits and nozzles), the entangled macromolecules are forced to give up their preferred (random-flight) conformation. This situation might be frozen-in into the solid polymer by cooling down so rapidly that there is no time for relaxation of the macromolecules via dis-entanglement processes. Alternatively, the deformation caused by the temporary external forces disappears as soon as the forces are removed, in particular when the temperature is moderately above T_g (above T_m for crystalline polymers); in other words, even the non-crosslinked polymer behaves elastically. However, this is not energy-elasticity (due to changes of bond angles and bond lengths) but so-called *entropy elasticity*. It results from the fact that macromolecular coils insist on their statistically most probable conformation (unperturbed coils, random walk conformation). The elasticity of polymer melts is, of course, not ideal since the elastic recovery after deformation is not complete ("residual deformation", hysteresis). This is because the internal stress in the polymer sample related to the deformation of the polymer coils can be relieved also by migration of the macromolecules. Thereby, they reduce the restoring force to some extent. This relaxation process occurs more quickly at higher temperatures since the increased macro-Brownian motion favors migration. In spite of this, a polymer melt is still elastic because of the efficient entanglement of the coiled macromolecules; the melt behaves in a *viscoelastic* way.

The rate of all these processes, of course, depends strongly on the temperature: in the vicinity of T_g the polymer chains are still relatively inflexible. Thus deformation requires considerable forces, and recovery occurs very slowly. Well above T_g the melt deforms more easily, but the tendency to flow as a result of increased macro-Brownian motion is still outweighed by the elastic recovery. The temperature range for pronounced elastic behavior of the polymer melt depends on the chain structure (e.g., whether it is branched) and in particular on the molecular weight and molecular weight distribution of the polymer.

Amorphous polymers whose glass transition temperatures are below room temperature are called *elastomers*; at room temperature they behave in a rubber-like fashion. However, as discussed above, non-crosslinked ("unvulcanized") elastomers do not return completely back to their original length after stretching: the larger the stretch is (and the longer the time is where the material is expanded), the larger is the residual extension due to relaxation of the internal tensions by migration of macromolecules. If such migration is prevented by crosslinking, the elastomer will reassume its original form even after high extension for long periods of time; such materials are known as *rubbers* and are described in Sect. 1.3.3.2.

1.3.3   Macromolecules in the Solid State

As explained previously, the thermal phase behavior of polymers differs markedly from that of common low-molecular-weight compounds. Even highly crystalline polymers have no sharp melting points but melt over a broader temperature range, T_m. This melting process, moreover, depends on the (thermal) history of the sample and on the heating rate. Amorphous polymers, on the other hand, do not melt but soften over an even broader temperature range. The characteristic thermal quantity characterizing the softening of amorphous phases in glassy or semicrystalline polymers is called the *glass transition temperature*, T_g.

Most polymers are applied either as elastomers or as solids. Here, their mechanical properties are the predominant characteristics: quantities like the elasticity modulus (Young modulus) E, the shear modulus G, and the temperature and frequency dependences thereof are of special interest when a material is selected for an application. The mechanical properties of polymers sometimes follow rules which are quite different from those of non-polymeric materials. For example, most polymers do not follow a sudden mechanical load immediately but rather yield slowly, i.e., the deformation increases with time ("retardation"). If the shape of a polymeric item is changed suddenly, the initially high internal stress decreases slowly ("relaxation"). Finally, when an external force (an enforced deformation) is applied to a polymeric material which changes over time with constant (sinus-like) frequency, a phase shift is observed between the force (deformation) and the deformation (internal stress). Therefore, mechanic modules of polymers have to be expressed as complex quantities (see Sect. 2.3.6).

The phenomenological ordering of polymers projected for use as constructing materials is not an easy matter. Sometimes the temperature stability is used as a criterion, i.e., the temperature up to which the mechanical properties remain more or less constant. Another attempt for classification, uses the E modulus or the shape of the curve of stress–strain measurements (see Sect. 2.3.6.1). In general, one can say that semicrystalline thermoplastics are stiff, tough, and impact-resistant while amorphous thermoplastics tend to be brittle. Their E modulus is of the order of 10^3–10^4 MPa. For amorphous thermoplastics, the application temperature is limited by T_g, while it is T_m in the case of semicrystalline materials. *Elastomers* show rubber-like properties at ambient temperature (E moduli are of the order of 10–10^2 MPa). They can be reversibly deformed without destruction, and they show very high elongations at break (>500%). Elastomers are obtained from polymers with very flexible main chains (rubbers, i.e. polymers with low T_g) via chemical (vulcanization) or physical (hydrogen bonds, phase separation, crystallization) crosslinking. Their static E modulus increases with increasing crosslink density. *Duromers* (thermosets, duroplasts) are polymers with extremely high crosslink densities. They show very high modulus, high strength, and high pressure resistance even at elevated temperatures. They can be processed only prior to crosslinking.

1.3.3.1 Macromolecules in the Elastomeric State

As was shown in the previous section, elasticity is a common phenomenon of all polymers above T_g. It is due to the extensive entanglement of the macromolecular chains and their efforts to maintain the preferred chain conformation. However, while non-crosslinked rubbers do not return completely to their original length after stretching, elastomeric networks do it almost completely. But even in crosslinked elastomers the recovery to the original state after extension is not ideal: in the stress–strain diagram, the stretching curve (extension at increasing force) does not coincide with the recovery curve (extension at decreasing force). There is thus a hysteresis loop whose width is a measure of the residual extension.

The properties of a rubber are determined essentially by the number of crosslinks (degree of crosslinking): assumed T_g is sufficiently low, weakly crosslinked rubbers are highly elastic and have a low elastic modulus. Upon increasing the crosslinking density, the elasticity decreases and the elastic modulus rises. Highly crosslinked rubbers lose their elasticity almost completely (hard rubbers, ebonite).

While elasticity of substances like steel or stones is determined by the tendency of the Gibbs free energy ΔG ($\Delta G = \Delta H - T\Delta S$) to strive towards a minimum as a consequence of a decrease in enthalpy ΔH, the situation in the case of elastomers is governed by the increase in entropy ΔS. This *entropy elasticity* provides an explanation of the fact that the tension of a rubber band (at constant length) increases with temperature while that of a steel wire decreases. This is because – upon elongation of the rubber band – the macromolecules are forced to change from their statistically most probable coil shape to the statistically less probable extended chain. The higher the temperature, the greater the restoring force since the change of the Gibbs free energy increases with both deformation and temperature ($T\Delta S$).

Orientations in elongated rubbers are sometimes regular to the extent that there is local crystallization of individual chain segments (e.g., in natural rubber). X-ray diffraction patterns of such samples are very similar to those obtained from stretched fibers. The following synthetic polymers are of technical relevance as rubbers: poly(acrylic ester)s, polybutadienes, polyisoprenes, polychloroprenes, butadiene/styrene copolymers, styrene/butadiene/styrene tri-block-copolymers (also hydrogenated), butadiene/acrylonitrile copolymers (also hydrogenated), ethylene/propylene co- and terpolymers (with non-conjugated dienes (e.g., ethylidene norbornene)), ethylene/vinyl acetate copolymers, ethylene/methacrylic acid copolymers (ionomers), polyisobutylene (and copolymers with isoprene), chlorinated polyethylenes, chlorosulfonated polyethylenes, polyurethanes, silicones, poly(fluoro alkylene)s, poly(alkylene sulfide)s.

1.3.3.2 Macromolecules in the Amorphous (Glassy) State

When polymer melts, rubbers, or elastomers are cooled down below T_g, they may freeze to glasses (noncrystalline amorphous phases). The rotational motions of the chain segments (micro-Brownian motions) are almost stopped now, and the transparent materials become stiff and (in most cases) brittle.

Polymer glasses are formed best when the macromolecular chains are irregular in structure (atactic, branched, crosslinked) so that crystallization is prevented.

Regular (isotactic, syndiotactic unbranched) polymer chains form glasses only if they are cooled down so fast that crystallization is prevented: such a quenching procedure freezes the material in the glassy state even if the polymer is able to crystallize.

The shape of macromolecules in the glassy state and their coil dimensions are almost identical with those in the melt and in solution under ϑ conditions. There is no long-range order but only a short-range order in the glassy state. Polymer glasses can be regarded, therefore, as isotropic frozen melts. This is indicated not only by theoretical considerations but also by numerous neutron scattering experiments: the chains are effectively entangled also in the glassy state, and the entanglements have a marked effect on the properties of amorphous polymers.

The attainment and maintenance of an ideally isotropic glassy state is not an easy matter. In general, macromolecular coils are oriented during their processing from the melt. Flow orientation assumed in the melt is often frozen after cooling, and different solid state properties are observed in the direction of flow and perpendicular to it. Orientation of the macromolecular coils can also occur when the polymer is already in the solid state. This may be achieved, for example, through the action of tensile or shear forces (stretching). It can also occur when casting films from solution, during the drying stage of the film as a result of shrinkage, and also when the thin film is pulled away from the support. Orientation causes anisotropy of various physical properties. Thus, orientation in a transparent polymer (e.g., polystyrene) is easily detected by the use of polarized light. Orientation also manifests itself in the dependence of mechanical properties on direction. This is of great importance during industrial processing of polymers. The complete prevention of orientation is not very easy: the manufactured polymer must be kept for some time at temperatures above T_g and then allowed to cool under conditions where no deformation by external forces can occur. In industry, however, orientation is deliberately caused in films and fibers, for example, by uniaxial or biaxial drawing in order to achieve or increase particular properties.

Amorphous polymers are characterized by the following properties: They are transparent and very often soluble in common organic solvents at room temperature. The following amorphous polymers have gained industrial importance as thermoplastic materials: poly(vinyl chloride), polystyrene, poly(methyl methacrylate), ABS-polymers, polycarbonate, cycloolefine copolymers, polysulfone, poly(ether sulfone), poly(ether imide).

1.3.3.3 Macromolecules in the Crystalline State

Macromolecules which have a chain constitution of sufficient uniformity (no branching, no statistic copolymers, regular substitution pattern, highly iso- or syndiotactic, etc.) are able to crystallize upon slow cooling of their melts or hot solutions. Ordered regions are formed where the chain segments are arranged regularly – on a three-dimensional lattice – over distances that are large in comparison to atomic dimensions, i.e., long-range order appears.

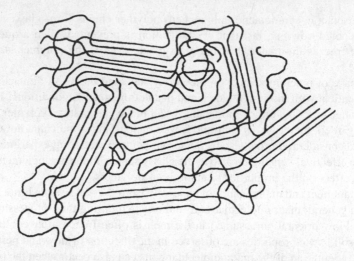

Fig. 1.10 Fringed micelles (schematic)

In most cases, however, polymers crystallize neither completely nor perfectly. Instead, they give semicrystalline materials, containing crystalline regions separated by adjacent amorphous phases. Moreover, the ordered crystalline regions may be disturbed to some extent by lattice defects. The crystalline regions thus embedded in an amorphous matrix typically extend over average distances of 10–40 nm. The fraction of crystalline material is termed the *degree of crystallinity*. This is an important parameter of semicrystalline materials.

The phenomenon that polymers generally crystallize only partially is attributable to the fact that they have difficulty in changing their shape from the coiled state in the melt into the ordered one necessary for their incorporation in a crystal. Bearing in mind these problems during the crystallization process, two models have been developed to describe the structure of the formed partially crystalline polymers: the fringed micelle model and the folded lamella model. According to the fringed micelle model (Fig. 1.10), the macromolecules lie parallel to one another like elongated threads (extended chain crystals called crystallites), thereby providing the order required for crystallization.

Since the length of the macromolecules exceeds by far that of the crystallites, each polymer chain traverses the crystalline and the amorphous regions several times. The individual crystallites are thus bound together by amorphous regions.

The fringed micelle picture is not particularly suitable for describing synthetic polymers crystallized from solution or melt. However, the fibrils of many natural substances, such as cellulose and proteins (collagen, silk), consist of bundles of macromolecules in a parallel alignment, compatible with the fringed micelle model. For synthetic polymers, however, it is more often found that they crystallize such that the macromolecules fold with an essentially constant length, leading to a lamellar-type crystallite structure (switchboard-model, Fig. 1.11).

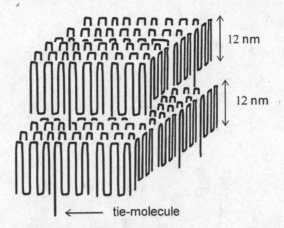

12 nm

12 nm

← tie-molecule

Fig. 1.11 Idealized picture of lamellae with back-folding (switchboard-model and tie molecules)

This is especially evident in the case of single crystals. Such single crystals can be formed by polymers upon careful cooling of highly dilute solutions (0.01–0.1%). Their formation starts from a nucleus which subsequently grows to a lamella through the back-and-forth folding of the macromolecules. Evidence for chain folding comes from X-ray diffraction and electron microscopy of single crystals of polymers like polyethylene and polypropylene. Moreover, there is rational evidence for this model: the lamella thickness is of the order of 10 nm. In the fully extended form, the macromolecules have a contour length of 10–1,000 μm (depending on the molecular weight). Hence, they must be folded in the crystallites, with the molecular chains aligned perpendicular to the lamellae. A further consequence of this model is that the crystals are generally not completely crystalline: not all the chains will be tightly folded, and chain ends may project from the lamellae, giving rise to a disordered portion. These disordered sections form the amorphous regions. The individual lamella stacks are bound together by so-called *tie molecules*. These tie molecules prevent movements of lamella stacks against each other except when mechanical forces are applied.

The shape of macromolecules within a folded lamella is not the same for all polymers. In crystalline polyethylene, for example, the chains assume a planar "zigzag" conformation, but in some other polymers like polypropylene and polyoxymethylene the chains prefer a helical shape, as in proteins. The helix might have three, four, or five monomer units per turn, i.e., the helices are three-, four-, or five-fold (Fig. 1.12)

Finally, it should be mentioned that polymers can exhibit polymorphism, i.e., they can crystallize in different types of lattices. The different crystal forms generally differ in their physical properties, e.g., crystallite melting point and density.

If the crystallization is carried out in a flowing solution in which a certain amount of so-called elongational flow is present, the morphology of a polymer like polyethylene may be completely different from that formed from the

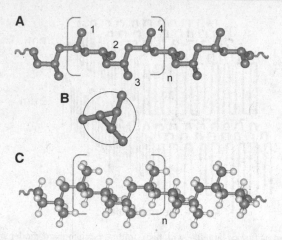

Fig. 1.12 The three-fold helix of isotactic polypropylene (A = carbon chain; B = cross-section; C = position of H atoms)

undisturbed solution. The elongational flow forces the coiled macromolecules to become extended to some degree so that crystallization from such a solution results in the formation of fibrillar crystals; the stretched molecules stack together side-by-side in the fibrils, somewhat like a piece of string. At regular distances the fibrils contain laterally arranged lamellar packets. The morphology resembles a shish kebab and is described by this term.

During crystallization from the melt, the situation is very complicated because of the mutual interactions of different macromolecules. Nevertheless, crystallites are formed according to similar principles. Ordered regions are formed containing lamellae in which there is a chain folding. However, the chain folding is not as regular as in single crystals since numerous chain molecules are built into different lamellae. These tie molecules substantially influence the mechanical properties of polymers. Compared to crystallization from solution, the overall shape of the chains can undergo relatively minor changes during crystallization from the melt.

Upon cooling of a polymer melt, the macromolecules generally crystallize as spherulites, containing both amorphous and crystalline regions. These morphological units are larger than the crystallites and can attain a diameter of several tenths of a millimeter, recognizable under the polarizing microscope by the characteristic "Maltese cross" shape (Fig. 1.13).

The structure of the spherulites is determined by the radial disposition and longitudinal extension of lamellae and amorphous interlayers. The circular banding around the center point of the spherulites can be attributed to a periodic twisting of the lamellae. The morphology also depends on the temperature at which the crystallization from the melt is carried out. This temperature is always below the melting point of the polymer crystal. Moreover, the number and size of the spherulites depend strongly on external conditions (pressure, orientation) and the number of nuclei. By the use of added nucleating agents, the number of spherulites per unit volume can be substantially raised and their resulting diameter diminished

Fig. 1.13 Spherulites of polypropylene

(heterogeneous nucleation). Some of the physical properties are thereby affected, especially optical properties such as transparency.

Whether, and to what degree a polymer crystallizes depends on various factors, notably on its structure, on the symmetry along the main chain, on the number and length of side chains, and on branching (e.g., in high-density and low-density polyethylene). Polar groups that lead to the formation of hydrogen bonds, and thus an association of chain segments in the molten state (e.g., amide groups in polyamides), substantially promote crystallization. Some polymers, such as poly-ethylene, can be obtained in the amorphous state only under special conditions; others, such as isotactic polystyrene, can be prepared in both amorphous and crystalline states. Even others, such as poly(ethylene terephthalate), can be brought into the amorphous state by rapid cooling to a temperature far below the crystalli-zation temperature; they remain stable if kept at low temperature, but begin to crystallize if warmed above the glass transition temperature. Some other polymers do not crystallize immediately, but must be induced to crystallize by appropriate treatment such as nucleating, long heating at a certain temperature (annealing), slow cooling of the molten polymer, or contact with a suitable swelling agent.

Crystalline polymers exhibit the following basic properties: They are opaque as long as the size of the crystallites or spherulites, respectively, lies above the wavelength of light. Their solubility is restricted to few organic solvents at elevated temperature. The following crystalline polymers have attained technical importance as thermoplastic materials: polyethylene, polypropylene, aliphatic polyamides, ali-phatic/aromatic polyamides, aliphatic/aromatic polyesters, polyoxymethylene, polytetrafluoroethylene, poly(phenylene sulfide), poly(arylene ether ketone)s.

1.3.4 Liquid-Crystalline Polymers (LCP)

Conventional liquids – those of low-molecular-weight materials as well as those formed by polymers – do not exhibit any order: they are amorphous or isotropic. However, in some specific cases so-called *liquid-crystalline mesophases* are

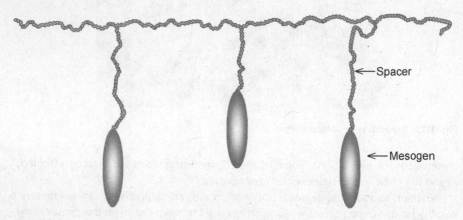

Fig. 1.14 The mesogen is built into the main chain

Fig. 1.15 The mesogen is attached as side chain to the mobile backbone, mostly via a flexible spacer

formed. The characteristic orientational order of such mesophases is between that of perfect crystals and that of amorphous liquids: while crystalline materials show long-range periodic order in three dimensions and an isotropic liquid has no orientational order, liquid-crystalline substances are less ordered than crystals in their mesophase state, yet having some degree of alignment, referred to as *liquid crystallinity*. These mesophases can be considered as ordered liquid states, i.e., anisotropic liquids. This fact can easily be recognized because these phases are liquid but not fully transparent. The opaqueness results from vectorial molecule assemblies forming neighboring domains which cause diffraction of visible light at the borderline of adjoining regions having different preferential orientations of their molecules. It is only on further heating that the liquid crystalline state is lost and an isotropic clear melt is achieved, provided no intermediate decomposition occurs. This final phase transition is called the *clearing point*. The phase transitions can be determined by DSC and by polarizing microscopy. X-ray diffraction provides information on the arrangement of the molecules in the LC-microphase (Figs. 1.14 and 1.15).

The term *mesophase* is used in reference to the liquid crystalline phase, and the molecular (sub)structures that give rise to this phase are termed *mesogens*. Macromolecules with rod-like or a flat, disc-like geometries tend to form such mesophases, in particular when the *mesogens* have strong dipole moments and/or are easily polarizable. Such polymers are called *liquid-crystalline polymers* (LCPs). Rod-like mesogens must be at least three times as long as they are wide (aspect ratio) in order to display the characteristics of liquid crystals. LCPs may have their rod- or disc-like mesogens either within their main chains or as lateral substituents.

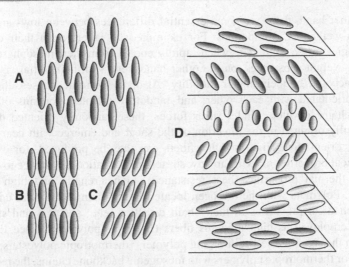

Fig. 1.16 Common mesophases of rodlike mesogens: **A** nematic, **B** smectic-A, **C** smectic-C, **D** cholesteric nematic

In main chain LCPs the mesogens are part of the polymer backbone, while side chain LCPs are formed when the mesogens are connected as lateral substituents to the polymer, in most cases via a flexible "spacer". Also, combinations of main chain and side chain LCPs are known.

LCP phases are subdivided into thermotropic or lyotropic. *Lyotropic* liquid crystals are formed by macromolecules that show liquid crystalline behavior in solution. This behavior is strongly concentration dependent. *Thermotropic* liquid crystals are molecules that show liquid crystalline behavior above the melting point of their crystallites.

The degree and the nature of order vary between the different liquid crystalline phases and are further classified accordingly (see Fig. 1.16). In the *nematic* phase all mesogens point in the same direction – at least locally – but there is no positional order with respect to each other. The *cholesteric* phase is similar to a nematic one but the mesogens are twisted about an axis perpendicular to the director in a periodic manner due to their chiral nature. Because chirality of the mesogens is required, this phase is also referred to as *chiral cholesteric*. The most ordered mesophase, however, is termed *smectic*. Here, the rods are all aligned towards some director, but in addition, the individual rods arrange themselves in a layered structure. The two most common types of smectic phases are smectic A and smectic C. The former occurs when the director is perpendicular to the ordering of the layers and the latter when the director is skewed from this layer.

Like their low-molecular-weight analogs, LCPs display characteristic colored textures in polarized light. Thus, the opalescence of lyotropic polyamides in solution can be viewed with the naked eye. For thermotropic polymers the birefringence of the anisotropic melt is often easily detectable in a polarizing microscope.

On the other hand, there are some essential differences between low- and high-molecular-weight LC compounds. For example, the latter retain their oriented morphologies in the solid state when rapidly cooled from their mesophase. Low-molecular-weight substances, on the other hand, return to their initial crystalline state on account of their higher mobility. Also, in lyotropic phases, the LCPs become orientated with each other, and randomly oriented domains of highly oriented chains develop. Under shear forces, these randomly oriented domains become fully oriented in the direction of the shear and emerge with near-perfect molecular orientation. High-modulus fibers can thus be produced, for example. Other typical properties of LCPs are low stretch or elongation, resistance to cutting, excellent thermal properties (heat resistance), high strength, and high impact resistance. Polymers with the mesogen located in the main chain yield unusually good mechanical properties in their chain direction, e.g., stiffness and strength. They are employed as high modulus fibers (lyotropic polyamides, see Example 4.12) or in the so-called self-reinforcing polymers (thermotropic polyesters, Example 4.6). For thermotropic polymers with mesogenic backbone chains, the relatively low viscosity of the nematic phase is of special practical interest. It affords the following peculiarities when a melt of these polymers is subjected to an extensional flow by, e.g., pressing it through a die: Not only the microscopic orientation is retained, but, in addition, there is a preferred transregional orientation of the respective domains in the direction of flow. One advantage is that this macroscopic order of the total system results in a considerable melt viscosity depression which means ease in processability. Moreover, the macroscopic orientation of the total system is largely retained on rapid cooling of the melt, i.e., also in the solid state. There will only be a minor volume dilatation and consequently a high precision of the molded (or extruded) part. Finally, the specific state of order is the reason for a number of unusual solid state properties. The physical properties that make LCPs unique include good barrier properties, high thermal stability, high modulus and strength, dimensional stability, and high chemical and solvent resistance.

If the mesogens are pendant to the polymer backbone, materials are obtained with special magnetic, electrical, and optical properties. They provide for nonlinear optics (NLOs) applications in numerous optoelectronic elements.

Bibliography

Textbooks

Eisele U (1990) Introduction to polymer physics. Springer, Berlin/Heidelberg/New York
Elias HG (2002a) An introduction to polymer science. Wiley, Weinheim
Elias HG (2002b) Makromoleküle, vol 1–4. Wiley-VCH, Weinheim
Flory PJ (1953) Principles of polymer chemistry. Cornell University Press, Ithaca
Furukawa Y (1998) Inventing polymer science Staudinger, Carothers and the emergence of macromolecular chemistry. University of Pennsylvania Press, Philadelphia
Morawetz H (1985) Polymers, the origins and growth of science. Willey, New York

Staudinger M, Hopf H, Kern W (eds) (1969–1976) Das wissenschaftliche Werk von Hermann Staudinger. Gesammelte Arbeiten nach Sachgebieten geordnet, vol 7. Hüthig Wepf Verlag, Basel/Heidelberg

Strobl GR (1999) The physics of polymers, 2nd edn. Springer, Berlin/Heidelberg/New York

Monographs and Handbooks

Allen G, Bevington J (eds) (1989) Comprehensive polymer science, vol 1–7. Pergamon, Oxford

Brandrup J, Immergut EH, Grulke EA (eds) (1999) Polymer handbook, 4th edn. Wiley, New York

Braun D (2003) Simple methods for identification of plastics, 4th edn. Hanser, Munich/Vienna

Collings PJ (ed) (1997) Handbook of liquid crystals. Oxford University Press, Oxford

Dautzenberg H, Jaeger W, Kötz J, Philipp B, Seidel C, Stscherbin D (1994) Polyelectrolytes – formation, characterization and application. Hanser, Munich

Flory PJ (1969) Statistical mechanics of chain molecules. Interscience, New York

Gelin BR (1994) Molecular modeling of polymer structures and properties. Hanser, Munich/Vienna

Glenz W (1993) A glossary of plastics terminology in five languages, 2nd edn. Hanser, Munich

Kricheldorf HR (ed) (1992) Handbook of polymer synthesis, parts A and B. Marcel Dekker, New York/Basel/Hong Kong

Lal J, Mark JE (eds) (1986) Advances in elastomers and rubber elasticity. Plenum, New York

Legge NR, Holden G, Schroeder HE (eds) (1987) Thermoplastic elastomers. Hanser, Munich

Lewis RJ (1999) Sax's dangerous properties of industrial materials, 10th edn. Wiley, New York

Mark JE, Bikales NM, Overberger CG, Menges G (eds) (2004) Encyclopedia of polymer science and engineering, vol 1–12. Wiley, New York

Northolt MG, Sikkema DJ (1990) Lyotropic main chain liquid crystal polymers. Adv Polym Sci 98:119

Rempp P, Merrill EW (1991) Polymer synthesis. Hüthig und Wepff Verlag, Basel/Heidelberg/New York

Rohn CL (1995) Analytical polymer rheology – structure, processing, property relationship. Hanser, Munich

Schlüter AD (2000) Synthesis of polymers. Wiley-VCH, Weinheim

van Krevelen DW (1990) Properties of polymers, their estimation and correlation with chemical structure, 3rd edn. Elsevier, Amsterdam

Ward IM (1983) Mechanical properties of solid polymers, 2nd edn. Wiley, New York

Woodward AE (1995) Understanding polymer morphology. Hanser, Munich/Vienna

Laboratory Manuals

Houben-Weyl (1987) Methoden der Organischen Chemie, vol E20. Makromolekulare Stoffe, Teil 1–3, Thieme, Stuttgart

Moore JA (ed) (1978) Polymer syntheses. Wiley, New York

Sandler SR, Karo W (eds) (1996) Polymer syntheses, vol 1–3, 2nd edn. Academic, New York

Sandler SR, Karo W (1998) Sourcebook of advanced polymers laboratory preparation. Academic, San Diego

Sandler SR, Karo W, Bonesteel EM, Pearce EM (1998) Polymer synthesis and characterization – a laboratory manual. Academic, San Diego

Sorenson WR, Sweeny F, Campbell TW (2001) Preparative methods of polymer chemistry. Wiley-VCH, Weinheim

Publications About Nomenclature

J. Kahovec, P. Kratochvil, A.D. Jenkins, I. Mita, I.M. Papisov, L.H. Sperling, R.F.T. Stepto (1997)
Source based names of non-linear polymers. Pure Appl Chem 69:2511

R.B. Fox, N.M. Bikales, K. Hatada, J. Kahovec (1994) Structure-based nomenclature for irregular
single-strand organic polymers (Recommendations 1993). Pure Appl Chem 66:873

Bareiss RE, Kahovec J, Kratochvil P (1994) Graphic representation (chemical formulae) of
macromolecules (recommendations 1994). Pure Appl Chem 66:2469

Jenkins AD, Loening KL (1989) Nomenclature. In: Allen G, Bevington JC, Booth C, Price C (eds)
Comprehensive polymer science, vol 1. Pergamon, Oxford, p 13

Jones RG, Jaroslav Kahovec, Robert Stepto, Wilks ES, Michael Hess (1994) Basic classifications
and definitions of polymerization reactions. Pure Appl Chem 66:2483

Metanomski WV (1999) Nomenclature of polymers. In: Brandrup J, Immergut EH, Grulke EA
(eds) Polymer handbook, 4th edn. Wiley, New York

Metanomski WV, Bareiss RE, Kahovec J, Loening KL (1993) Nomenclature of regular double-
strand (ladder and spiro) organic polymers (recommendations 1993). Pure Appl Chem 65:1561

Journals and Periodicals

In view of more than 150 journals and periodicals world wide which publish original papers on polymers, we refrained from listing these even in extracts. We recommend the internet – and Chemical Abstracts (CAS) – services.

Methods and Techniques for Synthesis, Characterization, Processing, and Modification of Polymers

2

In this chapter, the fundamentals and the most common methods and techniques for the synthesis, processing, characterization, and modification of macromolecular materials are described briefly, as an introduction to the special Chaps. 3, 4, and 5. The main emphasis is on the description of methods and techniques used in laboratories, but some examples from industrial practice are also mentioned.

2.1 Methods for Synthesis of Polymers

The formation of synthetic polymers is a process which occurs via chemical connection of many hundreds up to many thousands of monomer molecules. As a result, macromolecular chains are formed. They are, in general, linear, but can be branched, hyperbranched, or crosslinked as well. However, depending on the number of different monomers and how they are connected, homo- or one of the various kinds of copolymers can result. The chemical process of chain formation may be subdivided roughly into two classes, depending on whether it proceeds as a chain-growth or as a step-growth reaction.

2.1.1 Chain Growth Polymerizations

Chain growth polymerizations (also called addition polymerizations) are characterized by the occurrence of activated species (initiators)/active centers. They add one monomer molecule after the other in a way that at the terminus of each new species formed by a monomer addition step an activated center is created which again is able to add the next monomer molecule. Such species are formed from compounds which create radicals via homolytic bond scission, from metal complexes, or from ionic (or at least highly polarized) molecules in the initiating steps (2.1) and (2.2). From there the chain growth can start as a cascade reaction

D. Braun et al., *Polymer Synthesis: Theory and Practice*,
DOI 10.1007/978-3-642-28980-4_2, © Springer-Verlag Berlin Heidelberg 2013

(propagation; 2.3) upon manifold repetition of the monomer addition and reestab-lishment of the active center at the end of the respective new product:

$$I \longrightarrow I^* \tag{2.1}$$

$$I^* + M \longrightarrow I-M^* \tag{2.2}$$

$$I-M^* + n\,M \longrightarrow I\text{---}\!\!\left[\text{M}\right]_{n}\!\!\text{---}M^* \;(=P^*) \tag{2.3}$$

Finally, growth of an individual macromolecule is arrested in either a termina-tion or a transfer step (2.4) or (2.5)

$$I\text{---}\!\!\left[\text{M}\right]_{n}\!\!\text{---}M^* + S \longrightarrow I\text{---}\!\!\left[\text{M}\right]_{n}\!\!\text{---}M \;(+\,S') \tag{2.4}$$

$$I\text{---}\!\!\left[\text{M}\right]_{n}\!\!\text{---}M^* + T \longrightarrow I\text{---}\!\!\left[\text{M}\right]_{n}\!\!\text{---}M + T^* \searrow_{+\,n\,M} T\text{---}\!\!\left[\text{M}\right]_{n-1}\!\!\text{---}M^* \tag{2.5}$$

S: chain-terminating agent (stopper)
T: chain-transfer agent (solvent, monomer, initiator, polymer, regulator, ...)

While termination leads to the irreversible disappearance of an active center, chain transfer results in the growth of a second chain while the first one is terminated. Here, the active center is transferred to another molecule (solvent, initiator, monomer, ...) where it is able to initiate further chain growth. The resulting "dead" polymer, on the other hand, can continue its growth only when activated in a subsequent transfer step. Because this re-activation in general does not occur at the terminal monomer unit but somewhere in the chain, branched or cross-linked products will result:

$$I\text{---}\!\!\left[\text{M}\right]_{x}\!\!\text{---}M\text{---}M\!\left[\text{M}\right]_{y} \xrightarrow[-Z]{+Z^*} I\text{---}\!\!\left[\text{M}\right]_{x}\!\!\text{---}M^*\text{---}M\!\left[\text{M}\right]_{y} \xrightarrow{+\,n\,M} I\text{---}\!\!\left[\text{M}\right]_{x}\!\!\overset{\overset{\textstyle M\left[\text{M}\right]_{n-1}\!M^*}{|}}{M}\text{---}M\!\left[\text{M}\right]_{y} \tag{2.6}$$

In conclusion, chain-growth polymerizations are typical chain-reactions involv-ing a start-up step (initiation) followed by many identical chain-reaction steps (propagation) – stimulated by the product of the first start-up reactions. Transfer processes may continue until, finally, the active center disappears in a termination step.

Monomers appropriate for chain-growth polymerizations either contain double or triple bonds or are cyclic, having a sufficiently high ring strain

$$P^* + \ \overset{\diagup}{\diagdown}_R \ \longrightarrow \ P\overset{\diagup\diagdown}{\underset{R}{\ }}^* \tag{2.7}$$

$$P^* + \bigcirc \ \longrightarrow \ P\text{—}\wedge\wedge\wedge\wedge\wedge^* \tag{2.8}$$

Depending on the nature of the active center, chain-growth reactions are subdivided into radicalic, ionic (anionic, cationic), or transition-metal mediated (coordinative, insertion) polymerizations. Accordingly, they can be induced by different initiators or catalysts. Whether a monomer polymerizes via any of these chain-growth reactions – radical, ionic, coordinative – depends on its constitution and substitution pattern. Also, external parameters like solvent, temperature, and pressure may also have an effect. Monomers able to grow in chain-growth polymerizations are listed in Table 2.2 of Sect. 2.1.4.

2.1.2 Step Growth Polymerizations

In step growth reactions, on the other hand, neither are specific activated centers present to force the connection of the monomers, nor does the process occur as a cascade reaction. Instead, the monomers are tied together in discreet, independent steps via conventional organic reactions such as ester-, ether-, amide-, or urethane formation. Depending on whether small molecules are set free in the connection step, one distinguishes between polycondensations Eq. 2.9 and polyadditions Eq. 2.10

$$HOOC\text{—}R\text{—}COOH + HO\text{—}R'\text{—}OH \ \xrightarrow[-H_2O]{} \ HOOC\text{—}R\text{—}\underset{O}{\overset{\parallel}{C}}\text{—}O\text{—}R'\text{—}OH$$

$$\downarrow\downarrow \ -nH_2O$$

$$\left[\underset{O}{\overset{\parallel}{C}}\text{—}R\text{—}\underset{O}{\overset{\parallel}{C}}\text{—}O\text{—}R'\text{—}O\right]_n \tag{2.9}$$

$$OCN\text{—}R\text{—}NCO + HO\text{—}R'\text{—}OH \ \longrightarrow \ OCN\text{—}R\text{—}\underset{H}{\overset{\ }{N}}\text{—}\underset{O}{\overset{\parallel}{C}}\text{—}O\text{—}R'\text{—}OH$$

$$\downarrow\downarrow$$

$$\left[\underset{O}{\overset{\parallel}{C}}\text{—}\underset{H}{\overset{\ }{N}}\text{—}R\text{—}\underset{H}{\overset{\ }{N}}\text{—}\underset{O}{\overset{\parallel}{C}}\text{—}O\text{—}R'\text{—}O\right]_n \tag{2.10}$$

Table 2.1 Criteria for chain and stepwise reactions in polymer synthesis

Chain growth reactions	Step growth reactions
Start of the reaction (chain initiation) generally requires initiators or catalysts	Reaction often proceeds without the need for catalysts
Only active species (e.g., macro radicals or macro ions) can add further monomer molecules in the propagation process	Both monomer and polymer molecules with suitable functional end groups can react
Activation energy for chain initiation is higher than that for propagation	Activation energy is about the same for each reaction step
Monomer concentration decreases with reaction time	Monomer molecules disappear very quickly; more than 99% of monomer molecules have already reacted when the degree of polymerization is 10
Macromolecules are formed from the very beginning of the reaction	Monomer molecules first give oligomers; high polymer is formed only towards the end of the reaction
The average molecular weight of the poly-mer normally changes little with reaction time (exception: living polymerization)	The average molecular weight increases steadily with reaction time; long reaction times are usually necessary to produce high molecular weights

Evidently, monofunctional molecules cannot result in polymer chains via step growth polymerizations. Instead, each monomer molecule as well as all intermediates must possess two functional groups. When more than two reactive groups are present in a monomer, branched or crosslinked products will result. Moreover, step growth polymerizations are categorized according to how the functional groups are assigned to the monomers. When each monomer bears two identical functional units, the process is called AABB-type polycondensation/polyaddition. Here, mixtures of at least two different types of monomers are required, bearing the complementary functional groups (see Eqs. 2.9 and 2.10). If, on the other hand, each monomer molecule bears the two complementary functional groups required for step growth polymerization, the process is called AB-type polycondensation/polyaddition (see Eqs. 2.11 and 2.12).

$$n \; \text{HO-R-COOH} \xrightarrow[-n\,H_2O]{} \left[\text{O-R-}\underset{\underset{O}{\|}}{\text{C}} \right]_n \tag{2.11}$$

$$n \; H_2N\text{-R-COOH} \xrightarrow[-n\,H_2O]{} \left[\overset{\overset{H}{|}}{\text{N}}\text{-R-}\underset{\underset{O}{\|}}{\text{C}} \right]_n \tag{2.12}$$

As a consequence of the lack of special active centers, the chain formation in step growth polymerizations occurs via a sequence of accidental and independent reaction events. It proceeds via dimers, short and longer oligomers until, finally, at conversions higher than 99% long chains are formed which are called condensation polymers (polycondensates) or addition polymers, respectively. Apart from high degrees of conversion also a very precise 1:1 equivalence of the complementary functional groups is essential to achieve very high molar masses. A summary of the most significant differences between step-growth and chain-growth polymerizations are given in Table 2.1.

2.1.3   Modification of Polymers

The third possibility for synthesizing polymeric substances is the modification of existant natural or synthetic macromolecules (see Chap. 5). These processes can either be chemical or physical. Chemical modifications are reactions on macromolecules without degradation of the main chain (macromolecular substitution routes, "polymer-analogous reactions") like, for example, hydrolysis, esterification, and etherification of side groups. Physical modifications include addition of stabilizers as well as the addition of (inorganic) reinforcing agents and also the mixing of different polymers (polymer blends).

2.1.4   Polymer Recipes Reference List

The present book contains about 110 detailed polymer recipes. Yet, for quite a number of common polymers recipes are missing. The following Tables 2.2, 2.3, 2.4, and 2.5 attempt to fill this gap. The information provided includes the name of the monomer, the formula of the basic unit of the polymer, and references for detailed recipes. Table 2.2 lists polymers prepared by chain growth polymerization, Tables 2.3 and 2.4 those prepared by step growth polymerization, and Table 2.5 contains polymers obtained by chemical modifications of (natural) macromolecules.

2.2   Techniques for Manufacturing of Polymers

The processes for manufacturing macromolecules can be divided into three different categories (Scheme 2.1).

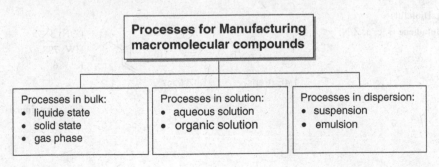

Scheme 2.1

Table 2.2 Synthesis of macromolecules by chain growth polymerization. *Note:* The citation of literature in the right-hand column of the following tables should be understood as follows: the number(s) in the first line(s) refer(s) to this book. For example, 3.2 means experiment number 2 in Chap. 3. The following line refers to: Houben Weyl, Methoden der Organischen Chemie, Vol. E20, Makromolekulare Stoffe Teil 1–3, Thieme Verlag, Stuttgart, New York, 1987 (abbreviated: HW). For example, the citation HW:135 refers to an experimental procedure described on page 135 thereof. The bottom line refers to H. R. Kricheldorf, Handbook of Polymer Synthesis, Vols. 1 and 2, Dekker, New York, 1992 (abbreviated: K). For example, K:34 refers to further references given on page 34 of that book

Monomer[a]	Basic unit of the polymer	Preparation examples, literature
1. Olefins		
Ethylene (r, c, Z/N)		(Z/N) 3–28
		(r+c) HW: 689
		K: 3
Propylene (c, Z/N)		(Z/N) 3–29/3–32
		(c) HW: 769
		K: 26
Isobutylene (c)		(c) 3–16
		HW: 769
		K: 53
Styrene (r, c, a, Z/N)		(r) 3–1/3–2/3–6/ 3–7
		(Z/N) 3–30
		HW: 762
		K: 77
α-Methylstyrene (r, c, a) (r only copolymerization)		(c) 3–18
		(a) 3–19
		HW: 1012
		K: 90
Tetrafluoroethylene (r)		HW: 1028
		K: 197
2. Diolefins		
Butadiene (r, c, a, Z/N)		(Z/N) 3–31
		HW: 798
		K: 386
	1,4E (trans) 1,4Z (cis) 1,2	

(continued)

Table 2.2   (continued)

Monomer[a]	Basic unit of the polymer	Preparation examples, literature
2-Methyl-1,3-butadiene (isoprene) (r, c, a, ZN)	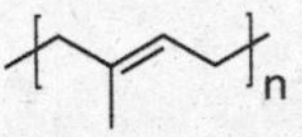 1,4E (trans) 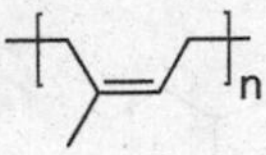 1,4Z (cis) 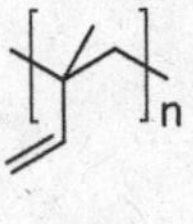 1,2 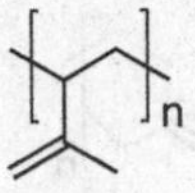 3,4	(r) 3–12 (a) 3–21 HW: 822 K: 393
2-Chloro-1,3-butadiene (chloroprene) (r)	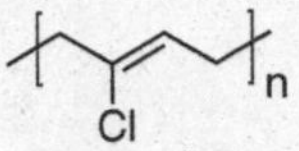 1,4Z (cis)	HW: 842 K: 396

3. Vinyl derivatives

Vinyl chloride (r, a)	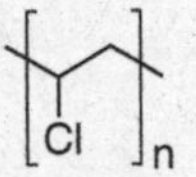	(a) HW: 1042 K: 172
Vinyl acetate (r)	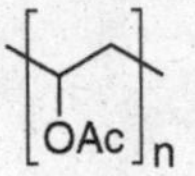	3–3/3–4 HW: 1115 K: 164
Isobutyl vinyl ether (r, c)	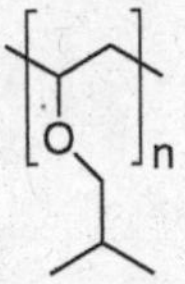	(c) 3–17 HW: 1072 K: 145
Vinylsulfonic acid (r)	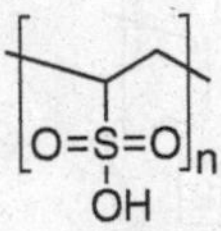	HW: 1259
Methyl vinyl ketone (r, c)	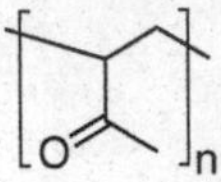	HW: 1138 K: 360
N-Vinylpyrrolidone (r, c)	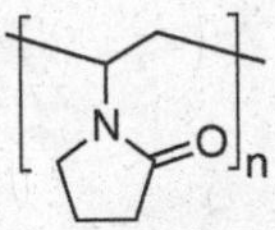	HW: 1267 K: 114

(continued)

Table 2.2 (continued)

Monomer[a]	Basic unit of the polymer	Preparation examples, literature
4. Acryl compounds		
Acrylic acid (r)	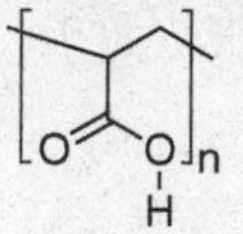	HW: 1148 K: 268
Methacrylic acid (r)	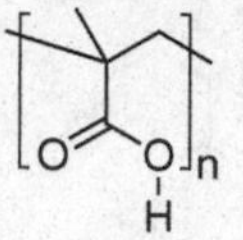	3–5 HW: 1148 K: 268
Methyl acrylate (r, a)	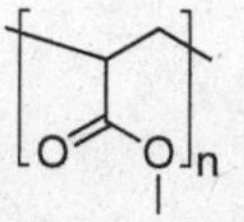	HW: 1144 K: 223
Methyl methacrylate (r, a)	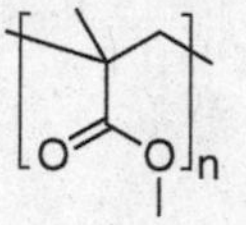	(r) 3–8 (a) 3–20 HW: 1144 K: 223
Acrylonitrile (r, a)	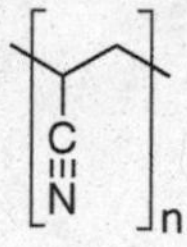	(r) 3–11 (a) HW: 1192 K: 280
Acrylamide (r, a)	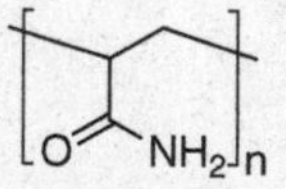	(r) 3–9 (a) HW: 1176 K: 256
Acrolein (1: r, c; 2: a)	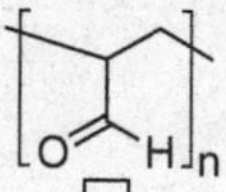 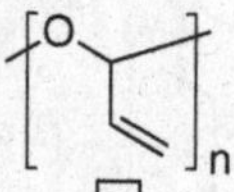	HW: 1127 K: 337
5. Allyl compounds		
Allyl alcohol (r)	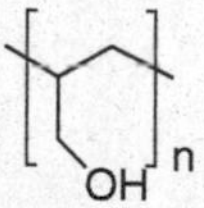	HW: 2013
Allyl chloride (r, c)	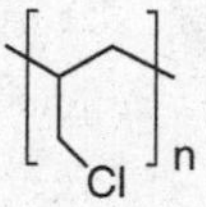	HW: 1013

(continued)

Table 2.2 (continued)

Monomer[a]	Basic unit of the polymer	Preparation examples, literature
Allyl acetate (r)	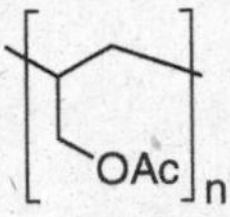 [structure]	HW: 1013

6. Monomers with C=O bonds

Formaldehyde (c, a)	[structure]	(a) 3–22
		(c) HW: 1388
		K: 618
Acetaldehyde (c, a)	[structure]	HW: 1380
		K: 632

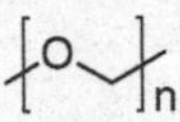

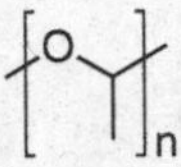

7. Ring-opening polymerization

(a) Cyclic ethers		K: 481
Ethylene oxide (c, a)	[structure]	HW: 1367
		K: 484
Propylene oxide (c, a)	[structure]	HW: 1367
		K: 501
Epichlorohydrin (c, a)	[structure]	HW: 1367
		K: 503
Tetrahydrofuran (c)	[structure]	3–23
		HW: 1367
		K: 521
Cyclosiloxanes (e.g., 2,4,6-hexamethyl-cyclotrisiloxane) (c, a)	[structure]	4–21/4–22
		HW: 2219
		K: 1144
(b) Cyclic acetals		K: 630
1,3,5-Trioxane (c)	[structure]	3–24
		HW: 1388
		K: 633
(c) Cyclic amides, oxazolines		K: 807, 876
β-Propiolactam (a)	[structure]	HW: 1505
		K: 890

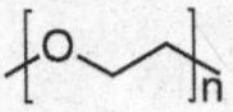

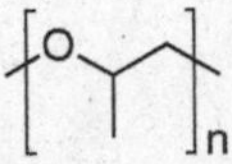

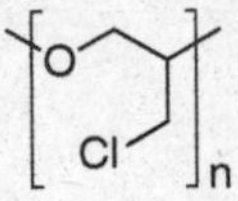

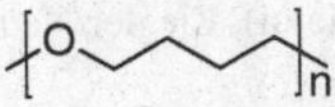

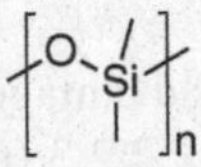

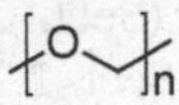

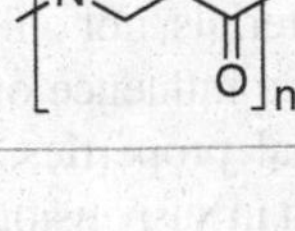

(continued)

Table 2.2   (continued)

Monomer[a]	Basic unit of the polymer	Preparation examples, literature
γ-Butyrolactam (a)		HW: 1505
		K: 891
ε-Caprolactam (c, a)		(a) 3–26
		HW: 1505
		K: 892
2-Alkyloxazolines (c)		3–27
		K: 747
(d) Cyclic esters		K: 652
β-Propiolactone (c, a)		HW: 448
		K: 653
δ-Valerolactone (c, a)		HW: 448
		K: 660
ε-Caprolactone (c, a)		HW: 448
		K: 661

[a]Polymerizability: radical (r), cationic (c), anionic (a), Ziegler/Natta (Z/N)

The basic characteristics as well as some advantages and disadvantages are illustrated in Sects. 2.2.2 (polyreactions in bulk), 2.2.3 (polyreactions in solution), and 2.2.4 (polyreactions in dispersion). Prior to this some special features that must be considered in the preparation of polymers (Sect. 2.2.1) and some suitable techniques for the preparation in the laboratory (Sect. 2.2.5) are described.

2.2.1   Particularities in the Preparation of Polymers

Especially in the case of macromolecular materials, not only the method of synthesis, but also the manufacturing process has a large influence on size and structure of the molecules, and consequently on the physical properties. For example, poly(vinyl chloride), produced by radical polymerization in suspension, differs in some practical properties from PVC obtained by radical polymerization in bulk. Changes in the

Table 2.3 Synthesis of macromolecules by step growth polymerization (polycondensation)

Monomer 1	Monomer 2	Polymer	Basic unit of the polymer	Preparation examples, literature
ω-Hydroxy-carboxylic acids		Linear polyesters		HW: 555 · K: 646
Diols	Dicarboxylic acids or derivatives	Linear polyesters		4-1 to 4-7 HW: 555 K: 649
Tri- or polyols	Di- or poly-carboxylic acids or their derivatives	Branched or crosslinked polyesters		HW: 555 K: 674, 677
ω-Amino carboxylic acids		Linear polyamides		4-9 HW: 1497 K: 810
Diamines	Dicarboxylic acids or derivatives	Linear polyamides		4-10 to 4-13 HW: 1497 K: 818, 829
Diamines	Bis-chloro-carboxylic acid esters	Linear polyurethanes		HW: 1561 K: 687
Diamines	Phosgene	Linear polyureas		HW: 1721 K: 715
Dicarboxylic acids		Poly-anhydrides		HW: 1400 K: 1668

(continued)

Table 2.3 (continued)

Monomer 1	Monomer 2	Polymer	Basic unit of the polymer	Preparation examples, literature
Dimethyl-silane-Diol		Polysiloxanes		HW: 2219 K: 1149
α,ω-Dihalo-alkanes	Sodium polysulfide	Polyalkylene sulfides		HW: 1458 K: 992
Phenols	Formaldehyde	Phenol/formaldehyde condensate		4-14 HW: 1764 K: 1484
Urea	Formaldehyde	Urea/formaldehyde condensate		4-15 HW: 1811 K: 1489
Melamine	Formaldehyde	Melamine/formaldehyde condensate		4-16 HW: 1811
2,6-Dimethyl-phenol		Poly(dimethyl phenylene ether)		4-17 HW: 1380 K: 546

			Structure	
Dichloro-benzene	Sodium sulfide	Poly(pheny-lene sulfide)		HW: 1463 K: 1397
Bisphenol	4,4'-Dichloro-diphenyl sulfone	Poly(arylene ether sulfone)		4-18 HW: 1475 K: 1038
Bisphenol	4,4'-Difluorobenzo-phenone	Poly(arylene ether ether ketone)		4-19
Bisanhydride	Aromatic diamines	Polyimides		4-20 HW: 2182 K: 943
Aromatic diphenyl carboxylate	Biphenyl-tetramine	Poly(benzimid azoles)		HW: 2186 K: 1247

Table 2.4 Synthesis of macromolecules by step growth addition polymerization (polyaddition)

Monomer 1	Monomer 2	Polymer	Basic unit of the polymer	Preparation examples, literature
Diols, e.g. polyether diols	Diisocyanates (X = arylene or alkylene)	Polyurethanes		4–23 HW: 1561 K: 685
Diamines	Diisocyanates (X = arylene or alkylene)	Polyureas		HW: 1721 K: 713
Di- or poly-epoxides	Amines or anhydrides	Epoxide resins		4–24 HW: 1891 K: 1496
Non-conjugated dienes	Dithioles	Polythioethers		HW: 1458 K: 1011

Table 2.5 Modification of macromolecules via polymer analogous reactions

Educt	Chemical reaction	Product	Preparation examples, literature
Cellulose	Esterification	Cellulose esters	5–5/5–7
			HW: 2042
			K: 1528
Cellulose	Etherification	Cellulose ethers	5–6
			HW: 2042
			K: 1518
Cellulose	Etherification with chloroacetic acid (carboxymethylation)	Carboxymethyl cellulose	5–6
			HW: 2042
			K: 1521
Poly(vinyl acetate)	Saponification	Poly(vinyl alcohol)	5–1
			HW: 2042
			K: 1504
Poly(vinyl alcohol)	Acetalization	Poly(vinyl butyral)	5–2
			HW: 2042
			K: 1506
Crosslinked polystyrene	Sulfonation	Cation exchanger	5–9
			HW: 1944
Polysulfone	Sulfonation	Sulfonated polysulfone	HW: 1944

properties of the products can be realized even within the same manufacturing process by changing the reaction conditions. Thus, in radical polymerization of styrene in bulk, polymers with the same molecular weight, but different distributions of molecular weight can be obtained solely by use of different reaction conditions. This is noticeable in the rheological characteristics of the molten polymer (processability) and in some properties of the solid material.

The reaction kinetics of polyreactions have their own characteristics. *Radical* and *ionic polymerizations* proceed as chain reactions, whereby the stationary concentration of the active ends of the chain, i.e., of the growing macro radicals or macro ions, is very small. In radical polymerizations, this concentration is about 10^{-8} mol/l. For this reason, chain polymerizations are remarkably sensitive towards impurities that can react with radicals or ions. Reactions of such impurities with the growing ends of the chains can cause a deceleration or interruption of the polymerization or a decrease of the molecular weight by chain transfer. A reaction with the initiator or catalyst can cause inhibition as well as acceleration (redox catalysis, cocatalysis). In addition, the steric arrangement of the monomers (*cis-trans* isomerism, tacticity) is strongly influenced by small amounts of certain substances. In *condensation polymerizations* and *stepwise addition polymerizations* of bifunctional monomers, conversion rates of more than 99% must be reached in order to obtain high molecular weights; this is only possible when monofunctional compounds are largely excluded. Therefore, the requirements in purity of the monomers and auxiliary materials are very high. In many cases, the tolerance limit for certain impurities is in the range of a few ppm. Methods for laboratory use which meet these requirements are described in Sect. 2.2.5.

One more feature is that polymer melts as well as polymer solutions have a very high viscosity due to the size and shape of the macromolecules. Thus, in polymerizations in the melt or in homogeneous solution, the viscosity of the reaction mixture increases, often by many orders of magnitude. Beside an impairment of mass transfer, high viscosity also causes difficulties in pumping, stirring, and mixing. In principle, the increase of the viscosity can be compensated by increasing the temperature, but this is strictly limited by the thermal stability of some polymers and, especially, by kinetic factors. An increase of the temperature in particular causes a strong decrease in molecular mass and an increase in side reactions (chain transfer, crosslinking) in many polymerizations. Finally, the high viscosity of the reaction mixture also substantially hinders heat transfer. Due to the fact that nearly all chain growth polymerizations and step growth polymerizations are strongly exothermic, the rapid removal of large amounts of heat creates severe difficulties, especially if one works in bulk or solution. Furthermore, polymers are poor heat conductors, thus the heat transfer coefficient decreases with increasing concentration of polymer. In addition, the heat transfer is reduced by the laminar boundary layer at the inner wall of the reactor, which is enlarged with increasing viscosity. In this respect, polymerizations in dispersion are distinctly more favorable because the reaction mixture retains a low viscosity and therefore turbulent stirring is possible and relatively high heat transfer coefficients can be reached. If the polyreaction is performed in aqueous medium, the high specific heat capacity of water additionally contributes to the heat dissipation. One more characteristic difference in comparison to the preparation of low-molecular-weight substances is the limitation in subsequent purification of the polymers. Thus, the complete removal of low-molecular-weight substances, such as solvents, oligomers, initiator residues, catalysts, and monomers can be very difficult. Because of this, purification is time- and cost-intensive in the laboratory and especially on an industrial scale. Special apparatus is often required (for laboratory techniques, see Sect. 2.2.5). Industrially, it is nearly impossible to separate the resulting polymer mixture into macromolecules of the same molecular size, structure, and composition, but for analytical use, some methods are available (see Sect. 2.3.3.4).

2.2.2   Polyreactions in Bulk

Polyreactions in bulk are carried out without solvents or diluents. In this way, high molecular weights can be obtained frequently with high rates of reaction. The resulting products are very pure, because only monomers and, if necessary, initiators and catalysts are added. Polyreactions in bulk are also advantageous for economical and ecological reasons because recycling and purification of solvents or dispersants as well as the disposal of liquid waste (as in the case of aqueous suspension or emulsion polymerizations) are not necessary. On the other hand, considerable difficulties in running the process are often caused by the removal of the heat of polymerization and the handling of the viscous reaction mixtures. Furthermore, polyreactions in highly viscous media show intensified side reactions such as chain transfer, thus influencing the properties of the resulting product.

Polyreactions in bulk are divided into homogeneous and heterogeneous reactions, depending on whether the polymer remains dissolved in its monomer or not or, respectively, whether the polyreaction is performed above or below the softening temperature of the polymer.

2.2.2.1 Homogeneous Polyreactions in Bulk

In homogeneous polyreactions in bulk, the polymer remains dissolved in the monomer. In some cases, this can only be achieved by carrying out the polyreaction above the softening temperature of the polymer. Chain polymerizations, condensation polymerizations, and stepwise addition polymerizations can be performed homogeneously in bulk. The number of *chain polymerizations* that can be performed in a homogeneous phase is very limited. On the one hand, only a few monomers are able to dissolve their own polymers (e.g., styrene and methyl methacrylate), on the other hand, temperatures above the softening point often cannot be used for kinetic and thermodynamic reasons (ceiling temperature, low molecular weights, increase of side reactions).

In contrast, *condensation polymerizations* and *stepwise addition polymerizations* in the homogeneous phase are the most frequently performed processes on an industrial scale. Because the resulting polymers usually have high softening points and the fact that the molecular weight – and therefore the melt viscosity – increases with progressing reaction, temperatures of above 250°C are often necessary, at least at the end of the reaction. Even so, the removal of the highly volatile reaction products (water, alcohol), which is necessary in order to obtain high molecular masses, is difficult. Therefore, evacuation and intensive mixing, which create larger surfaces and short diffusion paths, are necessary. Similarly, the conditions for high conversion rates and the equivalency of end groups in condensation polymerization and stepwise addition polymerization must be adjusted by suitable technical measures. Moreover, high thermal stabilities of the starting materials and the resultant polymers are necessary on account of the long reaction times at high temperatures. Thus, many high-melting polyamides are not accessible by this process. Hence, they must be prepared in solution or by interfacial condensation polymerization.

2.2.2.2 Heterogeneous Polyreactions in Bulk

In heterogeneous polymerizations in bulk, the formed polymer is insoluble in its monomer and the polyreaction is performed below the softening point of the polymer. On an industrial scale, this type of process is especially utilized for chain polymerizations, for example, the radical polymerization of liquid vinyl chloride, the polymerization of liquid propylene with Ziegler-Natta or with metallocene catalysts, and the polymerization of molten trioxane.

Besides in the liquid phase, some polyreactions are also performed in the *solid state*, for example, the polymerization of acrylamide or trioxane (see Example 3.24). The so-called post condensation, for example, in the case of polyesters (see Example 4.3), also proceeds in the solid phase. Finally, ring closure reactions on polymers with reactive heterocyclic rings in the main chain (e.g., polyimides, see Example 4.20) are also performed in the solid state.

A third possible variation of heterogeneous polyreactions in bulk is the polymerization in the *gaseous phase* or *gas phase*. The term gaseous phase is used when the reaction is performed with gaseous monomers. The polymerization itself does not occur in the gaseous phase, rather the gaseous monomer is adsorbed on the solid catalyst particles and polymerized in that state. With proceeding polymerization, the catalyst particles are encapsulated by the solid polymer layer. From there on, the monomer must diffuse from the gaseous phase through the polymer cover to the catalytically active centers. The reaction medium is now a gas/solid dispersion of a solid polymer in its gaseous monomer. Heterogeneous bulk polymerizations often show deviations in their kinetics in comparison to polymerizations in a homogeneous phase. Similar considerations hold true for processes on a technical scale due to lower viscosities and differences in material transport.

2.2.3 Polyreactions in Solution

There are two different ways for carrying out polyreactions in a solvent. When both the monomer and the resulting polymer are soluble in the solvent, one speaks of a homogeneous solution polymerization; on the other hand, if the polymer precipitates during the course of the reaction, it is called precipitation polymerization. By addition of a solvent, different effects are obtained: Basically, the viscosity of the reaction mixture is decreased in comparison to a bulk polyreaction; this facilitates heat transfer, mass transport, and handling.

In *chain polymerizations*, some additional aspects have to be considered. For example, the undesired Trommsdorff effect (gel effect, see Sect. 3.1) in radical polymerization can be completely or partially prevented by choosing the appropriate concentration and solvent. Furthermore, if solvents with a high chain transfer constant are used, the solvent molecules can undergo transfer reactions with the growing macroradicals; this reduces the degree of polymerization. In addition, terminal groups can be introduced which influence the thermal and chemical stability of the polymer. In ionic polymerizations, the influence of the solvent on the course of the reaction is even more pronounced. Besides transfer reactions, reactions with the initiator or solvation of the growing macroion may also occur. In certain cases, the solvent also influences the configurational sequence of the constitutional repeating units, as in the polymerization of dienes. Thus, the solvent for polymerization must be chosen very carefully and certain solvents must be avoided. If there are no side reactions, the same kinetic rules as in homogeneous polymerization in bulk can be applied to homogeneous polymerization in solution. On the other hand, if the polymer precipitates during the course of the reaction, abnormal kinetics are observed. This is because the typical chain termination via bimolecular reaction of macroradicals is hindered, whereas chain propagation is not affected. The solid concentrations obtainable in polymerizations in solution are mostly below 20% and therefore distinctly lower in comparison to polymerizations in suspension or emulsion.

Polycondensation and *polyaddition* reactions can also be conducted in solution. Condensation of diols with dicarboxylic acids in solution is advantageous for the preparation of polyesters that do not withstand the high temperatures necessary for melt polycondensations, or when molecular weights above 30,000 g/mol are required (see Example 4.2). For this purpose, the polycondensation is carried out with an approximately 20% solution of reactants in an inert solvent. Especially preferred are hydrophobic solvents, such as toluene, xylene, or chlorobenzene, which not only form an azeotrope with the liberated water, but also prevent a back reaction by providing a protective solvation shell for the ester linkages already formed. The low viscosity of the solution compared to that resulting from melt condensation allows the water formed to be removed much more easily; hence, solution condensation can be carried out at a relatively low temperature, controlled by the boiling point of the solvent. However, in order to obtain a sufficiently high esterification rate, a catalyst, usually an acidic compound such as p-toluenesulfonic acid, is necessary. If one of the starting components (diol or dicarboxylic acid) is insoluble in the desired solvent, it is possible first to carry out a pre-condensation in the melt at about 120–150°C and then to subject the resulting low-molecular-weight polyester to further condensation in solution.

Condensation polymers, especially polyamides, can also be prepared in solution by the Schotten-Baumann reaction at low temperature (see Example 4.11). For this purpose, two rapidly reacting monomers, for example, diamine and dicarboxylic acid dichloride, are mixed together with stirring in an inert solvent; the eliminated hydrogen chloride is trapped with an acid acceptor.

One usually works at room temperature in approximately 10% solution in toluene, methylene chloride, N-methylpyrrolidone (NMP), or dry tetrahydrofuran (THF). Tertiary amines (triethylamine) or dispersed calcium hydroxide are added as acid acceptor. This procedure has the following advantages: The polycondensation is carried out at low temperature (0–40°C); it is nevertheless very fast, the reaction usually being complete after a few minutes. At low temperatures practically no side reactions occur. Disadvantages are the following: relatively large amounts of solvent must be purified and handled and large amounts of salts are formed as by-products. Condensation in solution at low temperature is, therefore, above all a laboratory method, in which these disadvantages are not so significant.

The Schotten-Baumann reaction between dicarboxylic acid dichlorides and diamines can be performed not only in organic solvents, but also, by means of a special experimental technique known as interfacial polycondensation (see Examples 4.5 and 4.11). Both variants have the advantage of short reaction times at low temperature with simple equipment.

2.2.4   Polyreactions in Dispersion

Many polyreactions, especially chain polymerizations, can also be carried out under heterogeneous conditions. In this case, the liquid monomer is dispersed by stirring to small droplets in a liquid in which it is insoluble, and polymerized in that state. During the reaction, there is a change in the aggregation state of the dispersed phase, since the macromolecules formed are solids. Thus, the original liquid/liquid dispersion (emulsion) becomes a solid/liquid dispersion (suspension). If the polymer is insoluble in the monomer, e.g., polyacrylonitrile, this transition occurs early in the reaction; if, on the other hand, it is soluble in or swollen by the monomer, e.g., polystyrene, the state of the emulsion changes only at high conversion. The term "suspension polymerization" was chosen with respect to the final state, while the term "emulsion polymerization" refers to the initial state of the system. Despite this formal similarity, the two processes differ in some essential respects, for example, in the size of the resulting polymer particles (0.1–0.5 μm in emulsion polymerization, 0.5 μm–2 mm in suspension polymerization) and in the reaction kinetics. These methods offer the following advantages: The heat of polymerization is readily dissipated because of the segmentation of the monomer in small droplets and the high heat capacity of water. The reaction mixture remains very mobile even at high conversion, because the viscosity is strongly increased only in the monomer/polymer droplets, without larger changes in the overall viscosity. Therefore, solid contents of 50% are easy to handle. Polyreactions in dispersions can be conducted relatively simply in the laboratory. In some cases, they are also used in polycondensation and polyaddition reactions (e.g., interfacial polycondensation, see Example 4.11).

Suspension and emulsion techniques provide additional means for modifying the properties of the resulting polymers, for example:
- Variation of particle size and particle size distribution.
  This influences the rheology of the latex as well as the properties of the solid film (see Example 3.46).
- Incorporation of small amounts of ionic groups.
  This has an influence on the pigment load capacity and on the stability of high solid-containing latices (see Example 3.39).
- Preparation of polymer particles with core/shell-structure.
  Crosslinked rubbery polymers that are used as impact modifiers often do not have sufficient compatibility with the hard matrix of the surrounding thermoplast. One elegant method is to cover the rubbery modifier particle with a thin layer of a polymer that is compatible with both, the rubbery core and the thermoplastic matrix.

2.2.4.1 Polyreactions in Suspension

In radical polymerizations in aqueous suspension, the liquid monomer, usually containing a dissolved water-insoluble initiator (e.g., dibenzoyl peroxide), is finely dispersed by vigorous stirring and polymerized at increased temperature. Polymerization takes place in the monomer droplets, and hence it follows the kinetic laws of bulk polymerization. Normally, suspension aids (dispersants) must be added in order to facilitate the dispersion and, even more importantly, to prevent the coagulation of the polymer particles swollen by monomer in the later stage of polymerization. Therefore, either water-soluble macromolecules (so-called protective colloids) or fine-grained and insoluble inorganic compounds, the so-called Pickering emulsifiers (e.g., calcium carbonate) need to be added in amounts of 0.1–0.5%. Combinations of both types of dispersants are also used. The volume ratio of monomer to water phase is mostly between 25:75 and 50:50. If the size of the droplets is homogeneous, the ratio of 74:26 cannot be exceeded for steric reasons. In order to obtain high conversion, initiators with different decay constants are used and the temperature is increased towards the end of the polymerization.

The polymer particles accumulating in suspension polymerization of liquid monomers show almost the same size as the original monomer droplets when no secondary aggregation occurs, as for example, with PVC. When the resulting polymer is insoluble in the monomer, the polymer precipitates in the form of irregularly shaped particles, e.g., polyacrylonitrile. On the other hand, if the polymer is soluble in the monomer, the polymer is produced in the form of regular beads, e.g., polystyrene. This particular case of suspension polymerization is referred to as bead polymerization. In many technical suspension polymerizations (for example, expandable polystyrene or PVC), one is interested in definite particle size distributions and particle morphology. Both are influenced in a complex way by many factors, for example, by the volume ratio of the phases monomer/water, the type and concentration of the dispersant, the geometry of the reactor, the shape of the stirrer, and the intensity of stirring.

On the industrial scale, suspension polymerizations are not only carried out in the aqueous phase, but also in aliphatic hydrocarbons using Ziegler-Natta catalysts, as for example, in the polymerization of ethylene and propylene (see also Sect. 3.3.1).

2.2.4.2 Polyreactions in Emulsion

Working in emulsion is essentially limited to radical polymerization in water. Similar to suspension polymerization, the basic principle is to disperse a sparingly water-soluble monomer in water and bring about polymerization in this state. There are, however, some essential differences between the two procedures:

- At least a very low water solubility of the monomer is required. Extremely hydrophobic monomers, e.g., stearyl acrylate, do not polymerize under emulsion conditions.
- The emulsification of the monomer takes place in the presence of water-soluble emulsifiers that can form micelles. At the beginning of the polymerization, the monomer is present in form of monomer droplets as well as in the micelles.
- Water-soluble initiators (potassium peroxodisulfate; redox systems) are used except for a few special cases.
- The initiation step takes place preferentially in the aqueous phase.
- Polymerization does not occur in the monomer droplets but in the micelles which thereby slowly swell to latex particles.

- The rate of polymerization (at constant initiator concentration) depends on the number of micelles and therefore on the emulsifier concentration. The rate and degree of polymerization can be increased simultaneously.
- The size of the resulting polymer particles is much smaller than in suspension polymerization.

The ingredients for an emulsion polymerization consist essentially of four components:
- Water (demineralized),
- A monomer, sparingly soluble in water,
- A water-soluble, radical-generating initiator, and
- An emulsifier.

The course of an emulsion polymerization and the properties of the resulting polymer latex are strongly affected by the emulsifier. Emulsifier molecules consist of a hydrophilic and a hydrophobic moiety. According to the electrical charge on the hydrophilic part, one distinguishes between anionic, cationic, and nonionic emulsifiers. Examples of anionic emulsifiers are the K, Na, and NH_4 salts of fatty acids, sodium dodecyl sulfate, as well as salts of alkyl-substituted benzene- or naphthalenesulfonic acids. Examples of cationic emulsifiers are quaternary ammonium salts that possess at least one hydrophobic substituent. Typical non-ionic emulsifiers are ethoxylated phenols and *block* copolymers of ethylene oxide and propylene oxide. In very dilute aqueous solution, emulsifiers behave as isolated molecules or as electrolytes. With increasing emulsifier concentration, however, an abrupt change in some physical properties of the solution occurs, e.g., surface tension, viscosity, electrical conductance, and osmotic pressure. The concentration at which these abrupt changes are observed, is called the critical micelle concentration (CMC). It has a characteristic value for each emulsifier. Below the critical micelle concentration, the emulsifier is dissolved as individual molecule, but above the CMC, the emulsifier molecules cluster to form ordered molecular aggregates, the so-called micelles, in which the hydrophobic residues are turned inwards and the hydrophilic residues are turned outwards towards the aqueous phase. These micelles have a diameter of about 3.5 nm.

The size of the micelles is significantly increased by the addition of monomer up to a diameter of 4.5–5 nm. However, the size of the monomer droplets is still very much larger than that of the micelles (diameters up to 1 μm). In emulsion polymerization, one generally uses 0.5–5 wt.% of emulsifier relative to monomer. With the usual oil-in-water emulsions, the water content varies from half to four times the amount of monomer.

The micelles are present at a concentration of about 10^{18} per ml of liquor and each micelle contains around 100 monomer molecules. In contrast, the number of monomer droplets is only about 10^{10} per ml. Thus, despite the larger volume of monomer droplets, the micelles offer a very much larger surface area. A radical formed in the aqueous phase will thus encounter a monomer-filled micelle much more often than a monomer droplet. Therefore, the polymerization takes place practically only in the micelles and not in the monomer droplets. The monomer

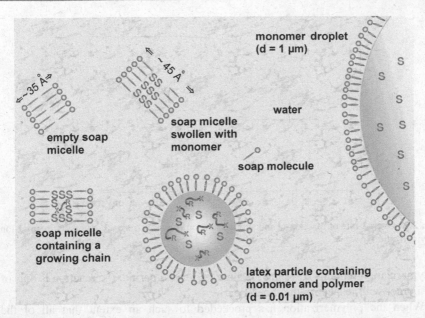

Fig. 2.1 Schematic presentation of species present in an emulsion polymerization (S = styrene monomer, R-X = growing polymer chain)

consumed in the micelles is replaced by diffusion from the monomer droplets through the aqueous phase. According to the theories of Harkins and of Smith and Ewart, the kinetic course of an emulsion polymerization is divided into three intervals: At first some of the micelles increase rapidly in size as the polymerization advances and are transformed into so-called latex particles, containing both monomer and polymer. These are still very much smaller than the monomer droplets and have an initial diameter of about 20–40 μm, corresponding to about 10^{14} particles per ml of liquor. The monomer used up is continuously replaced from the monomer droplets via diffusion through the aqueous phase. More and more emulsifier molecules are adsorbed from the aqueous phase onto the surface of the growing latex particles, assisting their stabilization; the micelles that still contain no polymer thereby slowly disappear. The concentration of free emulsifier finally falls below the critical micelle concentration; at this point, the surface tension increases significantly. From then on, practically no new latex particles can be formed. The first phase of the emulsion polymerization, the so-called particle-formation period, is completed after about 10–20% conversion.

From this point onwards, the polymerization occurs only in the latex particles, whose number, however, remains constant. The monomer droplets still have the function of a reservoir that delivers the monomer to the latex particles, so that the concentration of monomer in the latex particles and therefore the rate of

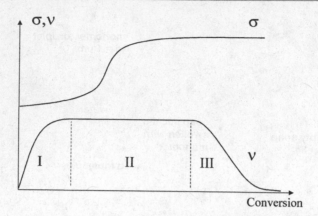

Fig. 2.2 Overall rate of reaction ν, and surface tension σ as function of conversion in emulsion polymerization

polymerization remains constant. During this second period, the reaction is thus of zero order with respect to the monomer.

When the polymerization has proceeded to such an extent that all of the monomer droplets have vanished, which occurs after 60–80% conversion, all of the residual monomer is located in the latex particles. The monomer concentration in the particles now declines as polymerization proceeds further, i.e., in this final period the reaction is first order. At the end of the polymerization, the emulsion consists of polymer particles with a size distribution between 50 and 150 μm, which is larger than the original micelles, but smaller than the original monomer droplets (Fig. 2.1). The changes of surface tension and overall rate of polymerization with conversion are schematically shown in Fig. 2.2.

A special procedure is reverse emulsion polymerization. In this case, a hydrophilic monomer (e.g., acrylamide) is dissolved in water and the resulting solution is emulsified using special water-in-oil emulsifiers in a water-immiscible organic liquid (petroleum ether). Then the polymerization is initiated with a water- or an oil-soluble radical generator. The resulting latex consists of very small water-swollen polymer particles, which are suspended in the oil phase. As in aqueous systems, the advantage of these latex systems lies in their low total viscosity, which is influenced by the nature of the oil phase. This is also the case when the water droplets contain extremely high-molecular-weight polymer products. The swollen polymer particles can be coagulated and isolated from the oil phase by adding special reverse emulsifiers, which abolish the effect of the water-in-oil emulsifiers.

With emulsion polymerization it is possible to prepare very high-molecular-weight polymers at high rates of polymerization. The required reaction temperatures are low and can even be below 20°C when redox systems are used for initiation (see Example 3.11). Polymer emulsions with solid contents of 50% and higher can be very stable. In many cases, e.g., poly(vinyl acetate), they are directly used as paints (paint latices), coatings, or adhesives (see Sect. 2.5.4).

2.2.5 General Laboratory Techniques for the Preparation of Polymers

2.2.5.1 Safety in the Laboratory
All persons being exposed to chemicals have to be instructed about the effects of dangerous substances (toxicity, point of ignition, etc.) as well as about preventive measures. Especially the following points are relevant:
- Wearing of suitable protective clothing (protective goggles, gloves, gas mask, laboratory coat, etc.).
- Knowledge about the safety devices (e.g., laboratory hood, fire extinguisher, emergency shower, first aid boxes, etc.).
- Avoidance or ban of unsafe techniques in the laboratory, e.g., oral suction of chemicals in pipettes, sucking of halogenated compounds with a water-jet pump because of effluent contamination.
- Controlled disposal of toxic substances in compliance with legal regulations.
- Strict ban on smoking and eating in the laboratory.

2.2.5.2 Working with Exclusion of Oxygen and Moisture
Molecular oxygen has an influence on the course of most polymerizations. In radical polymerizations this may occur through an effect on the initiation or termination reactions; in ionic polymerizations the initiator may be either activated or deactivated by oxygen. Oxygen may also cause oxidative degradation of macromolecules that have already been formed (especially in polycondensations at high temperatures). These effects are often already detectable at very low oxygen concentrations and it is therefore advisable to work under a nitrogen or argon atmosphere.

Charging of reaction vessels with nitrogen should always be done by repeated evacuation and admission of nitrogen rather than by simply passing nitrogen through the vessel, since one can never be sure when all the air has been displaced. The removal of oxygen in the vessels is easier if argon is used because the density of argon is higher than the density of air. Connections should be made with PVC tubing or glass tubing rather than long rubber tubing.

When conducting a reaction or distillation under a continuous flow of nitrogen or argon rather than in a closed system, a suitable outlet valve must be used in order to prevent back-diffusion of oxygen into the apparatus from the surrounding air. In the simplest cases, it is sufficient to use a Bunsen valve (a rubber tube with two lengthwise slits, closed at one end with a glass rod), or a paraffin oil valve.

If small volume experiments are carried out under exclusion of oxygen, an approved method is to use a balloon filled with inert gas attached to an otherwise closed apparatus. Likewise, using a balloon can easily compensate for changes in pressure. It is also recommended to use glove boxes when smaller amounts are to be handled under an inert gas.

When working with spontaneously inflammable substances (e.g., organometallic compounds) or with reactions on a larger scale, where there are rapid changes of pressure, it is advantageous to employ a backstop valve.

It is especially important to exclude oxygen and water in ionic polymerizations; the drying procedures normally used for preparative work are then generally inadequate. Glassware can be freed from water by drying in an oven at 150°C, but better by baking with a hot air ventilator under high vacuum. Gases can be dried by freezing out the moisture or by passage through columns filled with suitable solid reagents. Likewise, liquids can be dried by treatment with suitable drying agents at room temperature or by boiling under reflux or by azeotropic or extractive distillation.

Agents for the drying of liquids must, of course, be chemically inert, as otherwise undesired side reactions could occur (e.g., styrene polymerizes explosively on contact with concentrated sulfuric acid at room temperature).

Suitable drying agents are, for example, calcium chloride, silica gel, molecular sieves, calcium hydride, or phosphorus pentoxide; however, their use must be adapted to the requirements and conditions of each polymerization.

2.2.5.3 Purification and Storage of Monomers

In all polymerizations, the purity of the starting materials is of prime importance. Impurities present at concentrations of 10^{-2}–10^{-4} wt.% often have a considerable influence on the course of the reaction. With unsaturated monomers the impurities mentioned below may be encountered:

By-products formed during their preparation (e.g., ethylbenzene and divinylbenzenes in styrene; acetaldehyde in vinyl acetate); added stabilizers (inhibitors); autoxidation and decomposition products of the monomers (e.g., peroxides in dienes, benzaldehyde in styrene, hydrogen cyanide in acrylonitrile); impurities that derive from the method of storage of the monomer (e.g., traces of metal or alkali from the vessels, tap grease etc.); dimers, trimers, and polymers that are generally soluble in the monomer, but sometimes precipitate, for example, polyacrylonitrile from acrylonitrile. Likewise, in polycondensation reactions it is important to remove reactive impurities because they can cause considerable interference during the polyreaction.

The purification procedures to be applied depend on the monomer, the expected impurities, and especially on the purpose for which the monomer is to be employed, e.g., whether it is to be used for radical polymerization in aqueous emulsion or for ionic polymerization initiated with sodium naphthalene. It is not possible to devise a general purification scheme; instead the most suitable method must be chosen in each case from those given below. A prerequisite for successful purification is extreme cleanliness of all apparatus (if necessary, treating with hot nitrating acid and repeatedly thorough washing with distilled water).

The usual procedures of fractional, azeotropic, or extractive distillation under inert gases, crystallization, sublimation, and column chromatography, must be carried out very carefully. For liquid, water-insoluble monomers (e.g., styrene, Example 3.1), it is recommended that phenols or amines which may be present as stabilizers, should first be removed by shaking with dilute alkali or acid, respectively; the relatively high volatility of many of these kinds of stabilizers often makes it difficult to achieve their complete removal by distillation. Gaseous monomers (e.g., lower olefins, butadiene, ethylene oxide) can be purified and stored over molecular sieves in order to remove, for example, water or CO_2.

So-called prepolymerization is frequently used to achieve a very high purification: a monomer that has already been purified by the normal methods is polymerized to about 10–20% conversion by heating or irradiation, or if necessary by addition of initiator. It is then separated from the polymer by fractional distillation under nitrogen. Impurities that affect the initiation (e.g., by reaction with the initiator or its fragments) or react with the growing macromolecules (causing chain termination or chain transfer) are thereby removed.

Measurements of the common physical constants such as boiling point or refractive index are not sufficiently sensitive to determine the trace amounts of impurities in question. Besides the common spectroscopic methods, techniques like gas chromatography (GC), high-pressure liquid chromatography (HPLC), or thin-layer chromatography (TLC) are useful. The surest criterion for the absence of interfering foreign compounds lies in the polymerization itself: the purification is repeated until test polymerizations on the course of the reaction under standard conditions are reproducible (conversion-time curve, viscosity number of the polymers).

Storage of monomers and solvents also requires special precautions. The vessels must be specially constructed and be closed with a self-sealing cap through which the distilled contents can be transferred with a pipette or hypodermic syringe under inert gas. The contents should also not come in contact with tap grease.

Most monomers can be stored unchanged under nitrogen only for short times (hours or days), even in the dark at low temperature. For long-term storage, a suitable stabilizer is therefore indispensable. Effective stabilizers (inhibitors) of radical polymerization are quinones, phenols, amines, nitro compounds, and some metals or metal compounds. The addition of 0.1–1 wt.% of hydroquinone or 4-*tert*-butylpyrocatechol results in sufficient stabilization of many monomers.

Water, alcohols, ethers, or amines can cause inhibition of ionic polymerization. However, these substances can act in different ways according to their concentration. For example, in polymerizations initiated by Lewis acids (BF_3 with isobutylene) or organometallic compounds (aluminum alkyls), water in small concentrations behaves as a cocatalyst, but in larger concentrations as an inhibitor (reaction with the initiator or with the ionic propagating species).

2.2.5.4 Reaction Vessels for Polymerization Reactions
Numerous special experimental arrangements that are suitable for particular monomers and types of polymerization are described in the literature, but they may all be regarded as modifications of the equipment described below.

Radical polymerizations as well as polycondensation and polyaddition reactions of solid or liquid monomers are frequently conducted under inert gas in simple glass ampoules or vessels. Larger amounts are prepared in three or more necked flasks (flat-flanged flasks) equipped with adequate reflux condensers, tap funnels, stirrers and, if needed, a dip-in cooling coil and thermometer or temperature sensor for automatic temperature control. The flasks should be made from thick-walled glass in order to withstand any internal pressure built up during the reaction and should be sealed off as strain-free as possible. If strong exothermic reactions are expected, safety precautions must be taken against possible explosion (wrapping round with adhesive tape, protective shield, etc.). For safety reasons, ampoules with small volume should be used and should never be filled more than half full. In the filling process, it is necessary to ensure that none of the content remains adhering to the upper part of the vessel. Using a funnel can easily prevent this.

For many trial experiments it is sufficient to carry out the reaction on a gram scale in round-bottomed or Erlenmeyer flasks of 5–10 ml size. The reaction mixture should be layered with an inert gas using a pipette and should be stirred by using a magnetic bar. For closure of the reaction flask, a balloon filled with inert gas can be used. Flasks containing higher boiling reaction mixtures can be closed with a convenient taper-ground stopper, which should be protected with a metal clamp. If needed, a dropping funnel allows, for example, the addition of monomers under inert gas during polymerization.

For reactions in which only a slight pressure rise is expected (e.g., emulsion polymerization of gaseous monomers), thick-walled pressure bottles which are carefully placed behind a protective shield, can be used. For higher pressures or large-scale experiments, autoclaves are always to be preferred.

For polymerizations initiated by catalyst suspensions (e.g., alkali metals), a suitable high-speed stirrer is used. It should also be mentioned that high-speed mixers can be used as reaction vessels for polyreactions (interfacial polycondensations). Flasks with self-sealing septums made of silicon rubber, used in combination with hypodermic syringes, are very suitable as reaction vessels, especially for ionic polymerizations. They have the advantage that catalyst can be injected, or samples removed, with practically complete exclusion of air and moisture. It is recommended to puncture a second hypodermic needle through the septum for pressure compensation while injecting. Likewise, during sample taking, a second hypodermic needle is needed for volume compensation of the inert gas. If the polymerization is to be initiated by radiation, one generally uses ampoules or cells with well-defined dimensions. A proper geometrical relationship between the vessel and the radiation source is important.

In (living) anionic polymerizations (e.g., Example 3.19) in the laboratory, a vacuum line is often employed; all operations are then conducted under the same, rather high vacuum. Thus, optimum polymer yield and uniformity are secured.

For kinetic investigations of homogeneous polymerizations, a variety of methods and apparatus has been developed. The dilatometric method is especially worthy of mention on account of its simplicity and general applicability (Fig. 2.3).

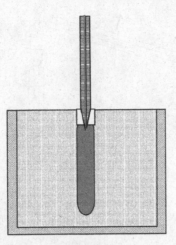

Fig. 2.3 Dilatometer in a thermostat. Conversion-time curves are obtained through volume contraction in the capillary

This procedure is based on the measurement of the contraction of volume that results from the different densities of the monomer and polymer. The conversion of the volume contraction to the yield of polymer can be made by means of a gravimetrically determined calibration curve or by calculation from the specific volumes (see Example 3.6).

With appropriate precautions, condensation and addition polymerization reactions can be carried out in the same apparatus as customarily used for organic preparative work (see Sects. 4.1 and 4.2). In order to obtain high molecular weights by polycondensation in solution, a special circulation apparatus can advantageously be used (Fig. 2.4).

Exact temperature control is very important for polymerization reactions, since, among other things, the rate and degree of polymerization are strongly dependent on temperature. For accurate work, for example, for kinetic analysis with a dilatometer, a thermostat filled with water or paraffin oil may be used instead of thermostatting in the normal way with the aid of a contact thermometer and an immersion heater.

2.2.5.5 Control and Termination of Polymerization Reactions

In chain growth polymerization reactions the average molecular weight, the molecular weight distribution and in some cases the type of terminal group of the polymer can be varied within certain limits by proper choice of reaction conditions and/or the addition of low-molecular-weight compounds (regulators, chain stoppers). It depends on the synthetic method which step is most suitable.

Thus, in a *radical polymerization* an increase of reaction temperature or amount of initiator causes an increase in the number of growing radicals. Since the rate of the propagation reaction is first order with respect to the concentration of growing radicals, while that of the termination reaction is second order, the average molecular weight is reduced (with simultaneous increase in rate of polymerization).

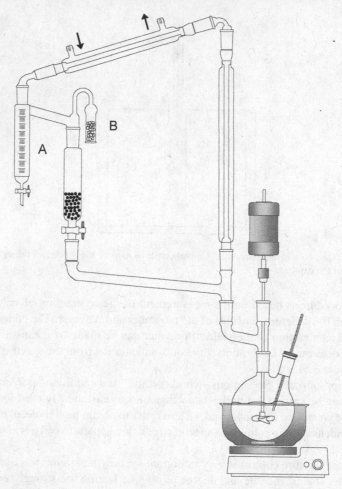

Fig. 2.4 Circulation apparatus for preparation of polyesters. A: water separator, B: drying tube

A decrease of monomer concentration also leads to lower molecular weights, but the rate of polymerization then decreases as well. Since side reactions may intervene at high temperatures or high initiator concentrations, the molecular weight is often better controlled by the addition of regulators, i.e., substances with high transfer constants (see Sect. 3.1 and Example 3.8b). Even at low concentrations, such compounds decrease the average molecular weight markedly by terminating the growth of polymeric chains; at the same time, a new chain is started so that, as a rule, the rate of polymerization is unaffected. The fragments of the regulator are built into the macromolecule as end groups. Especially suitable as regulators are thiols (1-butanethiol, 1-dodecanethiol) and other organic sulfur compounds, (e.g., diisopropylxanthogen disulfide).

In *ionic polymerizations,* the molecular weight can be regulated by temperature, type of catalyst and nature of solvent. In some cases also regulators can be used which, as in the case of cationic polymerization of trioxane, lead to the incorporation of special endgroups.

In polymerizations with *Ziegler-Natta* catalysts, molecular hydrogen is the preferred regulation agent for controlling the molecular weight.

In many cases (e.g., in kinetic investigations, in determination of the reactivity ratios, or in the preparation of unbranched polymers) it is not appropriate to allow polymerization to proceed to complete conversion of the monomer. Polymerization reactions can be stopped in different ways. Sometimes the reaction can be brought to a halt simply by cooling. In most cases pouring the reaction mixture, or a sample of it, into a sufficient quantity of precipitant can end the polymerization, whereby the polymer formed is precipitated and the residual monomer and initiator are highly diluted. Polymerizations are most effectively stopped by addition of an inhibitor or some compound that destroys the initiator; the details of this method depend on the type of polymerization. For radical polymerizations one uses, for example, hydroquinone or hydroquinone monoethers. Most ionic polymerizations can be stopped by addition of water, acids, or bases. When an ionic polymerization is carried out at low temperature, it is essential to terminate it by destruction of the initiator since otherwise warming to room temperature is likely to make the reaction go very quickly. Organometallic initiators (also Ziegler-Natta catalysts) can be destroyed with water or alcohols, and Lewis acids (BF_3) with amines.

Control of molecular weight is much simpler for those polymerizations where the macromolecules are built up stepwise (*condensation* and *stepwise addition*). In such cases, the reaction can be stopped at any time, for example by cooling, so that every step of molecular weight can be obtained. In principle, the molecular weight of the polymer formed in these reactions can also be controlled or the reaction can be interrupted by changing the molar ratio of the two bifunctional reactants or by adding an appropriate amount of monofunctional compound.

2.2.5.6 Isolation of Polymers

Isolation is simplest when the polymer is insoluble in the reaction mixture and precipitates during its formation. In these cases, the product can be separated by filtration or centrifugation. Aqueous solutions can be filtered through paper filters; solutions in organic solvents are better filtered through cloth or sintered glass discs. If the polymer remains dissolved in the reaction mixture, there are two possible ways of proceeding: either the excess monomer, solvent, and other volatiles can be distilled off under vacuum or the polymer can be precipitated by addition of a precipitant. The first procedure is only applied in exceptional cases, since it generally leads to a resinous polymer contaminated with initiator residues and especially with trapped monomer and solvent. The most usual procedure is, therefore, precipitation by means of a precipitant that should satisfy the following requirements: it must be miscible with the monomer and solvent, and dissolve all additives (e.g., initiator) as well as by-products (e.g., oligomers). Furthermore, the polymer should be nonswellable and should separate in flocculent form (not oily or

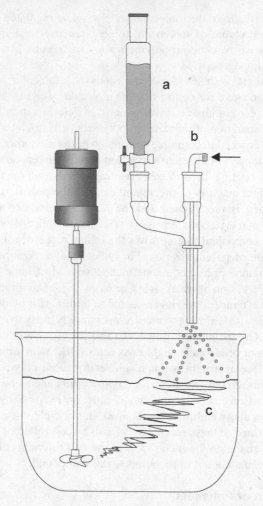

Fig. 2.5 Spray-type separator (**a** polymer solution; **b** compressed air; **c** precipitating agent)

resinous). Finally, the precipitant should be readily volatile and be absorbed or occluded by the polymer as little as possible.

The general procedure is to drop the reaction mixture or polymer solution into a 4- to 10-fold amount of precipitant under vigorous stirring. The concentration of the polymer solution (generally not above 10%) and the amount of precipitant are chosen so that the polymer precipitates in a flocculent, readily filterable form. It often happens that the precipitated polymer remains in colloidal suspension; in this case, it may help either to lower the temperature (external cooling or addition of dry ice) or to add electrolytes (solutions of sodium chloride or aluminum sulfate, dilute hydrochloric acid, acetic acid, or ammonia). Sometimes coagulation of the polymer can also be achieved by prolonged vigorous stirring or shaking (rheopexy). Polymers that tenaciously retain solvent and tend to resinify can be precipitated advantageously by the spray method (see Fig. 2.5). For this purpose, the polymer solution is sprayed in the form of a mist into the precipitant, thereby producing a fine floccular precipitate, the large surface area of which favors the outward diffusion of unpolymerized monomer and other compounds present.

Regarding industrial methods for isolation of polymers, the reader is referred to the relevant literature.

2.2.5.7 Purification and Drying of Polymers

Careful purification and drying of polymers is important not only for analytical characterization, but also because the mechanical, electrical, and optical properties are strongly influenced by impurities. Not the least important aspect of purification is the fact that even traces of impurities may cause or accelerate degradation or crosslinking reactions.

The conventional techniques for the purification of low-molecular-weight compounds, such as distillation, sublimation, and crystallization, are not applicable to polymers. In some cases, it is possible to remove the impurities by cold or hot extraction of the finely powdered polymer with suitable solvents or by steam distillation. Separation of low-molecular-weight components from water-soluble polymers (e.g., poly(acrylic acid), poly(vinyl alcohol), poly(acryl amide)) can be accomplished by dialysis or electrodialysis. However, the most widely used method of purification is by reprecipitation in which the solution of polymer (concentration less than 5–10 wt.%) is dropped into a 4- to 10-fold excess of precipitant, with stirring. If necessary, this operation is repeated with other solvent/precipitant pairs until the impurities are no longer detectable.

The drying of polymers often presents great difficulty, since many polymers tenaciously retain or trap solvent or precipitant; this phenomenon is termed "occlusion". The magnitude of this effect can be judged by the following examples. Cyclohexane is occluded so strongly by cellulose that, after drying for 2 days at 100°C and 0.1 mbar, one cyclohexane molecule is still retained for every six cellulose CRUs. A 0.2-mm-thick polystyrene film, prepared from a solution in THF by drying to constant weight in a stream of nitrogen at 75°C, still retains 13% THF.

There is no general rule for the prevention or avoidance of occlusion. In some cases, a change of solvent/precipitant system may help to achieve this goal. Raising the drying temperature is also beneficial.

An important prerequisite for successful drying is to subdivide the polymer as finely as possible (see Sect. 2.5.1).

2.3 Characterization of Macromolecules

For the unequivocal characterization of a low-molecular-weight compound, it is sufficient to specify a few physical or chemical properties, for example, boiling point, melting point, angle of optical rotation, refractive index, elemental analysis,

IR and NMR spectra. If two low-molecular-weight samples have the same characteristic properties, they may be considered as identical.

The characterization of macromolecular substances is considerably more difficult. Owing to the high intermolecular forces, macromolecules are not volatile without decomposition and so, no boiling point can be determined. The melting points of partially crystalline polymers are generally not sharp. Amorphous polymers frequently show only sintering or softening, often accompanied by decomposition. In addition to elemental analysis, other data must be determined, for example, solubility, viscosity of the solution, mean molecular weight, molecular weight distribution, and degree of crystallinity.

The fundamental difficulty is that polymeric substances cannot be obtained in a structurally and molecularly uniform state, unlike low-molecular-weight compounds. Thus, macromolecular materials of the same analytical composition may differ not only in their structure and configuration (see Sect. 1.2) but also in molecular size and molecular weight distribution; they are polydisperse, i.e., they consist of mixtures of molecules of different size. Hence, it is understandable that the expression "identical" is not, in practice, applicable to macromolecules. Up to the present time, there is no possibility of preparing macromolecules of absolutely uniform structure and size. It follows, therefore, that physical measurements on polymers can only yield average values. The aforementioned peculiarities of macromolecular substances mean that the methods of characterization suitable for low-molecular-weight compounds are frequently not applicable or only applicable in a substantially modified form; often completely new methods of investigation must be employed.

Since the properties of a polymer can be noticeably influenced by small variations in the molecular structure, and these in turn depend on the preparation conditions, it is necessary when reporting data to indicate not only the type of measurement (e.g., molecular weight by end group analysis; crystallinity by infrared measurement or by X-ray diffraction; etc.), but also the type of preparation (e.g., radical polymerization in bulk at 80°C; polymerization with a particular organometallic catalyst at 20°C).

The following Table 2.6 lists the most important characteristics of polymers together with appropriate analytical methods that can be applied for their determination.

2.3.1 Determination of Solubility

When studying a polymer, one should first determine its solubility. This is very characteristic for macromolecules and can serve as an early means of characterization. Important examples are the recognition of crosslinking, the separation and distinguishing between tactic and atactic macromolecules and the identification of copolymers. Moreover, solubility is a prerequisite for most physical measurements.

Table 2.6 Important polymer characteristics and analytical methods for their determination

Characteristics	Analytical methods
1. Composition and constitution	
(a) Chemical composition	Elemental analysis, UV, IR, NMR, pyrolysis-GC-mass spectrometry
(b) Endgroups	Elemental analysis, spectroscopy, titration
(c) Head/tail linkages	Spectroscopy
(d) Branching	Solution viscosity, melt viscosity, light scattering, NMR
(e) Crosslinking	Solution viscosity, melt viscosity, modulus of elasticity
(f) *cis/trans* Isomerism	Spectroscopy
(g) Stereoregularity (tacticity)	Spectroscopy
(h) Optical isomerism, optical activity	Polarimetry
(i) Refractive index	Refractometry
(j) Liquid crystalline domains	Polarization microscopy, X-ray scattering
(k) Composition of copolymers	Elemental analysis, spectroscopy, pyrolysis-GC-mass spectrometry
2. Molecular weight	*Absolute methods:* end group analysis, membrane osmometry, vapor pressure osmometry, static light scattering, mass spectrometry, sedimentation measurements
	Relative Methods: solution viscosity, melt viscosity, size-exclusion chromatography
3. Coil dimensions	Light scattering, sedimentation measurements, small angle X-ray, solution viscosity
4. Molecular weight distribution	Fractionation, size-exclusion chromatography
5. Bulk properties	
(a) Density	Pyknometer technique: flotation method
(b) Crystallinity	Birefringence, density, differential scanning calorimetry, X-ray diffraction
(c) Glass transition temperature	DTA, DSC, dynamic mechanical measurements
(d) Softening point	Vicat-method, Martens-method, heat distortion temperature
(e) Crystallite melting point	Polarization microscopy, DTA, DSC, dynamic mechanical measurements
(f) Melt viscosity	Capillary viscometry, rotational viscometry
(g) Tensile strength and elongation	Stress–strain measurements
(h) Modulus of elasticity	Dynamic-mechanical measurements
(i) Impact strength	Charpy-method, Izod-method
(k) Hardness	Ball hardness test
6. Morphology	
(a) Crystallites	Electron microscopy
(b) Spherulites	Polarization microscopy
(c) meso- and nano-Morphologies in copolymers and polymer blends	Electron microscopy
7. Surface properties	Scanning electron microscopy, scanning probe microscopy, ATR-IR spectroscopy, contact angle measurements

Table 2.7 Solubility of various polymers

Polymer	Solvents	Non-solvents
Polyethylene, poly-(1-butylene), isotactic polypropylene	p-Xylene[a], trichlorobenzene[a], decane[a], decalin[a]	Acetone, diethyl ether, lower alcohols
Atactic polypropylene	Hydrocarbons, isopentyl acetate	Ethyl acetate, propanol
Polyisobutylene	Hexane, toluene, carbon tetrachloride, THF	Acetone, methanol, methyl acetate
Polybutadiene, polyisoprene	Aliphatic and aromatic hydrocarbons	Acetone, diethyl ether, lower alcohols
Polystyrene	Toluene, methylene chloride, cyclohexanone, butyl acetate	Lower alcohols, diethyl ether, acetone
Poly(vinyl chloride)	THF, cyclohexanone, methyl ethyl ketone, DMF	Methanol, acetone, heptane
Poly(vinyl fluoride)	Cyclohexanone, DMF	Aliphatic hydrocarbons, methanol
Poly(tetrafluoroethylene)	Insoluble	
Poly(vinyl acetate)	Toluene, methylene chloride, methanol, acetone, butyl acetate	Diethyl ether, petroleum ether, butanol
Poly(isobutyl vinyl ether)	2-Propanol, methyl ethyl ketone, methylene chloride, aromatic hydrocarbons	Methanol, acetone
Poly(methyl vinyl ketone)	Acetone, methylene chloride	Water, aliphatic hydrocarbons
Polyacrylates and poly-methacrylates	Methylene chloride, acetone, ethyl acetate, THF, toluene	Methanol, diethyl ether, petroleum ether
Polyacrylonitrile	DMF, dimethyl sulfoxide, conc. sulfuric acid	Ethanol, diethyl ether, water, hydrocarbons
Polyacrylamide	Water	Methanol, acetone
Poly(acrylic acid)	Water, dil. alkali, methanol, DMF	Hydrocarbons, methyl acetate, acetone
Poly(vinylsulfonic acid)	Water, methanol, dimethyl sulfoxide	Hydrocarbons, acetone
Poly(vinyl alcohol)	Water, dimethylformamide[a], dimethyl sulfoxide[a]	Hydrocarbons, methanol, acetone, diethyl ether
Starch	Water, chloral hydrate, copper ethylenediamine	Acetone, methanol
Cellulose	N,N-Dimethyl acetamide (100 ml)/lithium chloride (5 g), aqueous tetraaminecopper(II) hydroxide, aqueous zinc chloride, aqueous calcium thiocyanate	Acetone, methanol, water
Cellulose triacetate	Acetone, chloroform, 1,4-dioxane	Methanol, diethyl ether
Cellulose trimethyl ether	Chloroform, toluene	Ethanol, diethyl ether, petroleum ether

(continued)

Table 2.7 (continued)

Polymer	Solvents	Non-solvents
Carboxymethyl cellulose	Water	Methanol
Aliphatic polyesters	Methylene chloride, formic acid, toluene	Methanol, diethyl ether, aliphatic hydrocarbons
Poly(ethylene terephthalate)	m-Cresol, 2-chlorophenol, nitrobenzene, trichloroacetic acid	Methanol, acetone, aliphatic hydrocarbons
Polyamides	Formic acid, m-cresol, γ-butyrolactone, DMF/calcium chloride, N-methylpyrrolidone/lithium fluoride	Methanol, acetone, aliphatic hydrocarbons
Polyurethanes (not cross-linked)	Formic acid, γ-butyrolactone, DMF, m-cresol	Methanol, diethyl ether, hydrocarbons
Polyoxymethylene	γ-Butyrolactone[a], DMF[a], benzyl alcohol[a]	Methanol, diethyl ether, aliphatic hydrocarbons
Poly(ethylene oxide)	Water, toluene, DMF	Methanol, diethyl ether, aliphatic hydrocarbons
Poly(tetrahydrofuran)	Toluene, methylene chloride, THF	Aliphatic hydrocarbons, diethyl ether
Poly(dimethylsiloxane)	Methylene chloride, heptane, toluene, diethyl ether	Methanol, ethanol
Poly(aryl ether sulfone)	Dimethyl sulfoxide, DMF	Alcohol, ether
Poly(aryl ether ketone)	Conc. sulfuric acid, trifluoromethanesulfonic acid, diphenyl sulfone[a], sulfolane[a]	Alcohol, ether, methylene chloride
Poly(phenylene sulfide)	Above 200°C: dichlorobiphenyl, 1-chloronaphthalene, N-ethylpyrrolidone, ε-caprolactam, diphenyl ether, diphenyl sulfone[a]	
Poly(phenylene oxide)	Chloroform, chlorobenzene, toluene	Methanol, ethanol

[a]Only soluble on heating

When determining the solubility it has to be remembered that macromolecules show extremes of behavior, like swelling and incompatibility (for details see Sect. 1.4.1.2). Although there are many thermodynamic theories for the description of polymer solutions, there is still no full understanding of these systems and quite often, one needs application of empirical rules and conclusions by analogy. As a guide, some solvents and non-solvents are listed in Table 2.7 for various polymers. However, not all combinations of solvent and non-solvent allow sufficient purification of a polymer via dissolution and reprecipitation, and trial experiments are often needed.

For the investigation of the solubility of a polymer one may proceed as follows:
- 30–50 mg samples of finely divided polymer are placed in small test tubes with 1 ml solvent and allowed to stand for several hours. The solution process is significantly influenced by the state of subdivision of the polymer.

- From time to time, the contents are stirred or shaken, and examined for the appearance of streaks. If no solution occurs after several hours at room temperature one can slowly raise the temperature, if necessary to the boiling point of the solvent. Any coloration or gas formation is indicative of decomposition of the polymer.
- If the polymer dissolves at higher temperature, which may require a long time, the solution should be allowed to cool slowly to check whether the polymer comes out of solution again and if so at what temperature (important for subsequent measurements).
- If the polymer simply swells without going into solution, the procedure is repeated with other solvents or solvent mixtures. If it swells in all solvents, without dissolving, one may assume that it is crosslinked.

2.3.2 Methods for Determination of Polymer Constitution

When a polymer synthesis is carried out, it is important to check whether this process was successful or not. This question has many facets. First of all, we may ask, for example,

- Whether the obtained chain molecules have the expected constitution
- How the monomer units are connected,
- What the structure of the repeating units is,
- What the composition of copolymers is,
- How complete a macromolecular substitution process was,
- What the configuration / tacticity of the repeating units is,
- Whether or not branches were formed,
- Which end groups the obtained macromolecules have.

To answer all these questions, spectroscopic and spectrometric techniques are required. Some of the most important techniques will be mentioned in the following.

2.3.2.1 High-Resolution NMR Spectroscopy

Chemical constitution, steric configuration and, in some cases, details about chain conformation, aggregation, association, and supramolecular self-organization behavior of macromolecular substances can be determined using high-resolution nuclear magnetic resonance (NMR) spectroscopy. This spectroscopic technique is sensitive towards nuclei with a nuclear spin different from zero. Identical nuclei (protons for example) incorporated at different places of a molecule – or bond to different molecules – have different shielding constants s and thus – at constant external field H_0 – different resonance frequencies n_1. This effect is called "chemical shift" d and is usually given relative to that of a standard compound like tetramethylsilane (TMS). Because of the smallness of this shielding constant the value of the chemical shift of a nucleus i is given in ppm (parts per million). For protons, the chemical shifts d are between 0 and approx. 12 ppm, for ^{13}C between 0 and approx. 220 ppm. Just by analyzing the chemical shifts of the signals found in an NMR spectrum a first rough analysis of the polymer constitution is possible. Moreover, the intensity of the absorptions of each nuclei is independent of the chemical environment but proportional to their relative concentration. This feature – together with the characteristic chemical shifts – is of special importance for qualitative as well as quantitative structural elucidation via NMR spectroscopy: position (δ/ppm) and intensity of an absorption give clear and direct information about constitutional, configurational, and other features of the material to be analyzed. And there is one more dominant effect which consolidates and deepens the structural information obtained from NMR investigations. This is the indirect spin-spin coupling of neighboring, nonequivalent nuclei of a molecule via the bond electrons. It leads to a fine-structure (muliplet structure) of the absorption signals which is caused by the generation of additional small magnetic fields at the locus of the observed nucleus. This is due to the relative orientation of the spins in the neighboring nuclei of the same molecule. In general, the number of absorptions of a multiplet, i.e., the "multiplicity" M, is given by

$$M = 2n\,I + 1 \tag{2.13}$$

where n is the number of equivalent neighboring atoms and I ($= 0, {}^1/_2, 1, 1\ {}^1/_2, 2, \ldots$) is their nuclear spin quantum number. Magnetically equivalent nuclei do not lead to any splitting due to indirect spin-spin coupling but give a common resonance line. The strength of the coupling is given by the spin-spin coupling constant J. The NMR spectra of dissolved polymers can be interpreted in the same way as those of low-molecular-weight compounds. Hence, it is a powerful tool for constitutional analyses: The chemical constitution of repeating units and end groups, the content of comonomers, or the steric configuration (tacticity; see below) of macromolecules can be determined in dilute solution using high-resolution NMR spectroscopy. Also, NMR spectra of linear polymers of low molar mass often show unique absorptions due to their end-groups. By referencing these absorptions to those of the nuclei in the repeating units it is possible to obtain the ratio of the number of end-groups to the number of repeating units. Thereby it is possible to evaluate the M_n of such a polymer. For branched polymers, NMR absorptions due to the branch point can be identified and reveal the chemical structure of those branch points, thus leading to a better understanding of the mechanism by which the branches form, and to information about the relative number of branch points within a macromolecule.

Using NMR spectroscopy, it is also possible to determine the polymer's tacticity. The constitutional repeating unit of polymethylmethacrylate (PMMA), for example, possesses three different types of (magnetically nonequivalent) protons with different chemical shifts d, i.e., the CH_2 protons (d $\approx$ 2 ppm), the α-CH_3 protons (d $\approx$ 1 ppm), and the OCH_3 protons (d $\approx$ 3.5 ppm).

Let us consider the –CH_2– group first. Depending on the tacticity, it can be the part of either a *racemic* or a *meso* diade (see Fig. 2.6).

It is evident that in the racemic diade both –CH_2– protons are imbedded into an identical microenvironment. Consequently, they are magnetically equivalent, absorb

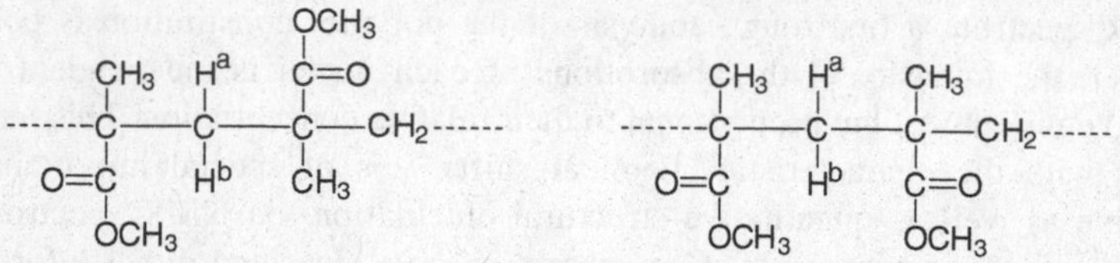

Fig. 2.6 Racemic (r; syndiotactic, *left*) and meso (m; isotactic, *right*) diade in PMMA

at the same resonance frequency ν (have the same value of δ), and do not couple with each other. Therefore, the proton in a racemic diade appears as a singlet in the NMR spectrum. For the meso diade, on the other hand, it is obvious that the two –CH$_2$– protons have a clearly different microenvironment: while H^a has two methyl groups as neighbors, there are the ester groups for proton H^b. Consequently, the two –CH$_2$– protons of the meso diade are magnetically nonequivalent, absorb at different resonance frequencies ν$_a$ and ν$_b$, respectively, and couple with each other. Therefore, these protons in a meso diade appear as a set of two doublets in the NMR spectrum.

The absorptions of the α-CH$_3$ group reflect the respective triade substructure (see Fig. 2.7):

One observes different values of δ for the isotactic, syndiotactic, and heterotactic triade, respectively, but all signals are singlets due to the lack of a proton at the carbon atom next to the CH$_3$ group (see Fig. 2.8). Via quantitative evaluation of the intensity of the respective absorptions information is available about the homogeneity of the chain configuration.

High magnetic fields and in particular ^{13}C-NMR spectroscopy allow the analysis of even longer configurational sequences (tetrads up to nonads). This proved to be important in particular for the analysis of polyolefins like polypropylene or cyclo-olefin copolymers (COC). These polymers are available via transition-metal mediated (Ziegler-Natta, metallocene) insertion polymerizations, and the configurational analysis provides deep insight into the respective polymerization mechanisms as well as into the structure–property relationships.

When high-resolution NMR spectra have to be recorded of a polymeric sample, one has to recognize that polymer solutions are in general highly viscous. To prevent excessive signal broadening caused by this restricted mobility of the solution, polymer solutions for NMR studies have to be highly diluted (approx. 1–2 mg · ml^{-1}). Accordingly, rather long acquisition times are required for readily resolved spectra, in particular for ^{13}C. Nevertheless, despite of high dilution, some polymer absorptions may remain broadened, especially those of atoms incorporated directly into the polymer backbone, while absorptions of lateral substituents tend to be well-resolved. This broadening even at high dilution is mainly due to the restricted mobility of the polymer backbone, preventing complete averaging of the dipolar environment within the time-window of the NMR experiment. Increase of temperature might sharpen some of these signals to a certain extent.

The limit of accuracy of ^{1}H-NMR experiments carried out in dilute solution is around 1–5%, depending on the resolution of the spectrum, and of approx. 10% for

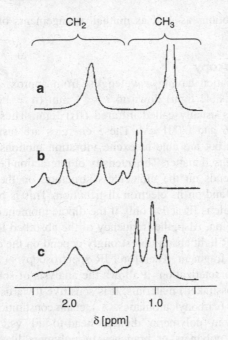

Fig. 2.7 Aliphatic regions of ^{1}H-NMR spectra of (**a**) syndiotactic, (**b**) isotactic, and (**c**) atactic PMMA

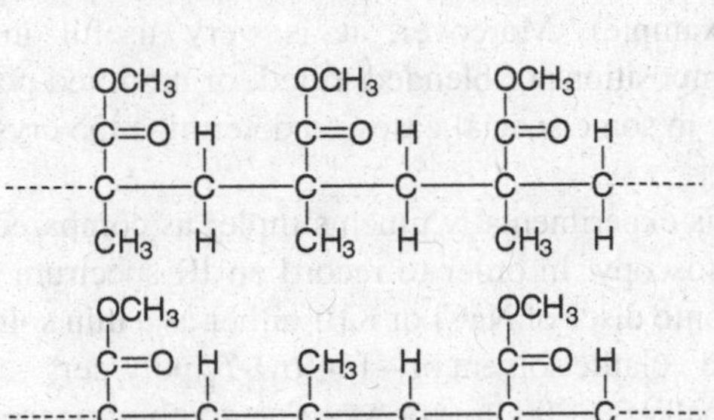

Fig. 2.8 Isotactic (*i, top*), syndiotactic (*s, middle*) and heterotactic (*h, bottom*) triades of PMMA

^{13}C NMR. If the polymer to be investigated proved to be insoluble, solid-state NMR techniques are available for further investigation. Solid-state NMR methods are also very useful for determining bulk properties of polymers such as relaxation behavior of local motions as well as mutual arrangements of chains and chain segments.

2.3.2.2 IR Spectroscopy

Electromagnetic radiation having wavelengths from approx. $\lambda = 760$ nm (n ≈ 13,000 cm^{-1}; near visible light) down to $\lambda \approx 1$ mm (n ≈ 10 cm^{-1}), where the microwaves begin, is usually called infrared (IR) light. Thus, IR photons have energies between 1.6 and 0.001 eV. These energies are insufficient to induce electronic transitions but are able to excite vibration motions of molecules and parts thereof in condensed matter. The intensity of interaction between IR radiation and a molecule depends on the molecule's structure, on the symmetry of the molecule's skeleton, and on its electron distribution. This is because a vibration transition of a molecule is IR active only if the dipole moment changes during the excited vibration motion. Also, the frequency of the absorbed IR radiation as well as the efficiency of IR light absorption strongly depend on the environment of the observed molecule's fragment. Therefore, IR spectroscopy is an important technique in polymer characterization. It allows the analysis of soluble polymers but also of insoluble (crosslinked) materials. It is sensitive towards structural features like functional groups (carbonyl, aromatics, . . .), chain constitution (1,2- vs. 1,4- and *cis-trans* isomerism in polymeric dienes, head-to-tail vs. head-head-tail-tail placements in vinyl polymers or branches in polymers like polyethylene, for example), end groups (M_n determination), and copolymer composition (in PE/PP copolymers, for example). Moreover, it is very useful in determining the components and compositions of blended, filled, or modified polymer compounds and composites and, in some special cases, to determine the crystallinity of a solid polymer sample.

IR spectroscopy is experimentally much simpler as compared to other methods of vibrational spectroscopy. In order to record an IR spectrum, in most cases the polymer is brought onto discs of NaCl or KBr either as a thin solid film (made from polymer solution in a volatile solvent or – for low-T_g polymers – from the melt; film thicknesses are typically 30–300 μm) or as a fine and homogeneous suspension in, for example, paraffin oil. Alternatively, solid polymers can be milled together with a large excess of KBr, and the resulting powder can be compressed to a (homogeneous, transparent) disc. Then, IR radiation is transmitted through the sample, and the absorbance (extinction) is measured as a function of the wavelength λ or of the wave number ν using a detector placed at the opposite site of the sample. In some other cases the ATR (attenuated total reflection) method is used. Here, the sample is placed as a thin film on the top of an ATR crystal, and the IR spectrum is recorded in reflection geometry. The IR spectra thus obtained provide information on what efficiency IR light is absorbed by the polymer sample at which wavelength λ or at which wave number ν. Characteristic IR spectra of PMMA and PE are shown in Fig. 2.9.

Despite the fact that a full assignment of all the observed absorptions to the respective macromolecule's natural frequencies is not possible in all cases – in particular for complex co- and terpolymers, stereoregular polymers, crosslinked

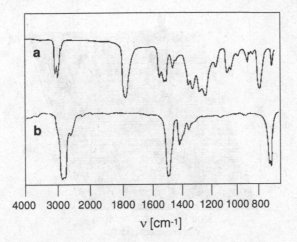

Fig. 2.9 IR spectra of (a) PMMA and (b) PE

systems, composites, compounds or blends this is very difficult – there are many bands caused by local group vibrations of a few atoms which can be interpreted very nicely. As an example, the C = O band (stretching vibration) is usually observed as an intense absorption between n $\approx$ 1850–1650 cm^{-1}. Because of the coupling with other vibrations of the molecule its frequency is characteristic for the constitution and the neighborhood of the observed atom group.

Hence, in principle, the identification of local atom groups of polymers proceeds in the same way as for low-molecular-weight materials, and the position of the respective bands is nearly unchanged. Also, IR spectra of oligomers are hardly different from those of high polymers if a minimum degree of polymerization is exceeded (P_n > 5–10). Moreover, characteristic absorption for chain end-groups might be observed in the spectra – in particular for strongly IR-active end-groups. Then, IR spectroscopy can be used to roughly estimate the degree of polymerization provided that the molecular weight is not too high (M < 10^4): While qualitative IR analysis is a rather simple technique, quantitative evaluation of the IR spectra is a more complicate matter. The samples have to be prepared very carefully (only measurements in transition are possible, using very homogeneous samples), and some further requirements have to be fulfilled in addition to this. The evaluation of the signal intensities is based on the Lambert-Beer law:

$$E = \log\left(\frac{I_0}{I}\right) = \varepsilon \cdot c \cdot d \qquad (2.14)$$

E being the optical density or absorbance, I_0 and I the intensities at a fixed wavelength λ (wave number ν) of the exciting beam and the beam after passing through the sample, ε is the decade molar extinction coefficient, c the polymer concentration, and d the layer thickness. E is determined directly by the

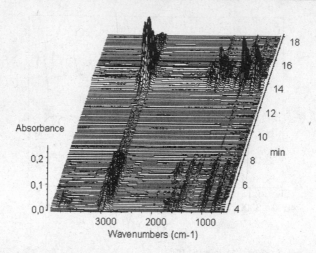

Fig. 2.10 Sequence of IR spectra recorded from the individual fractions of an SEC run taken at regular intervals

spectrometer but evaluation requires further considerations: in general, many bands in an IR spectrum at least partially overlap and are placed on a continuous background. A band shape analysis has to be carried out, and signal intensities caused by neighboring absorptions and the background have to be subtracted from the absorption of interest. Then, either the absorbance at maximum absorption, E_{max}, or the full signal intensity (obtained via integration over the whole signal) can be used for quantitative information.

Also, there might be absorptions in the IR spectra which are characteristic for syn- or isotacticity (such as in PMMA) or for branching points of nonlinear polymer chains (such as in polyethylene). Using data pools and programs which simulate IR spectra it is possible nowadays to characterize nearly all kinds of polymers very quickly using IR spectroscopy with respect to their constitution and their composition. Also, IR spectroscopy can be coupled with polymer chromatography (SEC, HPLC). Then it provides detailed chemical information on each individual chromatographic fraction (Fig. 2.10).

In the *near-IR range* (NIR, $\lambda = 0.76$–2.5 μm; $\nu = 13{,}200$–$4{,}000$ cm^{-1}) overtones and combinational vibrations are found. Because of the huge number of possible combinations of vibrations, this range usually shows a tremendous number of (rather weak) absorptions which overlap or even cover each other. The absorptions found in this range are very characteristic for the investigated material – and are thus of increasing importance – but computers are needed to evaluate these spectra. The *intermediate IR range* (MIR, $\lambda = 2.5$–50 μm; $\nu = 4{,}000$–200 cm^{-1}) is the most important range for structural elucidation using IR spectroscopy. Here, one can find the characteristic fingerprint bands which allow a fast and reliable first structural assignment for an unknown polymeric material, and many spectra are available as references in data pools. Finally, the range of the far-IR ($\lambda = 50\ \mu m$–1 mm; $\nu = 200$–10 cm^{-1}) is of only limited use for polymer analysis.

IR spectroscopy is not only useful for determining the chemical constitution of polymers. It additionally provides profound information on chain orientation and on the orientation of attached lateral substituents of polymers. In this case, polarized IR radiation is applied which is only absorbed by an IR-active bond if the plane in which the electrical field vector E of the IR beam oscillates is parallel to the transition dipole moment μ of the vibration to be excited. If, on the other hand, the transition dipole moment μ is perpendicular to the electrical field vector E of the IR beam no absorption is observed. Using this effect, the degree of orientation of a polymer sample (film, fiber) can be estimated by comparing the intensity at maximum $I(\|)$ and at minimum $I(\perp)$ absorption, i.e., the dichroic ratio.

2.3.2.3 UV–vis Spectroscopy

Alike IR spectroscopy, ultraviolet and visible light (UV–vis) spectroscopy are important optical methods for polymer characterization. For standard spectroscopic investigations in the UV–vis range, unpolarized light having wavelengths from 185 nm to 760 nm is used. The absorption spectra are usually recorded using double-beam spectrometers, and the transmissibility $T = I/I_0$ or the extinction, $\log I_0/I$, is recorded as a function of the wavelength. The UV–vis measurements are carried out on highly diluted solutions, but studies on (ultra)thin films are possible as well. This is because of the very high extinction coefficients of most chromophors and the fact that the extinction must not exceed a certain value in order to maintain the validity of the Lambert Beer law (see above). If this is the case, quantitative interpretation of the UV–vis spectra is possible. As indicated, however, UV–vis spectroscopy is limited to polymers that contain specific chromophores such as aromatic groups, conjugated double- or triple bonds, carbonyls or azo-subunits: When molecules absorb light in the UV–vis range, electronic transitions are induced in the molecules between electronic levels (usually $\pi \rightarrow \pi^*$ and $n \rightarrow \pi^*$ transitions in the most important range of wave length, i.e., between 200 nm and 600 nm). These transitions can be used for the quantitative determination of residual monomers in a polymer or of copolymer composition (e.g., copolymers of styrene). However, application of the Lambert-Beer's law might be affected here in some cases since the observed chromophors have a different next neighbors environment in these macromolecules. For example, in copolymers containing repeating units A and B with a chromophoric subunit in A, the extinction coefficient of the chromophor at maximum absorption (λ_{max}) might depend on whether it is the center of triades AAA, AAB, or BAB. Thus the possibility of quantitative evaluation has to be checked first by means of reference experiments in these cases. Also, configurational and constitutional features of macromolecules can be determined. In the case of conjugated p systems, for example, the $\pi \rightarrow \pi^*$ absorptions shift towards longer wavelengths at increasing conjugation length.

Last but not least, ageing and destruction processes can be monitored in polymers under application, and structural and quantitative analysis of unknown additives (stabilizers etc.) is possible in commercial polymers using UV–vis spectroscopy. Advantage can be taken here of the fact that the position of an electronic absorption in unsaturated systems depends only weakly on the surrounding medium. Even though UV–vis spectroscopy is not very specific in the absorption band, it is highly sensitive and therefore much better than NMR or IR spectroscopy to detect small amounts of chromophors.

2.3.2.4 Fluorescence Spectroscopy

Fluorescence spectroscopy is commonly used to characterize fluorescence effects in the UV and visual range of the electromagnetic spectrum. Such fluorescence is caused by the fact that the absorption of UV or visible light of specific wavelengths causes excitation of electrons within a molecule. If radiating relaxation occurs directly from the singlet S_1 state, the process is called fluorescence.

Readily measurable fluorescence intensities are found for molecules having aromatic and heteroaromatic rings, in particular when annulated rings are present, and in the case of conjugated π-electron systems. If the polymer molecules contain such fluorescence-active subunits they can be characterized by this technique, either directly via their fluorescence spectrum or via fluorescence quenching experiments (for polymers with appropriate quencher groups). It is also possible to introduce a very small amount of fluorescent-active groups (as comonomer units; "fluorescence marker") into a polymer. Inspection through a fluorescence microscope allows, for example, the monitoring of the molecule's movements (Brownian motions in solution, segment motions in the bulk above glass transition), or the determination of chain orientation in stretched samples. In polymer materials and composites, fluorescence spectroscopy can be used for the qualitative and quantitative determination of additives (stabilizers, for example) even when present at very high dilution only. This is possible due to the tremendous sensitivity of fluorescence spectroscopy.

2.3.2.5 Refractometry

The refractive index is an important quantity for characterizing the structure of polymers. This is because it depends sensitively on the chemical composition, on the tacticity, and – for oligomeric samples – also on the molecular weight of a macromolecular substance. The refractive indices n_D (determined using the sodium D line) of many polymers are collected in the literature. In order to characterize a molecule's constitution one requires knowledge of the mole refraction, R_g. For isotropic samples, it can be calculated in good approximation by the Lorentz-Lorenz equation:

$$R_g = \frac{n^2 - 1}{n^2 + 2} \cdot V_{g,mol}$$

(2.15)

where R_g is the mole refraction per monomer unit and $V_{g,mol}$ is the base mole volume. R_g can be calculated from well-known atom- and bond refraction increments. In the case of ideal mixtures the mole refractions behave additively like the specific volume of the components. For transparent samples, refractometry is therefore a well-appropriate technique for the determination of the polymerization kinetics – similar to dilatometry – and the characterization of mixtures (polymer solutions, compatibility etc.).

2.3.2.6 Elemental Analysis

Qualitative and quantitative elemental analysis of polymers can be carried out by the conventional methods used for low-molecular-weight compounds. So a detailed description is not needed here. Elemental analysis or determination of functional groups is especially valuable for copolymers or chemically modified polymers. For homopolymers where the elemental analysis should agree with that of the monomer, deviations from the theoretical values are an indication of side reactions during polymerization. However, they can also sometimes be caused by inclusion or adsorption of solvent or precipitant, or, in commercial polymers, to the presence of added stabilizers. The preparation of the sample for analysis must, therefore, be very carefully carried out (several reprecipitations, if necessary using various solvent/precipitant combinations; thorough drying).

2.3.2.7 Composition of Copolymers

In the characterization of copolymers one distinguishes between *qualitative* analysis, designed to test whether the material is a genuine copolymer or only a physical mixture of homopolymers, and *quantitative* analysis of the weight fraction of the incorporated comonomers.

The *qualitative* analysis is quite simple if the homopolymers differ in their solubility, for example, when one homopolymer is soluble in a solvent where the other is not. In this case a sample of the material is extracted with that solvent. The reprecipitated extracts and residues are examined for composition. The extraction must, however, be very carefully carried out and repeated several times since polymer mixtures are frequently quite difficult to separate by extraction. If no pure homopolymer is isolated in this way one can be sure that the sample is a genuine copolymer. If the solubility properties of the original homopolymers are insufficiently different it is sometimes possible to induce such differences through chemical transformation, for example, by oxidation and/or hydrolysis.

The qualitative investigation of copolymers is considerably more difficult when the homopolymers cannot be distinguished by their solubility. In this case other physical data of the supposed copolymer can be compared with the corresponding data for various physical mixtures of homopolymers, for example, softening point and melting range, density, and crystallinity. Copolymers can frequently be distinguished from physical mixtures of homopolymers by the qualitative and quantitative composition of their pyrolysis products.

Quantitative analysis of copolymers is relatively simple if one of the comonomers contains a readily determinable element or functional group. However, C,H elemental analyses are only of value when the difference between the carbon or hydrogen content of the two comonomers is sufficiently large. If the composition cannot be determined by elemental analysis or chemical means, the problem can be solved usually either by spectroscopic methods, for example, by UV measurements (e.g., styrene copolymers), by IR measurements (e.g., olefin copolymers), and by NMR measurements, or by gas chromatographic methods combined with mass spectroscopy after thermal or chemical decomposition of the samples.

In principle the composition of a copolymer may also be determined by analyzing the composition of the residual monomer by a suitable method after polymerization. It will usually be necessary first to separate the copolymer by precipitation, followed by careful recovery of the filtrate containing the residual monomer, but the direct method of analysis of the copolymer will generally be preferred. *Block* and *graft* copolymers can be characterized in the same manner. However, consideration must be taken of the fact that they usually contain large amounts of the homopolymers which must first be removed. The more refined characterization of a statistic copolymer involves the determination of the reactivity ratios r_1 and r_2 (copolymerization parameters), as well as the calculation of Q- and e-values (see Chap. 3.4, Example 3.36).

2.3.3 Determination of Molecular Weight and Molecular-Weight Distribution

The degree of polymerization, P, and the molecular weight, M, are some of the most important characteristics of a macromolecular substance. They indicate how many monomer units are linked to form the polymer chain and what their molecular weight is. In the case of homopolymers, the molecular weight of a macromolecule is given by:

$$M = P \times M_{\text{ru}}$$

with M_{ru} being the molecular weight of the constitutional repeating unit. However, while low-molecular-weight substances consist, by definition, of molecules of identical structure and size, this is generally not the case for polymers. Synthetic macromolecular substances are nearly always composed of macromolecules of similar structure but different molecular weight. These materials are therefore called *polydisperse*. As a consequence, chemical formulas of polymers are generally given in a way where the constitutional repeating unit is drawn in square brackets, bearing an index n indicating the average number of repeating units tied together to give the polymer chain. Full characterization of a macromolecular substance is not an easy task, therefore, and quite often statistical methods are required: because of polydispersity, the values of P and M are mean values only, and the macromolecular chain molecules of a synthetic polymer are characterized by a (more or less) well-defined chain-length distribution (or molecular-weight distribution). The respective molecular-weight distribution is the direct consequence of chain formation statistics and, moreover, in many cases very

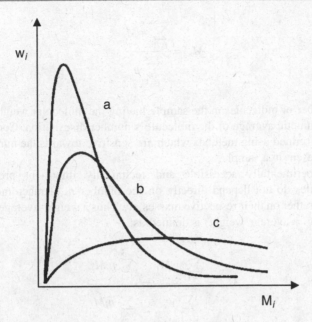

Fig. 2.11 Schematic presentation of the weight fractions of i-mers as a function of their molecular weight, M_i, for polycondensates of **a** low, **b** medium, and **c** high molecular weights

characteristic for the respective chain growth process. Let us ignore potentially different end groups which may also depend on the way of preparation of the respective polymer. Then macromolecules having the identical overall chemical constitution but different molecular weights represent so-called polymer-homologous series.

Depending on the selected polymerization reaction, the polymerization conditions, and potential side reactions one may obtain different molar mass distributions – even if only one single type of monomer is polymerized. As an example, Fig. 2.11 shows the plot of the overall masses w_i (w = weight) of all macromolecules i in the sample vs. their respective molecular weights M_i:

In order to normalize the molecular-weight distribution, mass fractions W_i are often used instead of the total masses w_i ($= n_i \cdot M_i$).

$$W_i = \frac{w_i}{\sum\limits_{i=1}^{i=\infty} w_i} \qquad (2.16)$$

Most molecular-weight distributions are sufficiently represented by a set of specific distribution parameters represented by different averages of the molecular weight. The most important averages of the molecular weight will be discussed in the following. The number-average molecular weight M_n (n = average number) is defined as:

$$M_n = \frac{\sum\limits_{i=1}^{i=\infty} n_i M_i}{\sum\limits_{i=1}^{i=\infty} n_i} \qquad (2.17)$$

n_i is the *number* of molecules in the sample having the molecular weight M_i. Thus M_i is the arithmetic average of the molecule's number distribution. Consequently, it can be determined using methods which are sensitive towards the number of the molecules present in a sample.

Many experimentally accessible and technically important properties of macromolecules do not depend directly on the number n_i of macromolecules in a sample but rather on their respective masses, w_i. Thus a weight-average molecular weight M_w (w = average weight) is defined as:

$$M_w = \frac{\sum\limits_{i=1}^{i=\infty} w_i M_i}{\sum\limits_{i=1}^{i=\infty} w_i} = \frac{\sum\limits_{i=1}^{i=\infty} n_i M_i^2}{\sum\limits_{i=1}^{i=\infty} n_i M_i} \qquad (2.18)$$

M_w corresponds to the first moment of the mass distribution of the molecular weight. Moments of the molecular weight distributions with other arguments do not have any descriptive meaning. For example:

$$M_z = \frac{\sum\limits_{i=1}^{i=\infty} z_i M_i}{\sum\limits_{i=1}^{i=\infty} z_i} = \frac{\sum\limits_{i=1}^{i=\infty} w_i M_i^2}{\sum\limits_{i=1}^{i=\infty} w_i M_i} \qquad (2.19)$$

is the z-average (centrifuge average) of the molecular weight, M_z. It can be determined by the measurement of sedimentation equilibria in an ultracentrifuge. A further important average of molecular weight is the viscosity-average molecular weight, M_h:

$$M_\eta = \left(\frac{\sum\limits_{i=1}^{i=\infty} w_i M_i^a}{\sum\limits_{i=1}^{i=\infty} w_i} \right) \qquad (2.20)$$

The exponent a can be determined experimentally from the relation between the intrinsic viscosity, $[\eta]$, and the molecular weight, M (Mark-Houwink-Kuhn relation):

$$[\eta] = K \cdots M^a \tag{2.21}$$

In most cases a is between 0.5 (ϑ solvent) and 0.9. In general, it is:

$$M_n \leq M_\eta \leq M_W \leq M_Z \tag{2.22}$$

Identity is only given for monodisperse samples, i.e., polymers whose macromolecules have all the same molar mass. Moreover, M_n might be equal to M_w if the exponent a in the $[\eta]$ to M relation is equal to 1.

As a simple measure of the width of a molecular-weight distribution the quotient of two averages is sufficient in many cases. The ratio of M_w and M_n is in particular important:

$$PDI = \frac{M_w}{M_n} \tag{2.23}$$

It is called polydispersity index (PDI). The value of PDI can range between approx. 1.01 (for anionically prepared polymers) up to more than 30 (high-pressure polyethylene). In general, it is between 2 and 5.

Averages of the degree of polymerization, P, are defined analogously to those of the molecular masses, M. As an example, it is for the weight-average degree of polymerization, P_w:

$$P_w = \frac{\sum\limits_{i=1}^{i=\infty} w_i P_i}{\sum\limits_{i=1}^{i=\infty} w_i} \tag{2.24}$$

where w_i is the mass of all molecules i having a degree of polymerization of P_i.

The number-average molecular weight, M_n, can be obtained by osmotic measurements or by determination of end-groups. The weight-average molecular weight, M_w, is measured by methods like light or X-ray scattering and – with limitations – viscosity measurements. Depending on the method of evaluation, ultracentrifuge analysis allows determination of M_n, M_w, and M_z. The latter one is characterized by a superproportional consideration of the larger macromolecules.

When the molecular-weight distribution needs to be described, the ratio of two different averages such as M_w/M_n is insufficient in some cases. This is because samples of identical values of M_n, M_w, and M_z might have completely different molecular-weight distribution curves. For the full description of a polymer sample with respect to the molecular weight it is, therefore, necessary to give the full molecular-weight distribution curve. Different ways are available to give this more profound information in a graphical diagram: quite often the molecular-weight distribution is represented as the integral (cumulative) mass distribution, $J_w(M)$, of the molecular weights. It represents the overall mass of the molecules having a molecular weight equal or smaller than $M = M_i$:

$$J_w(M) = \sum_{M=M_0}^{M=M_i} w_i \cdot \Delta M \qquad (2.25)$$

Because of the high numerical value of the argument M it is allowed to replace the summation (step function) by an integration (smooth curve), despite the discrete nature of the values of ΔM which are equal to the integer multiple of the molecular weight of the repeating unit, M_0, in the case of homopolymers:

$$J_w(M) = \int_{M_0}^{M_i} w(M)dM \qquad (2.26)$$

where $w(M)$ is the mass of all macromolecules having the molecular weight M, and $w(M)dM$ is the mass of macromolecules having a molecular weight ranging between M and $M + dM$. Since $w(0) = 0$, it is possible to set the lower integration limit equal to zero. For normalization reasons, moreover, the overall mass of the polymer, W, is set equal to 1:

$$W = \int_{0}^{\infty} w(M)dM = 1 \qquad (2.27)$$

It follows that $J_w(M) = 1$ as well for $M = 0 \rightarrow M = \infty$. Using the thus normalized integral mass distribution curves of the molecular weights – as can be determined by fractionated precipitation of a polymer – it is possible to calculate the averages of the molecular weights according to the above equations.

The thus obtained integral mass distribution curves of the molecular weight can be transformed into the differential mass distributions $w(M)$ of the molecular weight by differentiation with respect to M:

$$w(M) = \frac{dJ_w(M)}{dM} \qquad (2.28)$$

It shows us which mass fraction of the sample lies between M and $M + dM$.

2.3.3.1 Classification of the Methods for Molecular-Weight Determination

Knowledge of the molecular weight and of the molecular-weight distribution of a polymeric material is indispensable for scientific studies as well as for many technical applications of polymers. They effect the solution and melt viscosity, the processability, and the resulting mechanical properties tremendously. Therefore, we will give a short introduction into methods that allow us to determine the required information. Roughly, the methods developed for the determination of molecular weights are subdivided into absolute and relative methods:

Absolute methods provide the molecular weight and the degree of polymer-ization without any calibration. Their calculation from the experimental data requires only universal constants such as the gas constant and Avogadro's number, apart from readily determinable physical properties such as density, refractive index, etc. The most important methods in use today are mass spectrometry, osmometry, light scattering, and – to some extent – sedimentation and diffusion measurements. Also, some chemical and spectroscopic methods (determination of end-groups) are important because of their relative simplicity.

Relative methods measure properties that depend clearly on molecular weight, for example, the hydrodynamic volume of the polymer coils (GPC, viscosimetry) or their solubility as a function of chain length. However, these measurements can only be evaluated with respect to the molecular weight of the macromolecules if experimental calibration curves are available which were generated by comparison with an absolute method of molecular-weight determination.

A necessary prerequisite for application of the above methods is that the polymer is soluble in a suitable solvent. Moreover, one must ensure that the dissolved macromolecules exist as isolated species and do not form associates or aggregates. Proof of this can be obtained by carrying out reactions on functional groups of the polymer that do not lead to cleavage of the polymer chains. If the degree of polymerization of the original polymer agrees with that of the modified polymer, association can be excluded. Values of molecular weight determined in different solvents should also be in agreement if association is absent.

2.3.3.2 Absolute Methods
End-Group Analysis

If the macromolecules in a polymeric sample contain end groups which can be readily detected, analytically identified and quantified, and if the macromolecule's molecular weight is not too high ($<5 \times 10^4$), their number-average molecular weight, M_n, can be determined by chemical as well as by physical methods. Specific and very exact analytical methods must be applied here since the end groups to be estimated constitute only a small fraction of the macromolecule (less than 0.5%, depending on the molecular weight). Chemical methods are based mainly on titrations of the end groups. The most common procedure is potentiometric pH titration. Elemental analysis or trace analysis might be appropriate as well (halogen analysis, for example, when *p*-dibromobenzoyl peroxide has been used as the initiator in a radical polymerization). Physical methods are based on spectroscopic techniques such as IR, UV–vis (when azo compounds with characteristic absorption bands are used as initiator, for example), and NMR spectroscopy (especially for polymers made by step-growth polymerization). In the early days of polymer research, radiochemical analysis was used. This highly sensitive technique is based on the introduction of radioactive nuclei (^{3}H, ^{14}C) into the polymer chain ends using, for example, appropriately labeled per- or azo compounds as initiators.

The most important aspect for a reliable end-group analysis is that it must be absolutely clear what kind of end groups are present in a polymeric material and – if more than one kind of end group is present – how they are distributed over the material (one per chain, two per chain, more than two per macromolecule for branched systems etc.). In order to assure well-defined end groups, the macro-molecules formed by radical polymerization can be labeled by choosing a suitable initiator (or chain transfer agent) whose radical fragments become incorporated into the polymer. In this case it is also important to know the type of chain termination since this determines the number of labeled end groups per macromolecule (two for termination by combination, one for termination by disproportionation). Errors can occur if, for example, there is uncontrolled chain transfer to the monomer which reduces the number of labeled end groups per molecule. As a consequence, end-group analysis will lead to a too high apparent molecular weight. The molecular weights of macromolecules made by step-growth polymerization involving two compounds can also be obtained by end-group determination. In particular, the amino, hydroxyl, and carboxyl end-groups in polyesters and polyamides can be estimated very precisely both by potentiometric pH titration and by colorimetry. Hydroxy end groups (e.g., in polyoxymethylenes) can also be determined by acetylation or methylation.

The number-average molecular weight is calculated from the analytically deter-mined end-group content according to the following relationship:

$$M_n = \frac{100 \cdot z \cdot E}{e}$$

where E denotes the molecular weight of the end groups, z is their number per macromolecule, and e is the experimentally determined content of end-groups in grams per 100 g, i.e., wt%.

Membrane Osmometry

An important group of absolute methods allowing the determination of the mole-cular weight of macromolecules is based on the measurement of colligative properties. Here, the activity of the solvent is measured in a polymer solution via determination of the osmotic pressure Π_{os}. The value of Π_{os} required to determine the number-average molecular weight can be obtained using a membrane osmome-ter. Here, in a measuring cell having two chambers separated by a semipermeable membrane, one chamber contains the pure solvent and the second one the polymer solution in the same solvent (a membrane is called semipermeable if only the solvent can pass through but not the polymer molecules). Due to the lower activity (lower chemical potential) of the solvent in the polymer solution as compared to the pure solvent, solvent molecules migrate through the membrane from the solvent chamber into that of the polymer solution and dilute it. Therefore, the volume of the polymer solution increases until an equilibration is reached between the osmotic pressure Π_{os} and the hydrostatic pressure generated by the diluted polymer solution

$$\Pi_{os} = \varrho \, g \, \Delta h \qquad (2.29)$$

where ϱ is the density of the solvent and g is the acceleration of gravity. Following van't Hoff, it is

$$\Pi_{os} V = n RT \qquad (2.30)$$

for diluted solutions, with V being the volume of the polymer solution and n the number of moles of the dissolved polymer. Since $n = m/M_n$ (m is the mass (in g) of dissolved polymer) and $c = m/V$ it follows that:

$$\Pi_{os} = \frac{m}{V} \frac{RT}{M_n} = \frac{cRT}{M_n} \qquad (2.31)$$

Since van't Hoff's law is valid only for infinitely diluted solutions, one develops Π_{os}/c in power law series (break after the linear term in c)

$$\frac{\Pi_{os}}{c} = \frac{RT}{M_n} + A_2 \cdot c \qquad (2.32)$$

Thus, the osmotic pressure is first measured at different polymer concentrations, Π_{os}/c is then plotted vs. c, the values are linearly extrapolated to $c \to 0$, and the value of M_n is determined from the y axis intercept. A_2 is the second virial coefficient of the osmotic pressure. Solvents where $A_2 = 0$ are called "ideal" or ϑ solvents.

For membrane osmometry (as well as for all other techniques of molecular-weight determination via colligative properties) it is very important that the samples to be analyzed are very pure. In particular low-molecular-weight impurities have to be removed reliably. Otherwise, they will migrate through the semipermeable membrane and lower the chemical potential of the solvent in the reference chamber. An overestimation of the molecular weight will follow. The same effect applies when there are very small oligomers in the test sample. Therefore, the lower limit of M for application of membrane osmometry is approx. 10.000 – depending on the available membrane pore size. On the other hand, M should be below approx. 50.000 because of the limited sensitivity of this method. Moreover, complete dissolution and absence of aggregates is required for reliable measurements.

Vapor Pressure Osmometry
Not only is the osmotic pressure an appropriate quantity for the determination of the number-average of the molecular weight, M_n, of a polymer but also – at least in principle – all other colligative properties such as the lowering of the freezing point, the increase of the boiling point, or the lowering of the vapor pressure. While

ebullioscopy and cryoscopy are less common for M_n determination, vapor pressure osmometry is a well-established technique for this purpose. It is based on Raoult's law according to which – for the dilute solution of a compound 2 in a solvent 1 – the vapor pressure of the solvent, p_1, decreases proportionally with the mole fraction x_1 of the solvent:

$$\frac{p_1}{p_{1,0}} = x_1 = 1 - x_2 \qquad (2.33)$$

The relative decrease of the vapor pressure is:

$$\frac{\Delta p_1}{p_{1,0}} = 1 - \frac{p_1}{p_{1,0}} = x_2 = \frac{n_2}{n_1 + n_2} \approx \frac{n_2}{n_1} \qquad (2.34)$$

with $p_{1,0}$ the vapor pressure of the pure solvent and $\Delta p_1 = p_{1,0} - p_1$. So the measurement of the vapor pressure of a dilute polymer solution might lead to P_n and M_n. However, precise determination of the vapor pressure is not so easy as it should be for a standard method of polymer analysis. Therefore, the effect of vapor pressure lowering is measured indirectly by determining the increase of the solution temperature (due to the heat of condensation) when the solution is in contact with a saturated atmosphere of solvent vapor. Here, two adjusted thermistors are placed in a tempered cell containing a saturated solvent atmosphere. One of the thermistors bears a drop of the pure solvent, the other one a drop of the polymer solution (in the same solvent). The drop of the pure solvent assumes precisely the temperature of the measuring cell because condensation and evaporation of solvent molecules is balanced in the saturated atmosphere. In the polymer solution, however, the activity of the solvent molecules is decreased due to the presence of the polymer and thus their vapor pressure is lowered. Consequently, some more solvent condensates onto the solution drop and due to the condensation heat its temperature rises. Finally, its temperature is higher than the surrounding temperature by ΔT^*. This temperature difference is just enough to compensate the lowered vapor pressure of the solution, and the chemical potentials of the solvent are now identical in the solution and in the pure solvent. (In practice, however, this temperature difference ΔT^* is never reached completely because the experiment cannot be carried out adiabatically. Continuously, the solution drop loses some heat to the surrounding vapor phase. Therefore, a slightly lower temperature difference ΔT is measured instead of ΔT^*, and vapor pressure osmometry needs calibration despite the fact that it is – at least in principle – an absolute method). Introduction of the Clausius Claperon relation leads to:

$$\frac{\Delta T}{c} = K \cdot \frac{1}{M_n} \qquad (2.35)$$

with

$$K = \frac{const. \cdot RT^2}{L_1 \rho_s} \qquad (2.36)$$

K being a constant which is usually determined experimentally during cell calibration. L_1 is the heat of evaporation of the solvent, ρ_s the density of the solution, and c the polymer concentration. Finally, because the given deviation is valid only for ideal solutions but only real solutions can be studied in practice, the above equation is developed in a power law series with respect to c:

$$\frac{\Delta T}{Kc} = \frac{1}{M_n} + A_2 c + A_2 c^2 + \dots \qquad (2.37)$$

Experimentally, ΔT is determined for approx. five different polymer concentrations. After several minutes, a constant temperature difference ΔT of the two drops is reached which is proportional to their initial difference in vapor pressure and thus proportional to the number of dissolved macromolecules in the solution drop. ΔT can then be determined by measuring the difference in electric resistance of the two thermistors. Then, $\Delta T/Kc$ is plotted vs. c (thus the power law series is broken after the linear term in c) and the plotted values are extrapolated to $c \to 0$. M_n is finally calculated from the y axis intercept.

Vapor pressure osmometry is slightly less sensitive than membrane osmometry ($M_n < 2 \times 10^4$) but is not affected by very short chains in the polymer sample which migrate through the semipermeable membrane in the case of membrane osmometry. Therefore, it is in particular valuable for the analysis of oligomeric materials.

Static Light Scattering

Electromagnetic radiation excites electrons bound to atoms or molecules. If the energy of the radiation is insufficient – i.e., the wavelength is too long – to cause a transition from the electronic ground state into an electronically excited one, the excited electrons fall back immediately after excitation while they emit the absorbed radiation again in all directions in space. This light is observed as scattered light. In a different way of describing this process it is said that the electrons are excited by the light to vibrations which have the same frequency as the exciting light. These vibration of which positive (the nuclei) and negative charges (the electrons) are permanently shifted with respect to each other induce a dipole moment μ which is proportional to the absolute value E of the electric field vector of the light wave. The polarizability α is the constant of proportionality here. The oscillating dipoles again emit electromagnetic waves, i.e., light, with the same vibration frequency as the vibrating dipoles and thus as the incident light. This so-called ideal Rayleigh scattering is coherent and elastic.

The theory developed by Rayleigh and Debye for coherent light scattering shows that only sub-volume elements in a sample (whose size is determined by the wave length of the incident radiation) contribute to the scattering which are different in polarizability and thus refractive indices with respect to their surrounding: the scattering intensity is proportional to the square of the refractive index difference. In a pure solvent scattering is caused only by thermal density fluctuations and thus very weak for visible light. For solutions there is an additional contribution due to the dissolved material which causes concentration fluctuations. Here, the intensity of the scattered light is proportional to the square of the refractive index increment, dn/dc, of the dissolved substance in the solution. For much diluted solutions ($c \rightarrow 0$), the scattering intensity caused by the dissolved molecules, R_ϑ, is given by:

$$R_\vartheta = R_{\vartheta,solution} - R_{\vartheta,solvent} = \frac{I_\vartheta r^2}{I_0} = \frac{4\pi^2 n_0^2 f \cdot p}{\lambda^4 N_L} \left(\frac{dn}{dc}\right)^2 \cdot c \cdot M$$

$$= K \cdot c \cdot M \tag{2.38}$$

Here, I_0 is the intensity of the incident light, I_ϑ the scattered light intensity at a scattering angle ϑ, r is the distance between sample and detector, n_0 is the refractive index of the solvent, f a depolarization factor (≈ 1), p a polarization factor (≈ 1), and λ is the wavelength of the light. K can be calculated for known values of dn/dc, n and λ. Hence, the molecular weight M of the dissolved material can be determined by measuring R_ϑ at concentration c.

The above considerations are valid only for monodisperse samples of rather low molecular weight. When characterizing polydisperse samples, all components i having different molecular weights M_i and concentrations c_i, scatter independently from each other. Thus one obtains the following equation:

$$R_\vartheta = K \sum c_i M_i = K_c M_w \tag{2.39}$$

or

$$\frac{Kc}{R_\vartheta} = \frac{1}{M_w} \tag{2.40}$$

Thus light scattering delivers the weight-average molecular weight of a polydisperse sample. The usually high molecular weight of polymers enforces a further aspect: since the diameter of the polymer coil is usually larger than approx. $\lambda/20$, intramolecular interference effects become relevant in the scattered light. This interference is zero at $\vartheta = 0°$ and has a maximum at $\vartheta = 180°$.

This internal interference is described by the particle form factor, P_ϑ. It allows the direct calculation of the radius of gyration, $<s^2>^{1/2}$, of the macromolecules and thus provides information about their chain conformation:

$$P(\vartheta)^{-1} = 1 + \frac{16\pi^2 n_0^2 <s^2>}{3\lambda^2} \sin^2 \frac{\vartheta}{2} \tag{2.41}$$

Because all the above deductions are valid for infinite low polymer concentrations but practical measurements have to be carried out at finite values of c, it is necessary to include the second virial coefficient A_2. Thus the equation according to which evaluation of light-scattering experiments can be done is:

$$\frac{Kc}{R_\vartheta}\frac{1}{M_w} = \left(1 + \frac{16\pi^2 n_0^2 <s^2>}{3\lambda^2}\sin^2\frac{\vartheta}{2}\right) + 2A_2 c \qquad (2.42)$$

Light-scattering investigations are carried out in a way that the scattering intensity of several polymer solutions having different polymer concentrations c are measured at different scattering angles. Then, Kc/R_ϑ is plotted versus $\sin^2(\vartheta/2) + kc$ (with k an arbitrary constant) and then simultaneously extrapolated to $c \to 0$ and $\vartheta \to 0$ (Zimm plot). While M_w is obtained from the y axis intercept, the slope of the two extrapolated straight lines delivers $<s^2>$ and A_2.

The sensitivity of a light-scattering experiment is basically determined by the refractive index increment, dn/dc. Its value determines the lowest limit of M which is still accessible by light scattering. In general, the molecular weights of the polymers to be analyzed should be above approx. 30,000–50,000. Because light scattering is the more sensitive the larger the molecular weight of the scattering species is, special care has to be taken to remove all dust or other scattering particles from the solution. Therefore, special procedures are needed for the purification of the cuvettes, and the solvents and solutions have to be filtered carefully (preferentially using syringe filters). Here, however, one has to ensure that no polymer is filtered off during this process. This may happen in particular when the polymer tends to aggregate or not sufficient time was given for polymer dissolution. Last but not least, one has to assure that the polymer does not absorb the light used for the scattering studies.

Mass Spectrometry

Polymers are not easily converted to gas-phase ions but this is a requirement for compounds analyzed by mass spectrometry. Despite this difficulty, mass spectrometry has been utilized to study various aspects of polymers: polymers can be characterized – among others – with respect to their chemical composition, to their end groups, and to their molecular weight. Moreover, mass spectrometry can be used to study polymer surfaces.

Field desorption (FD) and fast atom bombardment (FAB) mass spectrometry provides mass spectral information about compounds that are not very volatile but these two techniques are not used often in polymer science since they have several disadvantages. Electrospray ionization (ESI) mass spectrometry can also be used to obtain the above information about polymers, but ESI spectra are generally complicated due to differences in charge state distributions. Static secondary ion mass spectrometry (static SIMS) is a surface-sensitive MS technique, which is suitable for studying the interfaces of polymers with respect to chemical structure and molecular weight as well as end groups and surface contaminants. Laser desorption mass spectrometry (LDMS) has been used for detection of polymer additives, characterization of polymer end groups, repeat units, and average molecular weights. Depth profiling experiments and film characterization have also been carried out using LDMS. Matrix-assisted laser desorption ionization mass spectrometry (MALDI-MS), however, is the most widely used MS technique for bulk polymer analysis. MALDI can be used to determine the average molecular weight for both high and low mass polymers, the mass of the end groups, and monitor polymerization reactions by characterizing the chemical structure of the products and the rate of polymerization. Also, MALDI can be used to examine the surface and bulk composition of biomaterials, whereas SIMS is used for examining monolayer and submonolayer coverage of polymers on surfaces. As a whole, the development of soft ionization methods has led to an increased use of mass spectrometry in characterizing polymers. MALDI appears to exhibit the best performance for reliable molecular weight determination at higher masses.

In MALDI, a polymer is dissolved in a solvent, and a special ingredient is added which has to absorb ultraviolet light very efficiently. Usually about 10^4 times more of the UV absorber are added to the solution than polymer. Then, the solution is placed in an airtight chamber, on the tip of the sample probe. The chamber is evacuated, and the solvent evaporates. Finally, a thin layer remains of the UV-absorbing compound together with a small amount of the polymer in it, i.e., the polymer is now dispersed in a matrix of the UV-absorbing compound. Then the sample is irradiated by a laser shoot (usually by an ultraviolet laser working in the 330–360-nm range). The UV-active matrix material absorbs the laser energy and reacts with the polymer in such a way that the macromolecules become charged ions. It is still unknown how this happens. Depending on what kind of polymer and what kind of matrix material is used, the polymers may be cations or anions. When the polymer forms cations, the positive cathode is placed right behind the sample, and the negative anode in front of it. Most of the time, there is only one single positive charge on each polymer molecule (ionized via alkali metal ion attachment). Moreover, the absorbed laser energy heats up the sample, causing evaporation of the matrix together with some of the polymer molecules at these high temperatures and low pressures. Now, the ionic polymers floating around in gaseous state between the electrodes are accelerated in the applied electric field (the positively charged polymers are travelling in the direction of the anode, attracted to its negative charge). The same electrical force is applied on each polymer molecule while it is being accelerated in the electric field between the two electrodes. But the more mass, the slower the acceleration. This means that the polymer will take longer to reach the detector at the end of the chamber. So the polymers will hit the detector, the small ones first, then the big ones. When a macromolecule reaches the detector, it registers a peak. The size of the peak is proportional to the number of molecules that hit at one time. So a series of peaks is obtained. Because the time a molecule needs to reach the detector is proportional to its mass, a plot of molecular weight on the x-axis and the number of molecules having given molecular weight on the y-axis is obtained, i.e., the molecular weight distribution. The Fig. 2.12 shows the MALDI spectrum of a poly(p-phenylene) derivate. It shows clearly the different chain lengths as well as the patterns of the end groups.

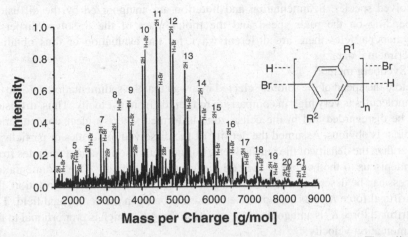

Fig. 2.12 MALDI mass spectrum of a side-chain substituted poly(*p*-phenylene). The splitting of the mass peaks of the respective *n*-mers is caused by the different combinations of end groups, i.e., H,H, H,Br, or Br,Br

MALDI measures the mass very accurately, and it gives an absolute measurement of mass. Still, sample and solution conditions must be optimized for the best performance of the matrix and therefore, it cannot yet be used as a routine method. Also, characterization of synthetic polymers by MALDI is sometimes limited by their solubility and mass discriminating desorption behavior, and the mass spectrum might be affected by the properties of the solvents used for polymer dissolution or by the matrix material.

There are also some further mass spectrometric applications in polymer science. Gas chromatography/mass spectrometry (GC/MS), for example, can be used to identify and characterize small volatile polymers and contaminants. GC/MS can also be used to characterize the degradation products of polymers. Pyrolysis-GC/MS can be used to determine the chemical structure of polymers and to examine their thermal degradation pathways. It is used to determine the chemical structure of analyte molecules by analyzing analyte fragmentation. This technique can also be used to monitor polymerization by identifying characteristic pyrolates. Usually, relatively low-mass fragments of the polymer can be analyzed; however, by controlling the temperature, pyrolysis-GC/MS can be used to analyze higher-mass pyrolates. Glow discharge mass spectrometry (GDMS) is another technique that is able to fingerprint polymer materials.

Ultracentrifuge Measurements

In a centrifugal field, dissolved molecules or suspended particles either sediment (if their density exceeds that of the pure solvent), or flotate for the opposite case (negative or inverse sedimentation). Under otherwise identical experimental conditions, the velocity of the molecules or particles depends on the viscosity of the solution or suspension and – very importantly – on the mass and shape of the dissolved species. Sedimentation and flotation are antagonized by the diffusion. Depending on the rotor speed and the molar mass of the dissolved/dispersed polymers/particles there are different ways for the evaluation of thus obtained experimental data.

(a) Svedberg method

Here the speed of the rotor is selected in a way that the sedimentation velocity of the molecules is very high in comparison to their diffusion velocity. Thus, diffusion can be disregarded and in the cell a zone is formed where a clear concentration gradient is obvious. Assumed the density of the dissolved molecules or particles is larger than the density of the pure solvent, this concentration gradient migrates from the meniscus to the bottom of the cell during the experiment. This sedimentation process can be described as follows: In a distance x from the center of rotation, the centrifugal force K_Z acts on a particle of mass m being in the centrifugal field. The centrifugal force K_z is antagonized by a friction force K_r which is proportional to the sedimentation velocity:

$$K_r = F \frac{dx}{dt} \tag{2.43}$$

The proportionality factor F is called friction factor and is identical for diffusion and sedimentation. Using the Einstein Sutherland equation

$$D = \frac{RT}{F \cdot N_L} \tag{2.44}$$

the Svedberg equation of sedimentation is obtained:

$$M_{s,D} = \frac{s \cdot RT}{D \cdot (1 - \bar{v}\rho_{solvent})} \tag{2.45}$$

If (1) the diffusion coefficient D of the polymer in the used solvent, (2) the specific volume $\bar{v}$ of the dissolved polymer, and (3) the density of the solvent $\rho_{solvent}$ are known, one can determine the molecular weight of the dissolved polymer according to the above equation by measuring the sedimentation coefficient (by measuring the maximum of the concentration gradient at regular time intervals).

The thus determined molecular weight is an apparent one since s and D depend on the polymer concentration. Therefore, extrapolation to concentration zero is required. The sedimentation coefficient obtained by extrapolating $c \rightarrow 0$ is called sedimentation constant s_0:

$$\frac{1}{s} = \frac{1}{s_0}(1 + k_s c) \tag{2.46}$$

Here k_s is a constant which depends on solvent and temperature. Thus the sedimentation constant can be calculated from the y-axis intercept when the reciprocal of the sedimentation coefficient s determined at different polymer concentrations c is plotted vs. the polymer concentration, c, and then c is extrapolated to zero. Using the thus determined sedimentation constant s_0 – which increases with growing molecular weight – and the known diffusion coefficient $D_{0(c \to 0)}$ ("diffusion constant") the Svedberg equation gives the molecular weight $M(s_0, D_0)$ of the polymer. This molecular weight is – in the case of polydisperse samples – in most cases somewhere between the viscosity average, M_η, and the number average, M_n, of the molecular weight.

When dispersions are analyzed where nonsolvated, sphere-like particles sediment, the sedimentation coefficients s are independent of concentration at low solid contents and, therefore, it is possible to determine the particle size distribution in dispersion from the distribution of the sedimentation coefficients.

(b) Sedimentation equilibrium

At low rotor revolution numbers an equilibrium state can be reached between sedimentation and diffusion. Now, a time-independent concentration gradient is established, i.e., $(dc/dt)_x = 0$. Under these conditions, the Svedberg equation becomes:

$$M = \frac{dc/dx}{\omega^2 xc} \frac{RT}{(1 - \bar{v}\rho_{solvent})} \tag{2.47}$$

If c and dc/dx are known as a function of x and the measurement is carried out in a theta solvent, the molecular weight M of monodisperse polymers can now be calculated precisely. If the solvent is not a theta solvent, the obtained value of M is an apparent molecular weight from which the true value can be calculated upon plotting $1/M$ vs. c and extrapolation to $c \to 0$. For polydisperse samples, one has to insert the average of dc/dx in the above equation, and the thus calculated molecular weight represents a weight-average, M_w.

An alternative approach for determining the molecular weight of a polymer in theta solvents includes the determination of the polymer's concentration at the meniscus (c_m) and at the bottom (c_b) (or alternatively at two other positions x_1 and x_2) in the cell. These two outstanding positions have a distance of $x_m(x_1)$ and $x_b(x_2)$, respectively, from the center of rotation. Then, one obtains the weight-average molecular weight of a polydisperse polymer sample via the equation:

$$M_w = \frac{c_b - c_m}{\omega^2 (x_b^2 - x_m^2) \cdot c_0} \frac{2RT}{(1 - \bar{v}\rho_{solvent})} \tag{2.48}$$

Here, c_0 is the polymer concentration of the original solution.

(c) Sedimentation in a gradient of density

When mixtures of solvents of different density are used for polymer dissolution, an equilibrium is established during ultracentrifugation where the concentration of the solvent of higher specific gravity is increased at the bottom of the cell while the specifically lighter solvent is enriched near the meniscus. Hence, the dissolved macromolecules encounter a density gradient in the cell, and they move to the place in the cell where the density of the solvent mixture is identical to their own density (one should take care that there is no preferential solvation in either one of the constituents of the solvent mixture!). However, due to Brownian motion monodisperse macromolecules do not collect at a single place x within the cell but rather in a certain zone. This zone has – for identical polymer molecules – the shape of a Gauss curve. The width of this curve decreases with increasing molecular weight:

$$\frac{c}{c_{x_0}} = \exp\left(-\frac{(x-x_0)^2}{2\sigma^2}\right) \tag{2.49}$$

with

$$\sigma^2 = \frac{RT}{M\bar{v}\left(\frac{d\rho}{dx}\right)_{x_0}\omega^2 x_0} \tag{2.50}$$

where x_0 is the distance between the center of the Gauss curve and the rotor axis, and $(d\rho/dx)_{x0}$ is the gradient of density at x_0. Chemically different molecules often have different densities and thus are enriched at different locations within the cell. For example, the chemical composition of (graft, block, statistic) copolymers and the tacticity of homopolymers can be characterized in the gradient of density. In the case of dispersions information is available about the density and the density distribution of the dispersed particles, and thus conclusions concerning their chemical composition are possible.

2.3.3.3 Relative Methods
Solution Viscosity
When a polymer is dissolved in a solvent, it makes the solution viscous. The caused thickening effect can be used to estimate a macromolecule's molecular weight because the higher the molecular weight, the more viscous the polymer solution will be. This is reasonable because the higher the molecular weight, the bigger the hydrodynamic volume is, and being bigger, the polymer molecule can block more motion of the solvent molecules. Also, the bigger a polymer is, the stronger its secondary forces are. So the higher the molecular weight, the more strongly solvent molecules will be bound to the polymer. This reduces even more the mobility of the solvent molecules.

For most polymers there is a definite relationship between molecular weight and solution viscosity. The viscosity method of molecular-weight determination was introduced by Staudinger. However, it is applicable only to linear and slightly branched molecules; it fails mostly for sphere-like or strongly branched molecules (globular proteins, glycogens). For the determination of the molecular weight of a polymer via solution viscosity measurements it is not necessary to determine absolute values of the solution viscosity. In principle, it is enough to measure the time t which a given volume of the polymer solution needs to flow through the capillary and to compare this with the time t_0 which is needed by the pure solvent. Then, to have a first measure of the viscosity-increasing effect of the polymer to be analyzed, the elution or flow time, t of the polymer solution at a given concentration c, is divided by t_0. This quotient is called the relative viscosity η_{rel}:

$$\frac{t}{t_0} = \eta_{rel} \tag{2.51}$$

However, the required information is the difference in the elution times of the solution and the pure solvent relative to the elution time of the pure solvent. Therefore, the elution time of the pure solvent t_0, is subtracted from the elution time of the solution t. The thus obtained result is divided by t_0. The resulting quantity is called the specific viscosity, η_{sp}, which is a dimensionless quantity:

$$\frac{t - t_0}{t_0} = \eta_{sp} \tag{2.52}$$

If the measurement is made in a capillary viscometer of specified dimensions and at low polymer concentration (so that the density of the solution is approximately the same as that of the solvent), the viscosities η and η_0 are represented in a good approximation by the elution times t and t_0. It follows that:

$$\frac{t - t_0}{t_0} = \eta_{sp} \approx \frac{\eta - \eta_0}{\eta_0} \tag{2.53}$$

If this value is divided by the concentration c of the polymer in solution, one obtains the reduced specific viscosity, η_{red}.

$$\frac{\eta_{sp}}{c} = \eta_{red} \tag{2.54}$$

Polymer solutions are never ideal since dissolved macromolecules influence each other even at very low concentration. On the other hand, a reliable correlation of solution viscosity and molecular weight is only possible if the dissolved macromolecules are not affected by mutual interactions: they must be actually independent of each other. Therefore, the viscosity of polymer solutions should be determined at infinite dilution. However, such measurements are impossible in practice. So one works at an as low as possible polymer concentration and extrapolates the obtained values to zero concentration. To do so, the elution time measurements are not only carried out for one single polymer concentration but for varying polymer concentrations (e.g., 10, 5, 2.5, 1.25 g/l). For each solution, the value of the reduced specific viscosity is figured out (the data will make evident that this quantity is clearly concentration-dependent even at the lowest possible polymer

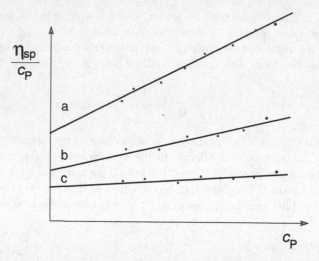

Fig. 2.13 Graphical evaluation of the limiting viscosity number (intrinsic viscosity) from viscosity measurements at different concentrations for (**a**) a high-molecular-weight, (**b**) a medium-molecular-weight, and (**c**) a low-molecular-weight polymer sample

concentrations). Then, the limiting value (intrinsic viscosity, Staudinger index or limiting viscosity number) [η] is determined as a reliable measure of the viscosity behavior of the isolated thread-like molecule at infinite dilution. It is defined by the following expression:

$$[\eta] = \lim_{c \to 0} \frac{\eta_{sp}}{c} \qquad (2.55)$$

Practically, the η_{sp}/c values are plotted against the concentration c, and linear extrapolation is done. [η] is obtained as the y-axis intersect (Fig. 2.13).

Since η_{sp} is dimensionless, [η] has units of reciprocal concentration (e.g., l/g or dl/g). Hence in viscosity measurements the concentration units must always be stated.

Since the intrinsic viscosity depends not only on the size of the macromolecule but also on its shape, on the solvent, and on the temperature, there is no simple relationship for the direct calculation of molecular weights from viscosity measurements. However, the Mark-Houwink-Kuhn equation gives a general description of how the molecular weight can be calculated from the intrinsic viscosity:

$$[\eta] = K \cdot M^a \qquad (2.56)$$

M is the viscosity average molecular weight, and K and a are the Mark-Houwink constants. There is a specific set of Mark-Houwink constants for every polymer-solvent combination. So one has to know these values for the applied

Table 2.8 K_m- and a-values for the calculation of molecular weights from viscosity measurements according to Eq. 2.56, $[\eta]$ in ml/g

Polymer	Solvent	Temperature (°C)	K_m (10^3 ml/g)	a
Polystyrene (atactic)	Toluene	25	7.5	0.75
	Cyclohexane	28	108.0	0.479
Poly(α-methylstyrene)	Toluene	25	7.06	0.744
	Cyclohexane	34.5	73.0	0.5
Polyisobutylene	Cyclohexane	25	40.0	0.72
Polybutadiene (98% *cis*)	Toluene	30	30.5	0.725
Polyisoprene	Toluene	25	50.2	0.667
	Cyclohexane	27	30.0	0.7
Poly(vinyl acetate)	Acetone	25	21.4	0.68
Poly(vinyl alcohol)	Water	25	20.0	0.76
Poly(methyl methacrylate)	Acetone	25	5.5	0.73
Polyacrylonitrile	DMF	20	17.7	0.78
Polyacrylamide	Water	30	6.31	0.8
Poly(ethylene glycol terephthalate)	*o*-chlorophenol	25	17.0	0.83
	Tetrachloroethane	50	13.8	0.87
Polycarbonate from bisphenol A	Chloroform	25	12.0	0.82
Nylon 6,6	*o*-chlorophenol	25	168.0	0.62
Poly(phenylene ether)	Toluene	25	28.5	0.68
Polysiloxane	Toluene	25	18.7	0.66
Cellulose triacetate	Dichloromethane	20	24.7	0.704

polymer-solvent combination in order to obtain an accurate measure of molecular weight. Therefore, for a new polymer for which no Mark-Houwink constants are available no good measure can be achieved. Under these conditions, one obtains only a qualitative idea of whether molecular weight is high or low. One is, therefore, always obliged to establish for each polymer a calibration curve or calibration function by comparison with an absolute method. This, however, is only valid for a given solvent and temperature (Table 2.8).

Mathematical evaluation of M is somewhat inconvenient. Graphical methods are preferable. The above equation can be expressed in logarithmic form:

$$\log[\eta] = \log K + a \cdot \log M \qquad (2.57)$$

so that a double logarithmic plot of $[\eta]$ versus M gives a straight line whose slope corresponds to the exponent a (see Fig. 2.14).

The exponent a depends on the shape of the macromolecules in solution. For rigid spheres $a = 0$; however, most macromolecules are present in solution as more or less expanded coils. Accordingly for most polymers a-values lie between 0.5 and 1.0, with 0.5 being the extreme value for nonexpanded ideal statistic coils (ϑ system) and 1.0 for fully expanded coils. Cases are also known where a is

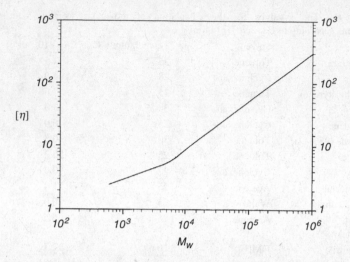

Fig. 2.14 Relation between viscosity and molecular weight for polystyrene in benzene at 20°C

greater than 1. This occurs with particularly stiff and elongated macromolecules, which approximate to the model of a rigid rod in solution, for which $a = 2$.

Since the degree of expansion of the polymer coils is directly dependent on the solvating power of the solvent, under otherwise comparable conditions, both a and $[\eta]$ provide a measure of the "goodness" of a solvent: high values of a and $[\eta]$ (at constant molecular weight and temperature) indicate remarkable coil expansion and therefore a good solvent. Low values of a and $[\eta]$ indicate a bad solvent. For example, the values a for poly(vinyl acetate) in methanol and acetone are 0.60 and 0.72, respectively.

The interactions between solvent and polymer depend not only on the nature of the polymer and type of solvent but also on the temperature. Increasing temperature usually favors solvation of the macromolecule by the solvent (the coil expands further and a becomes larger), while with decreasing temperature the association of like species, i.e., between segments of the polymer chains and between solvent molecules, is preferred. In principle, for a given polymer there is a temperature for every solvent at which the two sets of forces (solvation and association) are equally strong; this is designated the theta temperature. At this temperature the dissolved polymer exists in solution in the form of a nonexpanded coil, i.e., the exponent a has the value 0.5. This situation is found for numerous polymers; e.g., the theta temperature is 34°C for polystyrene in cyclohexane, and 14°C for polyisobutylene in benzene.

The following apparatus is needed to carry out viscosity measurements: a capillary viscometer with suitable mounting, a thermostatted bath, a stopwatch (0.1 s), several graduated 10-ml flasks, and graduated 5-ml and 3-ml pipettes. For the reasons already given the measurements are performed only on dilute solutions. The most commonly used capillary viscometer is the Ostwald viscometer

(Fig. 2.15a). The diameter of the capillary (in general between 0.3 and 0.4 mm) is chosen so that the flow time for the solvent is about 60–150 s. Special versions of the Ostwald viscometer have been developed for measurement of solution viscosity at higher temperatures. Since the viscosity of a solution depends strongly on the temperature, good thermostatting is necessary (accuracy within 0.05–0.1°C).

The viscosity measurements are conducted as follows: 100 mg of well-dried polymer are weighted into a 10-ml graduated flask and dissolved in somewhat less than 10 ml solvent. After the solution has been brought to the temperature of measurement, the solution is made up to the mark (polymer concentration 10 g/l). The polymer solution is now filtered through a glass frit in order to remove dust particles which would seriously disturb the measurement. It is filtered directly into bulb 4 of the viscometer.

The viscometer is suspended vertically in a thermostatted bath (Fig. 2.15.b). After temperature equilibration (about 5 min at 20°C), the surface of the polymer solution can be transferred from arm 2 to mark M_1 in arm 1 only by applying a slight pressure on the opening of arm 2 with a rubber bulb. The time required for the solution to flow from mark M_1 to mark M_2 is measured and the average of five measurements is taken as the flow time t. Depending on the total flow time, they should not deviate from one another by more than 0.2–0.4 s. The flow time of the solvent t_0 is likewise determined with a filtered 3-ml sample; this determination should be carried out each time before beginning the measurements on the solutions since it provides a simple and accurate check of the entire set-up (temperature control, cleanliness of the viscometer, etc.) The viscometer is now removed and the polymer solution is poured out through arm 2. After attaching the headpieces a and b, the viscometer is rinsed several times with pure solvent (application of slight vacuum at headpiece a) and then with purified acetone. It is finally dried by drawing air through the viscometer (the sintered glass filter should be covered with a piece of filter paper). The viscometer is then ready for the next measurement.

The viscosity behavior described so far is valid only for uncharged polymers. If polyelectrolytes are analyzed, a quite different viscosity behavior may be found in polar solvents (e.g., polymeric acids in water). The η_{sp}/c values at first fall off with decreasing concentration as for uncharged polymers but then climb steeply and may drop down later again (see Fig. 2.16). Addition of salt to the solution of polyelectrolytes (e.g., 1% and 5% sodium chloride in aqueous solution) restores, step by step, the normal behavior (see Fig. 2.16, curves b and c).

This is connected with the fact that in polyelectrolytes the shape and density of the macromolecular coils is affected by the degree of ionization and that the long-range intermolecular coulomb forces depend on the ionic strength of the solution. In the ionized state, the like charges distributed along the length of a macromolecule repel each other, leading to a marked coil expansion and hence a considerable increase in viscosity. Also, upon diluting the polymer solution the range of the intermolecular repulsion becomes larger, and mutual electrostatic interaction of the dissolved macromolecules becomes stronger despite of the increasing distance of the charged coils. Every factor that causes an increase in the degree of dissociation or a decrease in ionic strength, therefore, leads to a rise in the solution viscosity, and

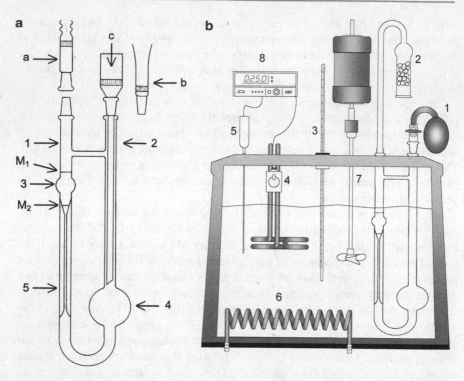

Fig. 2.15 (**a**) Ostwald viscometer: Total length: 25 cm; capillary length: 10 cm; bulb 3: diameter 1.3 cm, bulb 4: diameter 2.2 cm, filling level: 2 or 3 ml, flow volume: 0.5 ml; *a, b*: head pieces; *c*: sintered glass filter for filtration of solvent and polymer solution (**b**) Thermostatted bath and mounting for Ostwald viscometer: 1: rubber bulb; 2: drying tube; 3: thermometer; 4: heating coil; 5: thermo indicator; 6: cooling coil; 7: stirrer; 8: power and controlling unit

vice versa. The viscosity behavior of aqueous solutions of polymeric acids [e.g., poly(methacrylic acid)] of various concentrations can then be explained as follows: On dilution of the aqueous solution the normal effect is first observed, i.e., the viscosity decreases. With further dilution the increasing degree of dissociation of the carboxylic groups becomes noticeable and the ionic strength is as much lowered that intermolecular repulsion of the coils becomes relevant. This leads to an increase of the viscosity. The effect caused by the higher degree of dissociation and the high Debye length exceeds first that resulting from dilution; hence η_{sp}/c rises. Finally, the predominance is inverted, and the reduced viscosity drops down again. On addition of sodium chloride, the degree of ionization is essentially held steady, and therefore also the coil expansion; the rise in viscosity with decreasing concentration of polymer is thus suppressed.

Size-Exclusion (Gel Permeation) Chromatography

Synthetic polymers do not contain macromolecules of only one single molecular weight but are composed of macromolecules having a distribution of molecular

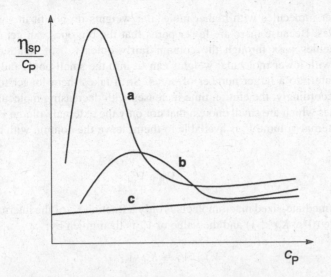

Fig. 2.16 Behavior of a polyelectrolyte in dilution viscosimetry in water: (**a**) without, (**b**) with a low quantity and (**c**) with a higher quantity of a low-molecular-weight electrolyte (salt) which screens the electrostatic interactions

weights. All the analytical techniques described so far (except MALDI-MS) represent the distribution curve by means of different averages of the molecular weight, i.e., M_n, M_w, M_η. Gel permeation chromatography (GPC, also called size-exclusion chromatography, SEC) does deliver such averages as well. However, it is moreover able to give the entire distribution curves. Thus GPC is a very powerful method of polymer fractionation and has become a standard method for determination of molecular-weight distribution and relative molar mass.

In a GPC experiment, the polymer is separated in a column which is filled with a swollen, uniformly packed resin ("gel", called stationary phase, while the solvent which passes through the column is called mobile phase). The gel beads are usually made of crosslinked polymers (in particular polystyrene but also various inorganic porous materials) with little holes and pores of different size where the pore diameter is of the dimension of the size of the solvated polymer coils, i.e., the pore-size distribution is approx. 10–10^5 nm.

A solution of the polydisperse polymer in the same solvent as was used to swell the resin is placed on the top of the column and eluted in the same manner as for standard column chromatography or high-pressure liquid chromatography (HPLC). In GPC, however, it is not the interaction of the dissolved analyte molecules with the stationary phase relevant for separation but the different (hydrodynamic) volumes of the polymers associated with their different molecular weights. Only solvent molecules and those macromolecules whose size is less than the prevailing pore size can diffuse into the pores of the swollen gel: their separation succeeds because the polymer molecules get caught up in the holes in the beads, then come out, pass on down the tube a little way, then get caught in another pore, and so on.

Big polymer molecules with higher molecular weights do not fit in some of the smaller holes. Because there are fewer pores that the big ones can get caught in, these molecules pass through the column fairly quickly. But smaller polymer molecules with lower molecular weights can fit into the small pores and therefore will penetrate into a larger number of pores. So it takes them longer to pass the column. Accordingly, the elution time increases with decreasing molecular size.

Molecules which are small enough that not only the external volume V_0 but also the total internal volume V_i is available to them, leave the column with an elution volume, V_e:

$$V_e = V_0 + V_i \qquad (2.58)$$

For intermediate-sized macromolecules only a fraction K_d of the internal volume is accessible ($0 < K_d < 1$) and the value of V_e is then given by:

$$V_e = V_0 + K_d \cdot V_i \qquad (2.59)$$

The constant K_d is the apparent distribution coefficient for the distribution of a substance between the swelling medium inside and outside the gel particles. K_d depends mainly on the molecular size and to a lesser extent on the shape of the molecule in solution.

Very large macromolecules cannot penetrate into the pores of the gel. Hence, such large molecules cannot be separated from one another. The so-called exclusion limit gives an approximate indication of the limiting molecular weight up to which the macromolecules of the polymer to be fractionated can penetrate the network and therefore be separated. Network structure and exclusion limit are closely related: the tighter the network, the smaller the exclusion limit.

The efficiency of fractionation by gel chromatography not only depends on the type of gel but also on the dimensions of the column. The internal volume V_i of the gel pores is determined by the amount of dry resin used and by its swellability, which in turn depends upon the eluting agent. The total volume of the gel bed V_t is thus made up of the volume of the gel framework, the internal volume V_i of the gel, and the external volume V_0 between the gel particles. The external volume V_0 is identical with the elution volume V_e of a substance with a molecular weight above the exclusion limit. Macromolecules of this size cannot penetrate the network but pass through the column unimpeded. V_0 can thus be readily determined.

In order to detect the macromolecules that elute from the column, detectors are needed that can count how many polymer molecules are coming out of the end of the column at a given time interval. The polymer concentration in the eluate can be determined discontinuously by precipitation and weighing of the dry polymer. The commercially available GPC equipment measures continuously the refractive index or the difference in refractive index between the solution and the pure solvent. The polymer concentration can also be determined spectroscopically (e.g., by UV–vis) providing the macromolecules possess relevant absorption bands. Using the thus collected data, a plot of time can be made on the x-axis and the number of polymer molecules coming out at a given time on the y-axis. As GPC is not an absolute method, calibration is required. For this, one usually takes samples of very narrow and well-known molecular-weight distribution. Calibration of the GPC column(s) in use delivers a calibration curve which correlates the elution time (or volume) with the logarithm of the polymer's molecular weight, log M. Using this calibration curve, the molecular weight can be calculated from elution time. This results in a plot of molecular weight on the x-axis and the number of molecules with a particular weight on the y-axis. On this plot, molecular weight *decreases* from left to right.

When a size-exclusion chromatograph is calibrated correctly, one can know the molecular weight of a polymer just based on the time it takes to pass, or *elute* through the column. From Fox and Flory's theory of solution viscosity one can learn that the size of a solvated macromolecular coil is directly correlated with its solution viscosity. The correlation is:

$$[\eta]M = \Phi \left(<r^2>_0 \right)^{3/2} \cdot a^3 \tag{2.60}$$

A universal calibration is therefore possible for SEC by plotting log ($[\eta] \cdot M$) vs. V_e when a viscosity detector is used. Absolute molar masses can be obtained using a light-scattering detector.

SEC became the most widely used method for molar mass and molar mass distribution determination due to its broad applicability, easy sample preparation, and the large amount of information resulting from the full distribution curve. The commercially available SEC systems work automatically with small sample amounts and even at elevated temperatures. In addition, chromatographic systems coupled with spectroscopic methods giving chemical information on the separated fractions gain more and more importance for analysis of complex polymer systems and mixtures.

2.3.3.4 Determination of Molecular-Weight Distribution by Fractionation

Polymer syntheses nearly always result in polydisperse products, i.e., are composed of macromolecules of different molecular weights. Since many physical properties depend not only on the average molecular weight but also on the broadness and the shape of the molecular-weight distribution (MWD) curve, it is an important technique to determine (and perhaps to modify) the MWD by fractionation. While there is no separation procedure which provides truly monodisperse samples from the polydisperse starting material, nevertheless, one can obtain fractions whose MWD is really small. These fractionation methods are based on the (slightly) decreasing solubility of polymers with increasing molecular weight: simplisticly, phase separation into a polymer-rich gel phase and a solvent-rich sol phase occurs in an originally homogeneous polymer solution when the Flory-Huggins interaction parameter χ exceeds a critical value χ_c. The critical value of χ_c decreases with increasing molecular weight of the polymer:

$$\chi_c = \frac{1}{2}\left(1 + m^{-\frac{1}{2}}\right)^2 \approx \frac{1}{2} + m^{-\frac{1}{2}} \qquad (2.61)$$

with $m = V_p/V_{solvent}$ and V_p and $V_{solvent}$ being the mole volume of the polymer and the solvent, respectively. In dilute solutions the second virial coefficient of the osmotic pressure, A_2, depends on c as follows:

$$A_2 = \left(\frac{1}{2} - \chi\right)RT \cdot \rho_p^{-2} \cdot V_{solvent}^{-1} \qquad (2.62)$$

with ρ_p being the density of the dissolved polymer. Phase separation starts as soon as the value of A_2 drops below a critical value $A_{2,c}$:

$$A_{2,c} \approx -m^{-\frac{1}{2}}RT \cdot \rho_p^{-2} \cdot V_{solvent}^{-1} \qquad (2.63)$$

For $m \to \infty$, the critical value is identical with that in a ϑ solvent, i.e., $A_2 = 0$ and $\chi = 0.5$. Since the solubility of macromolecules decreases with increasing molecular weight, it is possible to separate these materials with respect to their molecular weights by changing the composition of the solvent and/or the temperature. In general, one roughly distinguishes between two methods, namely fractional precipitation and fractional extraction.

To fractionate a polymer by precipitation, a precipitant is slowly added to the polymer solution (concentration of polymer 0.1–1 wt.%) at constant temperature until a persistent cloudiness appears. After some time the droplets separate as a second liquid (or swollen gel) phase. This fraction contains the highest molecular weight components and is separated by decantation or centrifugation. Further precipitant is then added to the majority layer until further phase separation is observed. Then the above procedure is repeated several times until all polymer is separated off. A disadvantage of fractional precipitation is that the residual solutions become more and more dilute so that separation of the late fractions might be difficult. Furthermore, the method is rather time consuming since the formation of the gel phase occurs very slowly.

These disadvantages can be circumvented to some extent by means of the Meyerhoff's triangular fractionation technique. Here, precipitant is added to the dilute polymer solution until about approximately half the polymer material is separated off in the gel phase. After separating the gel it is redissolved. Now, one has two solutions with which one proceeds as for the original solution, i.e., enough precipitant is added to bring about separation of approximately half the polymer in each case, and so on. In this way large volumes of solution can be avoided and there is a considerable saving in time, since several fractions can be worked up simultaneously.

In some cases the molecular-weight distribution can be determined by turbidimetric titration, a technique which is based on the fractional precipitation. A precipitant is added to a very dilute solution of the polymer, and the resulting

turbidity is measured as a function of the amount of added precipitant; the preparative separation of the fractions is thereby avoided. If the polymer is chemically
homogeneous, the mass distribution function can then be calculated. Turbidimetric
titration is also suitable as a means for establishing the best fractionation conditions
(e.g., choice of solvent/precipitant combination, size of fractions, etc.), before
carrying out a full-scale fractionation by precipitation.

Fractional extraction is free from the disadvantages encountered in fractional
precipitation. Here, the technique consists in extracting the polymer with a series of
solvent/precipitant mixtures, the proportion of solvent being increased stepwise.
Since one begins with the poorest solvent mixture – in contrast to fractional
precipitation – the first fraction contains the low-molecular-weight components
and the final fraction the high-molecular-weight components. The isolation of large
fractions is possible by fractional extraction using a modified technique. The
physical state of the polymer is very important for the efficiency of fractional
extraction.

Once the amounts and molecular weights of the fractions have been determined,
the molecular-weight distribution of a polydisperse material can be expressed
graphically in the form of a distribution curve. The mass distribution function is
written as:

$$m_P = H(P) \tag{2.64}$$

where m_p is the mass fraction (in gram or in%) of macromolecules with a degree of
polymerization of P. At first sight it seems reasonable to present the fractionation
results in the form of a histogram in which the amount in percentage is plotted
directly vs. the degree of polymerization P. However, this representation only gives
a realistic picture if each individual fraction deviates from its molecular weight
average by the same amount, i.e., if the width of the steps are all equal. Unfortunately this condition is rarely fulfilled in practice. Therefore, the amount of the
fraction should not be represented by a line (height of step), but by an area, i.e., so
that the mass distribution function should be applied in integral form:

$$I(P) = \int_1^P H(P) \cdot dP \tag{2.65}$$

The integral distribution function is obtained from the experimental data as
follows: The mass distribution within a given fraction is approximately symmetri-
cal. Assumed the degree of polymerization of the mth fraction is P_m one can expect
that half the fraction has a smaller degree of polymerization and the other half
a larger one. Moreover, since fractions 1 to $(m-1)$ all have a degree of polymerization smaller than P_m one obtains – by summing the amounts of all fractions from 1
to $(m-1)$ and adding half the amount of the mth fraction – the mass fraction of all

Table 2.9 Fractionation of polystyrene of average degree of polymerization P = 800

Fraction No.	Amount [%]	$I(P)$ [%]	$m_p \cdot 10^2$ [%]	P
8	3.4	1.7	1.10	170
7	3.7	5.25	3.45	360
6	7.3	10.75	4.35	430
5	16.8	22.8	6.95	680
4	24.9	43.7	7.55	900
3	9.9	61.6	5.30	1,300
2	26.5	79.25	4.25	1,470
1	7.5	96.25	1.40	2,240

degrees of polymerization from zero to P_m and hence a pair of values of the above integral equation.

For example, to obtain the value of I(P) for fraction 5 in Table 2.9 one adds the percentage amounts of fractions 8, 7, and 6 (3.4 + 3.7 + 7.3 = 14.4) and adds half the amount of fraction 5 (14.4 + 8.4 = 22.8). The integral distribution curve (see Fig. 2.17) is obtained by plotting $I(P)$ values obtained in this way, versus the corresponding degree of polymerization.

A polymer is the more uniform with respect to the molecular weight the steeper the integral distribution curve is. The differential mass distribution function

$$dm_P = H(P) \cdot dP \qquad (2.66)$$

in which dm_p is the mass fraction (in gram or %) with degrees of polymerization between P and $P + dP$, can be most simply obtained by graphical differentiation of the integral curve. This gives the values of dm_p/dP. An inflection point in the integral curve corresponds to a maximum in the differential curve. This procedure gives information about the distribution that is more detailed, the sharper the fractions and the greater their number. This kind of differential distribution curve tells us how many macromolecules there are with a given degree of polymerization P in the polymer sample. It normally has a single maximum and resembles a Gaussian bell-shaped curve. Distribution curves with two or more maxima are an indication of side reactions during the preparation, or degradation reactions during fractionation.

Proper fractionation according to molecular weight can only be expected if the macromolecules are chemically uniform. For polymers that are nonuniformly branched, or in which partial chemical conversions were carried out, or which are statistic, graft, or block copolymers, the solubility also depends on factors other than the molecular weight. Fractionation can then lead to separation according to the chemical composition of the polymer. These results will, therefore, only lead to a reliable molecular-weight distribution if there is convincing evidence for the chemical uniformity of the single fractions. This can be achieved by elemental analysis, spectroscopy, or pyrolysis gas chromatography. By changing the solvent,

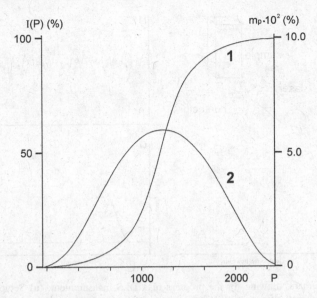

Fig. 2.17 Integral (1) and differential (2) mass distribution function of a polystyrene sample

precipitant, or eluent, it is sometimes possible to choose between fractionation according to molecular size and fractionation according to chemical composition.

2.3.4   Determination of size distribution of Polymers Dynamic Light Scattering (DLS)

Dynamic Light Scattering (DLS) is a fast method for the determination of size distributions of polymers in solution. A clear distinction must be drawn between DLS and static light scattering, discussed in Sect. 2.3.3.2.4, which is an absolute method for the determination of $\overline{M}_w$. Instead of angle dependent light scattering intensity (R_ϑ), *fluctuation* of light scattering intensity (I) of a polymer in solution is observed over a certain period of time by the DLS-method.

Brownian motion causes dispersed macromolecules to diffuse in and out of a certain volume of the solution. Hence, the number of scattering centres and therefore the light scattering intensity changes permanently, which can be observed in a short time scale. In case of small molecules, intensity fluctuations of scattered light are fast but low in their amplitude and for large molecules vice versa. Figure 2.18 shows the observable measurement signal for two different sizes.

If the obtained intensity curve is correlated with itself by a small time shift (τ), curves for larger molecules resemble themselves over a longer period as small molecules. Plotting the similarity versus applied time shift results in an exponential decaying curve. This curve is called the correlogram (chart 1) and can be fitted by the autocorrelation function (1). This equation depends on the translational

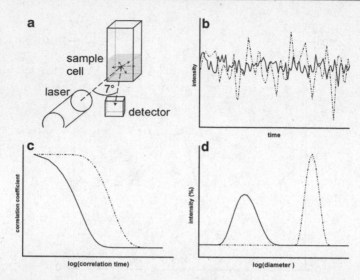

Fig. 2.18 Schematic drawing for the progress of a DLS measurement: (**a**) Setup of a DLS-apparatus in backscattering mode. (**b**) Typical intensity fluctuation for a large (•••••) and a small (—) molecule. (**c**) Correlogram for a large (•••••) and a small (—) molecule. As can be seen, correlograms for small molecules decay more rapidly. (**d**) Obtained size distributions by intensity for a large (•••••) and a small (—) molecule

diffusion coefficient (D), which is easy to understand regarding the explanation given above. Other parameters, which affect the autocorrelation function are: the wavelength of the laser λ_0, the scattering angle θ and the refractive index of the dispersant $\bar{n}$:

$$G(t) = \int_0^\infty I(t)I(t+\tau)dt = B + Ae^{-2\frac{4\pi\bar{n}}{\lambda_0}\sin\left(\frac{\theta}{2}\right)D\tau} \qquad (2.67)$$

By knowledge of the measurement temperature (T) and the system viscosity (η), the averaged size of observed molecules can be calculated via the Stokes-Einstein equation (Eq. 2.68):

$$d_h = \frac{kT}{3\pi\eta D} \qquad (2.68)$$

It should be noted that it is the *hydrodynamic* diameter (d_h), which is determined by DLS. d_h represents a sphere consisting of the dispersed macromolecule as well as surrounding molecules, which have the same value of D as the macromolecule. Particularly the solvent shell and polar interactions with ions should be considered. Moreover, aggregation of macromolecules may occur, which lead to bigger spheres.

A further parameter, which is of relevance beside d_h, is the width of size distribution. This parameter, referenced as polydispersity index (PDI), is calculated from the deviation between the autocorrelation function and values actually measured. For narrowly dispersed samples PDI is below 0.1. PDI values above 0.3 point at widely size dispersed samples, which should be fitted by a more complex model. Several algorithms have been developed, which all base on a sum of autocorrelation functions. As each autocorrelation function results in one Gaussian distribution, the obtained size distribution turns out to be a sum of optionally superimposed curves. Thus, these complex algorithms reproduce reality better but are also more affected by measurement errors.

Special attention should be given to the measurement parameter I. So, the resulting distribution received from DLS measurement is an intensity distribution. In fact, number distributions are more interestingly. Though, for small molecules ($d_h \leq \lambda/10$), Rayleigh scattering theory is applicable involving $R_\vartheta \propto d_h^6$! Hence, DLS is much more sensitive to larger molecules or aggregates entailing the calculation of number distributions is strongly affected by small measurement errors in the region of small values of D. Thus, DLS results should always be created from a couple of measurements regarding carefully their standard deviation. In addition samples have to be prepared accurately as it is described for static light scattering in Sect. 2.3.3.2.4.

2.3.5 Polymer Characterization in Bulk

2.3.5.1 Determination of Density

The densities of polymers can be determined by the pyknometer technique or by the flotation method. In the pyknometer technique the liquid volume displaced by the polymer sample is determined by weighing. Most polymers have a density larger than that of water, which can, therefore, be used as the liquid. Polymers in the form of powders or pressed discs tend to adsorb or occlude air bubbles, which can lead to serious errors. This can be largely prevented by careful degassing of the pyknometer and polymer sample under vacuum before filling with liquid, and/or by addition of a small amount (0.1%) of commercial detergent to lower the surface tension of the water.

The flotation method is especially suitable for powdered polymers. A liquid of higher density is added to a liquid of lower density until the test particles neither sink nor rise to the surface. The densities of the solid and liquid mixture are then equal and it remains only to determine the density of the latter. The experiment is conducted as follows: The powdered polymer is placed in a small beaker with a certain amount of less dense liquid. The heavier liquid is then run in from a burette with gentle stirring until the state of suspension is attained.

The density of the liquid mixture can be determined using a pyknometer or it can be derived from a previously determined calibration curve. For the determination of densities less than unity (e.g., polyethylene), ethanol/water mixtures are suitable; for densities larger than unity one may use mixtures of water with aqueous salt solutions (40% $CaCl_2$ solution: $d_4^{20} = 1.40$ g/ml; 72% $ZnCl_2$ solution: $d_4^{20} = 1.95$ g/ml). The density gradient method, which is an elegant variation of the flotation method, should also be mentioned.

2.3.5.2 Determination of Crystallinity

The simplest way of establishing qualitatively the crystallinity of a polymer is by the observation of birefringence under a suitable microscope, taking care to exclude the possibility of orientation birefringence. Also thermotropic liquid crystalline polymers can show birefringence combined with relatively low viscosity. X-ray diffraction allows a quantitative determination of the degree of crystallinity as well as the usual crystallographic data.

In some cases crystalline polymers show additional absorption bands in the infrared spectrum, as in polyethylene ("crystalline" band at 730 cm^{-1}, "amorphous" band at 1300 cm^{-1}) and polystyrene (bands at 982, 1318, and 1368 cm^{-1}). By determining the intensity of these bands it is possible to follow in a simple way the changes of degree of crystallinity caused, for example, by heating or by changes in the conditions of preparation.

The degree of crystallinity is also reflected in the density of the polymer so that the determination of density provides at least a relative measure for crystallinity. Differential scanning calorimetry (DSC) is frequently applied to determine the crystallinity from the heat of crystallization or melting (see Sect. 2.3.5.8).

2.3.5.3 Glass Transition Temperature

Below a certain temperature any amorphous polymer behaves as a hard glass. When heated above this temperature individual segments of the macromolecules achieve larger mobility (micro-Brownian motion); as a result the polymer becomes soft and elastomeric. The temperature at which this change sets in is called the glass transition temperature T_g. This temperature is very important in technological applications and depends, amongst other things, on the chemical nature of the polymer, on the configuration, on the degree of crystallinity, on the length of the side chains, and on the degree of branching. The glass transition temperature is generally determined with specialized equipment and depends to some extent on the method used. One measures the temperature dependence of a particular physical quantity such as refractive index, elastic modulus (or torsional modulus), dielectric constant, heat capacity, expansion coefficient, or specific volume; all these change abruptly at the glass transition temperature. In amorphous polymers the glass transition temperature T_g frequently coincides with the softening point. In crystalline polymers, on the other hand, the crystallite melting point lies considerably above T_g. For crystalline polymers the glass transition temperature can be estimated rather well by rule of thumb (Boyer-Beaman rule): it is about two-thirds of the crystallite melting point, expressed in Kelvin.

The measurements of Young's modulus in dependence of the temperature (dynamic-mechanical measurements, see Sect. 2.3.6.2) and the differential thermal analysis (DTA or DSC) are the most frequently used methods for determination of

Table 2.10 T_g- and T_m-values of some polymers (Source: Polymer Handbook 3rd edition, 1989, Wiley & Sons, Inc., pp VI/393)

Polymer	T_g [°C]	T_m [°C]
cis-1,4-Polyisoprene	−73	–
Poly(n-butyl acrylate)	−55	–
Poly(ethyl acrylate)	−24	–
Poly(isobutyl acrylate)	−24	–
Poly(methyl acrylate)	6	–
Poly(n-butyl methacrylate)	20	–
Poly(vinyl acetate)	31	–
Poly(ethyl methacrylate)	65	–
Polystyrene	100	–
Poly(methyl methacrylate)	105	–
Polycarbonate	145	–
Poly(α-methylstyrene)	168	–
cis-1,4-Polybutadiene	−102	11[a]
trans-1,4-Polybutadiene	−93	83[a]
Polyoxymethylene	−83	183
Polyisobutylene	−73	44[a]
Poly(tertrafluoroethylene)	−33	332
Polyethylene	−21	142
Polypropylene	−13	188
Polyamide-6 (Nylon-6)	40	260
Polyamide-6,6 (Nylon-6,6)	50	280
Poly(ethylene terephthalate)	69	280
Poly(vinyl chloride)	81	273[a]
Polyacrylonitrile	105	320[a]

[a]A few percent crystallinity can be reached after special post-treatment, e.g., stretching or by influencing the chain structure

the glass transition temperature. In Table 2.10 are listed T_g and T_m values for several amorphous and crystalline polymers.

2.3.5.4 Softening Point

The softening point is generally determined by slowly heating a test piece under constant load until it experiences a certain deformation. The temperature at which this occurs is known as the softening point. Since the methods used involve empirical and arbitrarily chosen test parameters, the softening point is physically less well-defined than the glass transition temperature. In practice, the softening point is generally determined by one of three relatively simple methods.

The most widely used method is that of Vicat in which a blunt steel needle (area of point 1 mm^2) is applied vertically to the surface of a test piece (about 1 cm^2 and 3–4 mm thick) under a load of 49 N. The oven temperature is then raised at 50°C/h and the temperature determined at which the needle has sunk 1 mm into the test piece; this is taken as the softening point (Vicat temperature).

In the method of Martens a test rod is mounted upright in a support and the upper free end is put under a bending stress via a small weighted lever. The rod is slowly heated in an oven until a specified deflection is attained. The softening point determined in this way is called the Martens temperature.

Finally, in the English-speaking countries, there is another extensively used method in which a rod is supported at its two ends and a load is placed on the center; this is slowly heated in a liquid bath until a certain distortion is attained; the temperature at which this occurs is called the heat distortion temperature (HDT). These methods are applicable both to thermoplastics and to crosslinked polymers. Interconversion of the softening temperatures determined by the different techniques is not possible.

2.3.5.5 Crystallite Melting Point

The melting point of a low-molecular-weight compound is an important quantity, which, through determination of the mixed melting point with a reference sample, can serve for its identification. On the other hand, macromolecular substances seldom have a precise melting point, but melt instead over a certain temperature range that is influenced by various factors. The melting point or the melting range of a polymer can thus assist its characterization, but not its identification in the sense used for low-molecular-weight compounds.

Recognition of the melting range is relatively simple if the polymer is partly crystalline because the change of birefringence can then be observed using a hot-stage polarizing microscope. First, the approximate melting temperature of the polymer is determined on a Kofler hot-block. A small amount of polymeric substance is then melted on a microscope slide using the hot-block at a temperature somewhat above the melting range. A normal coverslip, heated to about the same temperature, is finally pressed onto the molten polymer with a cork, so that a thin homogeneous film is produced. The slide is now placed on the hot stage of a melting-point microscope at a temperature about 20°C below the melting point. The sample is observed under polarized light and slowly heated. The temperature range from the first noticeable change to the final disappearance of birefringence is noted. The mean value of the first and last readings may be taken as the crystallite melting point. Many polymers recrystallize on cooling so that duplicate measurements are possible. The temperature at which crystallization commences on cooling from the melt is called the *crystallization temperature*. It is not identical with the melting point of the crystallites but lies somewhat below it, on account of the hindered motion of the macromolecules in the molten state.

The determination of the melting range according to the method described can be upset by the fact that a noncrystalline polymer can also show birefringence if its macromolecules have been oriented by the action of external forces. For example, orientation can occur when the coverslip is pressed onto the molten polymer, or, if a film is used, during the preparation or cutting of the test piece. The disappearance of birefringence due to orientation does not then occur at the melting point of the sample. Such errors can be avoided by duplicate determinations on the same sample, since once the polymer has been melted it is unlikely to undergo any reorientation in the absence of external forces. Today's usual standard procedure to characterize the melting process of polymers is differential scanning calorimetry (DSC; see Sect. 2.3.5.8). From the thus obtained thermogram (see Fig. 2.19, for example), the melting temperature T_m corresponds to the temperature at the peak maximum. In addition, the heat of melting can be obtained through integration of the curve in reference to an imaginary baseline. Conducting the measurement by cooling the molten probe, one can observe and record the crystallization process.

For amorphous polymers the melting range depends, amongst other things, on the molecular weight distribution and the degree of branching. An effect of molecular weight is only observed in the oligomeric region, e.g., for poly(oxymethylene dimethyl ethers) or for oligoamides and oligoesters. This is similar to the usual increase in melting point with molecular size observed in a homologous series of low-molecular-weight compounds. The glass transition temperatures (T_g) and the melting points (T_m) of some selected polymers are listed in Table 2.10. Crosslinked polymers do not melt, but possess a softening range that generally lies in the vicinity of the decomposition temperature. There are also some uncrosslinked polymers that likewise have no melting point or range (e.g., polyacrylonitrile). They begin to decompose above a certain temperature.

2.3.5.6 Melt Viscosity (Melt Flow Index)

Polymer melts are in most cases viscoelastic, meaning the viscosity is dependent on measurement conditions, like frequency or shear rate. Most polymer melts show a viscosity plateau at low shear rates, called Newtonian plateau and the viscosity drops with increasing shear rate.

Two main types of rheometers are suitable for the determination of the viscosity of polymer melts: Rotational rheometers (working in rotation or oscillation mode, geometries used are Couette-geometry, cone and plate, or plate and plate) and capillary rheometers where a stamp is pressed with a certain velocity trough a capillary filled with the melt. Whereas rotational rheometers have a limited shear rate range and are suitable for rheological investigations characterizing structural changes, capillary rheometers and especially high pressure capillary rheometers give information about the processability of the polymer at high shear rates, as they apply in extrusion or injection molding.

Simplified versions of capillary rheometers are melt flow testers, which are especially suitable for laboratory use since they are relatively easy to handle and represent a fast measure. However, they only give information about the fluidity at a certain shear rate which however for chemists in most cases is enough to judge the material's fluidity. Using a melt flow tester the measure of fluidity is not expressed in terms of the melt viscosity but as the amount of material extruded in a given time (10 min). The amount of extrudate per unit of time is called the melt index or melt flow index i (MFI). It is also necessary to specify the temperature and the shearing stress or load. Thus $MFI/2$ (190 C) $= 9.2$ g/10 min means that at 190 C and 2 kg load, 9.2 g of polymer melt are extruded through a standard nozzle in 10 min. Today the expression melt flow index is often replaced by the melt flow rate (MFR)

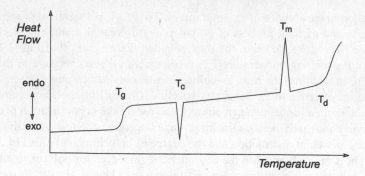

Fig. 2.19 Schematic DSC thermogram of a semicrystalline polymer: T_g = glass transition temperature; T_c = crystallization temperature; T_m = melting temperature; T_d = decomposition temperature

indicating the volume extruded under the above mentioned conditions (unit $cm^3/10$ min).

The melt flow tester consists of a heatable cylinder fitted with a standard nozzle at the lower end, the upper opening being closed by a piston that can be loaded with different weights. The cylinder is heated to the required temperature and then filled with powdered or granulated polymer; the piston is loaded with the desired weight. After a specified initial melting period (about 5 min) the piston catch is released so that the polymer melt is forced through the nozzle. At intervals of 1 min (longer for highly viscous melts) the extruded polymer is cut off at the nozzle and weighed. The time is measured with a stopwatch. The melt index is calculated from the mean of at least five measurements.

The frequency dependent viscosity of a polymer melt reflects even small changes in molecular weight or molecular weight distribution as well as branching and crosslinking or chemical variation, especially when using rotational rheometers in oscillation mode and regarding low frequencies/shear rates. Therefore, measuring of melt viscosity curve delivers not only valuable data for the selection of methods and conditions for the processing of thermoplastic polymers; moreover, it is an easy analytical tool for characterizing macromolecular substances. In this respect it is superior to the measurement of solution viscosity.

2.3.5.7 Thermogravimetry

Prior to any other bulk characterization of a polymer it is necessary to know more about its thermal stability under inert conditions (spontaneous chain degradation, carbonization, dehydration, etc.) as well as in the presence of oxygen (oxidation). Also, it might be of interest whether there is still monomer, solvent, or other volatile material present in the sample. The activity of stabilizers (against depolymerization or oxidation, for example) and the content of inorganic fillers and reinforcing fibers is, moreover, often required information. All these aspects can be studied by thermogravimetric analysis (TG or TGA). This technique monitors the weight loss of a sample in a chosen atmosphere as a function of temperature. The temperature program is in general a linear increase, but isothermal studies can be carried out as well.

The equipment required for TGA is a thermobalance. It is composed of a recording balance, a furnace, a temperature programer, a sample holder, an enclosure for establishing the required atmosphere, and a means of recording and displaying the data. Balance sensitivity is usually approx. one microgram, with a total capacity of a few hundred milligrams. A typical operating range for the furnace is ambient to 1000°C, heating rates are up to 100 K · min^{-1}. The sample temperature is measured by a thermocouple close to the sample. Sample holder materials commonly available include aluminum, platinum, silica, and alumina. Many factors affect the TGA curve. The primary factors are heating rate and sample size. The particle size of the sample material, its morphology, and the gas flow rate can affect the progress of the thermal reaction.

Gas chromatography (GC) and mass spectrometry (MS) can be coupled to the TGA instrument for online identification of the evolved gases during heating: pyrolysis-GC/MS is a popular technique for the evaluation of the mechanism and the kinetics of thermal decomposition of polymers and rubbers. Moreover, it allows a reliable detection and (semi)quantitative analysis of volatile additives present in an unknown polymer sample.

2.3.5.8 Differential Scanning Calorimetry (DSC)

Differential scanning calorimetry (DSC) analyzes thermal transitions occurring in polymer samples when they are cooled down or heated up under inert atmosphere. Melting and glass transition temperatures can be determined as well as the various transitions in liquid crystalline mesophases. In a typical DSC experiment, two pans are placed on a pair of identically positioned platforms connected to a furnace by a common heat flow path. One pan contains the polymer, the other one is empty (reference pan). Then the two pans are heated up at a specific rate (approx. 10 K · min^{-1}). The computer guarantees that the two pans heat at exactly the same rate – despite the fact that one pan contains polymer and the other one is empty. Since the polymer sample is extra material, it will take more heat to keep the temperature of the sample pan increasing at the same rate as the reference pan. A plot is created where the difference in heat flow between the sample and reference is plotted as a function of temperature. When there is no phase transition in the polymer, the plot parallels the x-axis, and the heat flow is given in units of heat, q, supplied per unit time, t:

$$\frac{q}{t} = heat\ flow \tag{2.69}$$

With the heating rate defined as rise of temperature, ΔT, per unit time, t

$$\frac{\Delta T}{t} = heating\ rate \tag{2.70}$$

the heat capacity, C_p, follows as:

$$\frac{\frac{q}{t}}{\frac{\Delta T}{t}} = \frac{q}{\Delta T} = C_p = heat\ capacity \tag{2.71}$$

When the glass transition temperature of the polymer sample is reached in the DSC experiment, the plot will show an incline. It is obvious that the heat capacity increases at T_g, and therefore DSC can monitor the T_g of a polymer. Usually the middle of the incline is taken to be the T_g. Above T_g, the polymer chains are much more mobile and thus might move into a more ordered arrangement: they may assume crystalline or liquid-crystalline order. When polymers self-organize in that way, they give off heat which can be seen as an exothermal peak in the DSC plot. The temperature at the highest point is usually taken as the crystallization temperature, T_c. The area of the peak corresponds to the latent energy of crystallization. When the thus obtained (semi)crystalline polymer is heated further in the DSC pan, its melting temperature, T_m, can be reached. Here, either thermotropic liquid-crystalline phase might be formed or the material changes into its amorphous melt. The sample absorbs heat during melting, and thus an endothermal peak appears in the DSC plot. The heat of melting is obtained by measuring the peak area, and the temperature at the apex of the peak is taken to be the melting point. When all transitions are compiled in one single plot, an idealized DSC plot of a semicrystalline polymer is obtained (see Fig. 2.19).

Amorphous polymers will not show crystallization or melting peaks but a glass transition; liquid-crystalline polymers show some additional endotherms between T_m and isotropization. Semicrystalline polymers show also a defined exothermic recrystallization peak in a cooling cycle after reaching full isotropization (T_m). Because glass transition is associated with a change in heat capacity but no latent heat is involved with this process, the glass transition is called a second-order transition. Transitions like melting and crystallization, which do have latent heats, are called first-order transitions.

The observation and the exact position of T_g, and especially of T_c and T_m are strongly dependent on the thermal history of the sample. Therefore, often several DSC temperature cycles are carried out (maximum temperature must be clearly below decomposition temperature) with analyzing the thermal transitions usually in the second and not in the first run after a defined cooling cycle.

DSC measurements can be used, moreover, to determine how much of a polymer sample is crystalline. For that purpose, the area of the melting peak has to be measured: the DSC plot gives the heat flow per gram of material as a function of temperature. Heat flow is heat given off per second. So the area of the peak is given as:

$$peak\ area = \frac{heat\ \times\ temperature}{time\ \times\ mass}\ \left[\frac{J\ \cdot\ K}{s\ \cdot\ g}\right] \tag{2.72}$$

When the peak area is divided by the heating rate

$$\frac{peak\ area}{heating\ rate} \equiv \frac{\frac{J \cdot K}{s \cdot g}}{\frac{K}{s}} = \frac{J}{g} \tag{2.73}$$

and multiplied with the mass of the sample

$$\frac{J}{g} \cdot g = J \tag{2.74}$$

the total heat is obtained that is given off when the polymer melts. The same calculation can be done for the peak of crystallization, leading to the total heat evolved during the crystallization. With $H_{m,total}$ the total heat given off during melting and $H_{c,total}$ the heat of crystallization it follows that:

$$H_{m,total} - H_{c,total} = H' \tag{2.75}$$

H' is the heat given off by that part of the polymer sample which was already in the crystalline state before the polymer was heated above T_c. With this number H' it is possible to figure out the percent crystallinity: it is divided by the specific heat of melting, H_c^*, which is the amount of heat given off by a certain amount of the polymer. So the mass of crystalline material, m_c, follows as:

$$\frac{H'}{H_C^*} = m_c \tag{2.76}$$

This is the total amount of grams of polymer that were crystalline below T_c. Now, if this number is divided by the weight of the sample, m_{total}, the fraction of the sample is obtained that was crystalline:

$$\frac{m_c}{m_{total}} = crystalline\ fraction \tag{2.77}$$

and

$$crystalline\ fraction \times 100 = \%crystallinity \tag{2.78}$$

Last but not least, DSC is a powerful technique for many other polymer-relevant aspects such as monitoring of curing reactions, detection of degradation, determination of heat capacity of chemical conversions, monitoring of initiator decomposition, etc.

2.3.5.9 Small- and Wide-Angle X-Ray Scattering (SAXS and WAXS)

Scattering of electromagnetic radiation can be used to obtain information about structures within materials having dimensions of the same order as the radiation wavelength. In X-ray scattering experiments, a sample is irradiated with X-rays, and the resulting scattering pattern is recorded, i.e., the intensity of the scattered radiation as a function of the scattering angle, θ. Then, the structure that caused the observed pattern is calculated. Scattering of X-rays is caused by differences in electron density.

Small-angle X-ray scattering (SAXS) is typically done at angles in the vicinity of the primary beam (extending to less than 2° for standard wavelengths). The scattering features at these angles correspond to structures ranging from tens to thousands of Ångstroms and are used to explore microstructure on the colloidal length scale as found, e.g., in phase-separated block copolymers. On the other hand – since the larger the diffraction angle, the smaller the length scale probed – wide-angle X-ray scattering (WAXS) is used to determine crystal structure on the atomic length scale. A typical WAXS pattern produced by a crystalline polymer looks as follows: The sharp peaks are caused by reflections from the various crystalline planes in the sample. They are observed at different angles according to Bragg's law:

$$2d \sin \theta = n\lambda \tag{2.79}$$

where θ is the spacing between crystalline planes, θ is the angle that those planes make with the incoming X-ray beam, l is the wavelength of the X-ray beam, and n is an integer. Since the crystalline parts of polymers are randomly oriented with respect to each other, the scattering is the same in all directions. So only a one-dimensional slice of the scattering pattern is needed to be sampled. However, only part of the polymer is crystalline, and the thickness of the crystalline lamellae is typically about 10–50 nm. For $d = 250$ Å, for example, the angle at which we observe the scattering caused by these lamellae is 0.17° for a fixed wavelength of 1.5 Å. This is an order of magnitude smaller than the angles at which the crystalline peaks are observed in the WAXS. Hence the structure of the lamellae has to be determined by SAXS, while the crystalline part is studied by WAXS: regular WAXS tends to focus on the location, width, shifts, etc. of Bragg peaks which arise from crystalline lattice structures. It is also a characteristic feature of WAXS patterns of semicrystalline polymers that the reflexes sit on the top of an amorphous halo. Upon graphic separation of the scattering intensities contributing to the sharp reflexes, A_c, and to the amorphous halo, A_a, respectively, it is possible to estimate the weight fraction of the crystalline phase, w_c:

$$w_c = \frac{A_c}{A_c + A_a} \tag{2.80}$$

For SAXS investigations, the primary experimental requirement is a well-collimated X-ray beam with a small cross-section. Synchrotron radiation sources with their intense brightness and natural collimation are ideal because most polymers are very poor scatterers. In most typical experiments, SAXS is performed in transmission mode. In this mode, polymer samples are typically 1–3 mm thick, offering about 63% absorption of the incident X-ray beam. In situations where transmission mode operation is not a feasible option, such as when the sample of interest is a thin film on an opaque substrate or when only the surface microstructure is of interest, one must use a combination of Grazing Incidence Diffraction geometry and SAXS, known as GISAXS. An extended sample-detector distance is usually required for SAXS to give the barely scattered photons room to spread out from the main beam, and also to reduce the detected X-ray background. A position-sensitive detector is required to measure the scattered intensity.

Interpreting SAXS data can be a very difficult task. At very small angles, the shape of the scattering in the so-called Guinier region can be used to give an idea of the radius of gyration of any distinct structures that are on this lengthscale. At higher angles, if a diluted system of relatively identical particles is considered, one might be able to see broad peaks that would also give information on the shape of the particles. In a system of strongly interacting particles, on the other hand, Bragg peaks may be observed which will obscure single-particle information but give structural information. At still higher angles, in the Porod region, the shape of the curve is useful in obtaining information on the surface-to-volume ratio of the scattering objects and on the dimensions of the scattering particles. This regime is also important for the analysis of regular structures shown by, for example, block copolymers.

2.3.5.10 Phase Contrast Microscopy

Phase contrast microscopy is a contrast-enhancing optical technique. It can be used to produce high-contrast images of transparent specimens such as lithographic patterns, fibers, multiphasic polymer samples, polymer morphologies, and latex dispersions. It is moreover useful for examination of surfaces, including integrated circuits, crystal dislocations, defects, and lithography.

In effect, the phase contrast technique employs an optical mechanism to translate minute variations in phase into corresponding changes in amplitude, which can be visualized as differences in image contrast. Light waves that are diffracted and shifted in phase by the specimen (termed a phase object) are transformed into amplitude differences that are observable in the eyepieces.

In a phase contrast microscope, an incident wavefront present in an illuminating beam of light becomes divided into two components upon passing through a phase specimen. The primary component is an undeviated (or undiffracted) planar wavefront, commonly referred to as the surround (S) wave. It passes through and around the specimen but does not interact with it. In addition, a deviated or diffracted spherical wavefront (D-wave) is also produced. It becomes scattered in many directions. After leaving the specimen plane, surround and diffracted light waves enter the objective front lens element and are focused at the intermediate image plane where they combine through interference to produce a resultant particle wave (often referred to as a P-wave). The mathematical relationship between the various light waves generated in phase contrast microscopy can be described simply as:

$$P = S + D \tag{2.81}$$

Detection of the specimen image depends on the relative intensity differences, and therefore on the amplitudes, of the particle and surround (P and S) waves. If the amplitudes of the particle and surround waves are significantly different in the intermediate image plane, then the specimen acquires a considerable amount of contrast and is easily visualized in the microscope eyepieces. Otherwise, the specimen remains transparent and appears as in an ordinary brightfield microscope.

Images produced by phase contrast microscopy are relatively simple to interpret when the specimen is thin and distributed evenly on the substrate. When thin specimens are examined using positive phase contrast optics, they appear darker than the surrounding medium when the refractive index of the specimen exceeds that of the medium. It should also be noted that numerous optical artifacts are present in all phase contrast images, and large extended specimens often present significant fluctuations in contrast and image intensity. Symmetry can also be an important factor in determining how both large and small specimens appear in the phase contrast microscope. Sensible interpretation of phase contrast images requires careful examination to ensure that artifacts are not incorrectly assigned to important structural features.

2.3.5.11 Polarization Microscopy

Polarized light microscopy provides all benefits of brightfield microscopy but offers a wealth of further information not available with any other optical technique. For example, it can distinguish between isotropic and anisotropic materials. This is because isotropic materials have the same optical properties in all directions, only one refractive index, and no restriction on the vibration direction of light passing through them. In anisotropic materials, in contrast, optical properties vary with the orientation of incident light with the crystallographic axes. They demonstrate a range of refractive indices, depending both on the propagation direction of light through the substance and on the vibrational plane coordinates. More importantly, anisotropic materials act as beam splitters and divide light rays into two parts. The technique of polarizing microscopy exploits the interference of the split light rays, as they are re-united along the same optical path, to extract information about these materials.

The wave model of light describes light waves vibrating at right angles to the direction of travel of light, with all vibration directions being equally probable. This is "common" light. It can be converted into plane-polarized light by passing through a polarizing filter. Here only light with one specific vibration plane passes through. There are two polarizing filters in a polarizing microscope – the polarizer and analyzer. The polarizer is situated below the specimen stage – usually with its permitted vibration direction fixed in the East–west direction. The analyzer, usually aligned North–south, is placed above the objectives and can be moved in and out of the light path. When both the analyzer and polarizer are in the optical path, their permitted vibration directions are positioned at right angles to each other. In this configuration, the polarizer and analyzer are said to be crossed, with no light passing through the system and a dark field of view present in the eyepieces. But if there is a birefringent material on the specimen stage, the plane of the polarized light passing through the polarizer is turned to a certain extent. Consequently, there arises some light having a polarization plane which again can pass through the analyzer. A (partially) bright and (potentially) colored picture of the sample can be seen in the field of view.

Polarization colors result from the interference of the two components of light split by the anisotropic specimen. The two components of light travel at different speeds through the specimen and have different refractive indices, or refringences. The faster beam emerges first from the specimen with an optical path difference (OPD). It may be regarded as a "winning margin" over the slower one. The analyzer recombines only components of the two beams traveling in the same direction and vibrating in the same plane. The polarizer ensures that the two beams have the same amplitude at the time of recombination for maximum contrast: there is constructive and destructive interference of light in the analyzer, depending on the OPD on the specimen and the wavelength of the light, which can be determined from the order of polarization color(s).

Superimposed on the polarization color information is an intensity component. As the specimen is rotated relative to the polarizers, the intensity of the polarization colors varies cyclically, from zero (extinction) up to a maximum after 45° and back down to zero after a 90° rotation. Whenever the specimen is in extinction, the permitted vibration directions of light passing through are parallel with those of either the polarizer or analyzer. This can be related to geometrical features of the specimen, such as fiber length, film extrusion direction, and crystal faces. In crossed polarizers, isotropic materials can be easily distinguished from anisotropic materials as they remain permanently in extinction (remain dark) when the stage is rotated through 360°.

Polarized light microscopy is thus used to study composites such as cements, ceramics, mineral fibers, and polymers, and crystalline or highly ordered biological molecules such as DNA, starch, and wood. The technique can be used both qualitatively and quantitatively and is an outstanding tool for materials science. During the solidification of polymer melts there may be some organization of the polymer chains, a process that is often dependent upon the annealing conditions. When nucleation occurs, the synthetic polymer chains often arrange themselves tangentially and the solidified regions grow radially. These can be seen in crossed polarized illumination as white regions with the black extinction crosses. When these spherulites impinge, their boundaries become polygonal. In other cases, polymers can undergo lyotropic or thermotropic liquid crystalline phase transitions, which can often be observed and recorded in a polarized light microscope.

In summary, polarizing microscopy provides a vast amount of information about the composition and three-dimensional structure of a variety of samples. The technique can reveal information about thermal history and the stresses and strains to which a specimen was subjected during formation. Polarizing microscopy is a relatively inexpensive and well accessible investigative and quality control tool.

2.3.5.12 Scanning Electron Microscopy (SEM)

In the scanning electron microscope (SEM), an image is formed by a very fine electron beam which is scanned across the surface of a sample in a series of lines and frames called a raster: at any given moment, the specimen is bombarded with electrons over a very small area. These electrons may be elastically reflected with no loss of energy (backscattered electrons), they may be absorbed and give rise to secondary electrons of very low energy (together with X-rays), they may be absorbed and give rise to the emission of visible light, and they may give rise to electric currents within the specimen. All these effects can be used to produce images. The contrast in the image is determined by the sample morphology. A high-resolution image can be obtained because of the small diameter of the primary electron beam.

For backscattered electrons, the contrast in the produced image is determined by the atomic number of the elements in the sample. The image will therefore show the distribution of different chemical phases in the sample. Because these electrons are emitted from a depth in the sample, the resolution in the image is not as good as for secondary electrons. Interaction of the primary beam with atoms in the sample causes shell transitions which result in the emission of an X-ray. The emitted X-ray has an energy characteristic of the parent element. Detection and measurement of the energy permits elemental analysis (energy dispersive X-ray spectroscopy or EDS). EDS can provide rapid qualitative or even quantitative analysis of elemental composition with a sampling depth of 1–2 μm. X-rays may also be used to form maps or line profiles, showing the elemental distribution in a sample surface.

By far the most common technique in SEM, however, is image formation by means of the low-energy secondary electrons. The primary electrons enter a surface with an energy of 0.5–30 keV, and generate many low-energy secondary electrons. The intensity of these secondary electrons is largely governed by the surface topography of the sample. An image of the sample surface can thus be constructed by measuring secondary electron intensity as a function of the position of the scanning primary electron beam. High spatial resolution is possible because the primary electron beam can be focused to a very small spot (<10 nm). The secondary electrons are selectively attracted to a grid held at a low (50 V) positive potential with respect to the specimen. Behind the grid is a disc held at about 10 kV positive with respect to the specimen. The disc consists of a layer of scintillant coated with a thin layer of aluminum. The secondary electrons pass through the grid and strike the disc, causing the emission of light from the scintillant. The light is led down a light pipe to a photomultiplier tube which converts the photons of light into a voltage. The strength of this voltage depends on the number of secondary electrons that are striking the disc. Thus, the secondary electrons produced from a small area of the specimen give rise to a voltage signal of a particular strength. The voltage is led out of the microscope column where it is processed and amplified to generate a point of brightness on a cathode ray tube (or television) screen. An image is built up simply by scanning the electron beam across the specimen in exact synchrony with the scan of the electron beam in the cathode ray tube.

The SEM does not contain objective, intermediate and projector lenses to magnify the image as in the optical microscope. Instead magnification results from the ratio of the area scanned on the specimen to the area of the television screen. Increasing the magnification in an SEM is therefore achieved quite simply by scanning the electron beam over a smaller area of the specimen. Scanning electron microscopy (SEM) is the best known and most widely-used of the surface analytical techniques. High-resolution images of surface topography, with excellent depth of field are produced using a highly-focused electron beam.

Samples suitable for SEM measurements include most solids which are stable under vacuum (metals, ceramics, polymers, minerals). Samples must be less than 2 cm in diameter. Non-conducting samples are usually coated with a thin layer of carbon or gold in order to prevent electrostatic charging.

2.3.5.13 Scanning Transmission Electron Microscopy (STEM)

In a scanning transmission electron microscope (STEM) the objective lens focuses the electron beam onto an atomic-scale sample volume. All scattered electrons can then be collected by a variety of detectors placed behind the specimen. An image is generated simply by moving the focused beam step by step over the specimen. Hence a STEM image may be considered as a collection of individual scattering experiments. Various types of signals discriminated in scattering angle and/or energy loss yield different structural and chemical information and may be captured simultaneously in different channels. This simultaneous and controlled acquisition of information lends itself to quantitative analyses that are difficult to realize with other instruments. In addition, as there is no limitation of the solid angle and the energy loss interval over which the scattered electrons may be collected, 60–100% of them contribute to the image. This provides a unique opportunity to image even beam-sensitive biomolecules at low dose.

Elastically scattered electrons are collected by an annular detector and provide the elastic dark-field signal. A spectrometer deflects those electrons that have lost energy at a larger angle than the unscattered electrons, thus facilitating the acquisition of the inelastic dark-field signal. The coherent bright-field signal arising from unscattered and low-angle elastically scattered electrons is collected through a small aperture placed on the optical axis. The various signals can be collected in parallel and processed online as the probe is scanned over the sample.

2.3.5.14 Transmission Electron Microscopy (TEM)

In transmission electron microscopy (TEM), a beam of highly focused and highly energetic electrons is directed toward a thin sample (<200 nm) which might be prepared from solution as thin film (often cast on water) or by cryocutting of a solid sample. The incident electrons interact with the atoms in the sample, producing characteristic radiation. Information is obtained from both deflected and non-deflected transmitted electrons, backscattered and secondary electrons, and emitted photons.

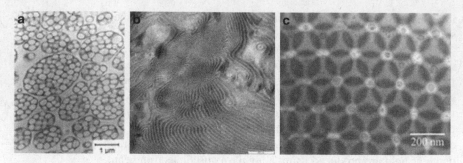

Fig. 2.20 TEM-pictures of multiphasic materials (**a**) Micromorphology of high impact polysty-rene. White: Polystyrene, black: Polybutadiene (see Sect. 5.5.2.4). (**b**) Lamellar micromorphology of an AB diblock copolymer (see Sect. 3.4.2.1). (**c**) Quasi crystalline order of polystyrene particles in an "artificial opal"

Because the electron beam passes through the sample, transmission electron microscopy reveals the interior of the specimen. It is sensitive toward the internal structure of the material (size, shape, and distribution of phases within the material), its composition (distribution of elements, including segregation if present), and the crystalline structure of the phases and the character of crystal defects.

A typical electron microscope at an accelerating voltage of 75 kV works with electrons having a wavelength less than 5 pm. This makes the theoretical resolution about hundred thousand times better than that of light. Unfortunately, this theoreti-cal resolution has never come even close to being attained. The basic drawback is that magnetic fields cannot be manipulated, shaped, and grouped the way it is possible with glass lenses: as a result, electron microscopes must use very small apertures which seriously attenuates the resolution, about 100 times. Nevertheless, excellent information can be achieved in polymer science, for example, for multi-phasic polymer blends and composites (Fig. 2.20).

2.3.5.15 Scanning Probe Microscopy

Around 1980 a new method of microscopy known as scanning probe microscopy (SPM) was invented. Within the past 10 years, applications have been increasing exponentially in fields like surface physics and chemistry, biology and optics. SPM is also beginning to emerge as a useful and popular technique for R & D and quality control in several industries.

Probe microscopes are characterized by two common features. On the one hand, a sharp, tiny probe gets very close to the sample and feels the surface by monitoring some kind of interaction between the probe and the surface, which is very sensitive to distance. On the other hand, the sample is scanned in a raster fashion with near atomic accuracy, and the variation in the interaction is translated into a topographic map of the surface.

Among the family of SPMs the two most commonly used are Scanning Tunneling Microscopy (STM) and Atomic Force Microscopy (AFM). In STM, a sharp metallic probe and a conducting sample are brought together until their electronic wave functions overlap. By applying a potential bias between them, a tunneling current is produced. The probe is mounted on a piezoelectric drive that scans the surface. Combining the piezoelectric drive with a feedback loop allows imaging of the surface in either a constant current or a constant-height mode. In AFM, the probe tip is attached to a cantilever with a small spring constant. The probe is much like a spring which changes dimension upon experiencing a force and the interaction that will be monitored is the repulsion between two atoms when they are brought extremely close to each other. The forces acting on the probe tip deflect the cantilever and the tip displacement is proportional to the force between the surface and the tip. The resultant bending of the AFM cantilever is measured optically by the deflection of the reflected laser beam. The most important advantage that AFM has over STM is that the former is not limited to conducting samples, so materials can often be imaged "as is" with essentially no sample preparation as long as the surface roughness is not too high. While the technique of AFM is maturing, many AFM modes have appeared for special purposes. The most common techniques are contact mode, non-contact mode, and tapping mode.

In the *contact mode* the tip scans the sample in close contact with the surface. The force on the tip is repulsive with a mean value of 10^{-9} N. This force is set by pushing the cantilever against the sample surface with a piezoelectric positioning element. In contact mode AFM the deflection of the cantilever is sensed and compared in a DC feedback amplifier to some desired value of deflection. If the measured deflection is different from the desired value, the feedback amplifier applies a voltage to the piezo to raise or lower the sample relative to the cantilever in order to restore the desired value of deflection. The voltage that the feedback amplifier applies to the piezo is a measure of the height of features on the sample surface. It is displayed as a function of the lateral position of the sample.

Problems with contact mode are caused by excessive tracking forces applied by the probe to the sample. The effects can be reduced by minimizing tracking force of the probe on the sample. Moreover, under ambient conditions, sample surfaces are covered by a layer of adsorbed gases consisting primarily of water vapor and nitrogen which is 10–30 monolayers thick. When the probe touches this contaminant layer, a meniscus forms and the cantilever is pulled by surface tension toward the sample surface. The magnitude of the force depends on the details of the probe geometry, but is typically on the order of 100 nN. This meniscus force and other attractive forces may be neutralized by operating with the probe and part or the entire sample totally immersed in liquid.

An attempt to avoid these problems is the *non-contact mode*. This technique is used in situations where tip contact might alter the sample in subtle ways. In this mode the tip hovers 50–150 Ångstrom above the sample surface. Attractive Van der Waals forces acting between the tip and the sample are detected, and topographic images are constructed by scanning the tip above the surface. Unfortunately the attractive forces from the sample are substantially weaker than the forces used by contact mode. Therefore the tip must be given a small oscillation so that AC detection methods can be used to detect the small forces between the tip and the sample by measuring the change in amplitude, phase, or frequency of the oscillating cantilever in response to force gradients from the sample. For highest resolution, it is necessary to measure force gradients from Van der Waals forces which may extend only a nanometer from the sample surface. In general, the fluid contaminant layer is substantially thicker than the range of the Van der Waals force gradient and therefore, attempts to image the true surface with non-contact AFM fails as the oscillating probe becomes trapped in the fluid layer or hovers beyond the effective range of the forces it attempts to measure.

Tapping mode is a key advance in AFM. This potent technique allows high-resolution topographic imaging of sample surfaces that are easily damaged, loosely hold to their substrate, or difficult to image by other AFM techniques. Tapping mode overcomes problems associated with friction, adhesion, electrostatic forces, and other difficulties that plague conventional AFM scanning methods by alternately placing the tip in contact with the surface to provide high resolution and then lifting the tip off the surface to avoid dragging the tip across the surface. Tapping mode imaging is implemented in ambient air by oscillating the cantilever assembly at or near the cantilever's resonant frequency using a piezoelectric crystal. The piezo motion causes the cantilever to oscillate with a high amplitude (typically greater than 20 nm) when the tip is not in contact with the surface. The oscillating tip is then moved toward the surface until it begins to lightly touch, or tap the surface. During scanning, the vertically oscillating tip alternately contacts the surface and lifts off, generally at a frequency of 50,000–500,000 cycles per second. As the oscillating cantilever begins to intermittently contact the surface, the cantilever oscillation is necessarily reduced due to energy loss caused by the tip contacting the surface. The reduction in oscillation amplitude is used to identify and measure surface features.

In addition to these topographic measurements, AFM can also provide much more information. The AFM can also record the amount of force felt by the cantilever as the probe tip is brought close to – and even indented into – a sample surface and then pulled away. This technique can be used to measure the long-range attractive or repulsive forces between the probe tip and the sample surface, elucidating local chemical and mechanical properties like adhesion and elasticity, and even thickness of adsorbed molecular layers or bond rupture lengths. Phase images allow to distinguish between softer and harder domains on the sample surface.

Studies on fundamental interactions between surfaces extend across physics, chemistry, materials science, and a variety of other disciplines. With a force sensitivity on the order of a few pico-Newtons, AFMs are excellent tools for probing these fundamental force interactions. Force measurements in water revealed the benefits of AFM imaging in this environment due to the lower tip-sample forces. Some of the most interesting force measurements have also been performed with samples under liquids where the environment can be quickly changed to adjust the concentration of various chemical components. In liquids, electrostatic forces between dissolved ions and other charged groups play an important role in determining the forces sensed by an AFM cantilever.

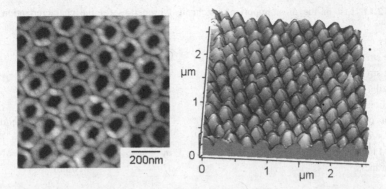

Fig. 2.21 AFM pictures of (*left*) a porous crystal made from a core-shell latex: the crystalline lattice of the SiO$_2$ appears *grey* while the nanopores are black; (*right*) surface of a dried core-shell latex

Insulators, organic materials, biological macromolecules, polymers, ceramics, and glasses are some of the many materials which can be imaged in different environments, such as liquids, vacuum, and low temperatures. The materials being investigated include thin and thick films, coatings, ceramics, composites, glasses, synthetic and biological membranes, metals, polymers, and semiconductors (Fig. 2.21). AFM is being applied to studies of phenomena such as abrasion, adhesion, cleaning, corrosion, etching, friction, lubrication, plating, polishing, and any nano- and mesostructures in multiphase systems. By using AFM one can not only image the surface in atomic resolution but also measure the force at nano-Newton scale.

2.3.6 Mechanical Measurements

The determination of the mechanical characteristics of a polymer serves ultimately to establish its usefulness and applicability as an industrial material. Although at first sight such measurements are of a purely applied character, some methods of investigation yield data that are not only useful for engineering practice, but also allow deductions about composition, structure, and state of aggregation of the polymeric material. Thus, they supplement the methods of characterization of polymers discussed in Sects. 2.3.1, 2.3.2, and 2.3.3. The following mechanical properties can, in this sense, serve for the physical characterization of a polymer in the solid state: strength and elongation, stiffness (modulus of elasticity), brittleness and toughness, and hardness. They are collected in Table 2.10 and discussed separately in Sects. 2.3.6.1, 2.3.6.2, 2.3.6.3, and 2.3.6.4.

For analytical purposes and an initial characterization, quick tests (duration minutes to few hours) are sufficient. However, the estimation of the usefulness as an industrial material needs long-term testing (months to years) in different environments (air, water, solvents, etc.). The numerous other tests employed in

Table 2.11 List of the most important mechanical properties for the characterization of a polymer

Property	Symbol	Units
Tensile strength (yield strength)	σ_B	N/mm^2
Tensile strength at break	σ_R	N/mm^2
Elongation at yield	ε_B	%
Elongation at break	ε_R	%
Modulus of elasticity or Young's modulus	E	$(J/m)/mm^2$ or N/mm^2
Impact strength	a_n	$N \cdot mm/mm^2 = N/mm$
Notched impact strength	a_k	$N \cdot mm/mm^2 = N/mm$
Ball hardness	H	N/mm^2

engineering practice to determine mechanical (and other) properties, as well as the special methods for testing rubbers, films, fibers, foams, coatings, and adhesives, will not be dealt with here.

The results of mechanical tests on polymers can obviously only be compared with one another when they are obtained at the same temperature, since the physical properties of polymers change markedly with temperature (see Sect. 1.4.3). Furthermore, the manner of preparation and pretreatment (conditioning) of the test specimen is decisive for the reliability and reproducibility of mechanical tests on polymers. For the determination of constants that are characteristic of the material, it should be as isotropic as possible, that is, exhibit the same characteristics in all directions, and therefore be free of internal stress; in addition, the temperature and humidity should be held constant for all measurements. While the latter conditions are relatively easy to fulfill, the preparation of completely stress-free and therefore isotropic test specimens is quite difficult to achieve for the reasons already given (see Sect. 1.4.3). Flow orientation in test specimens is especially liable to occur in thermoplastic materials (less so in thermosetting polymers), and in extreme cases can lead to values of tensile strength or impact strength that are two or three times as high when measured in the flow direction compared with those perpendicular to this direction. Possible ways of suppressing the anisotropy have been suggested in the literature. Test specimens can be fabricated either directly by injection or compression molding or casting (thermosetting plastics) in suitable forms, or by milling, sawing or punching of sheets of the polymer.

Because of reasons mentioned above the preparation, the dimensions, and the pretreatment of the test specimen have been standardized (DIN standard or ASTM standard). The important characteristic values of technical polymers are listed in databases (Campus, Polymat) (Table 2.11).

2.3.6.1 Stress–Strain Measurements

In a stretching experiment a test specimen is placed under tension, causing the length to increase and the cross-section to decrease, until finally it breaks. For these stress–strain measurements the test specimen has shoulders at both ends, such that the break occurs in the desired place, namely at the position of lowest cross-section.

The specimen is held at its broader parts in the clamps of the testing machine. The machine then pulls the clamps apart at constant speed, whereby a force is transmitted to the test specimen. The latter is plotted continuously against the change of length by means of a coupled recorder. The maximum tension P_{max} during the experiment is not always the same as the tension at break.

The prevailing tension divided by the smallest cross-section F_0 of the test specimen at the beginning of the experiment gives the corresponding stress σ, which is thus the tension per unit cross-section (1 mm^2). The ultimate tensile strength σ_B is obtained by dividing the maximum load P_{max} by the initial cross-section F_0 measured in N/mm^2 or MPa:

$$\sigma_B = \frac{P_{max}}{F_0} \tag{2.82}$$

The elongation e is generally understood to be the extension with respect to the original length. The elongation at yield is accordingly the extension, $\Delta l = l - l_0$, at maximum load P_{max} divided by the initial length l_0:

$$\varepsilon_B = \frac{\Delta l}{l_0} \cdot 100 \, [\%] \tag{2.83}$$

From the values of the yield strength and the corresponding elongation one thus obtains a measure of the ultimate load that can be carried by the material. However, it is very much more informative for the characterization of a polymer to observe not just the values of yield strength σ_B and elongation at yield ε_B, but the whole stress–strain experiment shown graphically as a plot of stress s_B against ε (stress–strain diagram). The point in the stress–strain curve where deviation from linearity sets in is called the proportionality limit. This means that Hooke's law is no longer obeyed beyond this point. The position of the proportionality limit depends very much on temperature. Polymers may be divided into three main categories according to the shape of such a stress–strain diagram (see Fig. 2.22).

The first group comprises materials whose stress–strain curve is very steep and almost linear, and flattens only slightly near the break point (curve I in Fig. 2.22). Like metals, these materials deform only to a small extent at relatively high loads. Amongst these may be numbered all thermosets, some thermoplastics such as polystyrene and poly(methyl methacrylate) as well as high modulus/high strength fibers, i.e., substances that are only slightly elastic and rather brittle.

The second group exhibits the phenomenon of drawability. This manifests itself in the stress–strain behavior (curve II in Fig. 2.22) as follows: At first these materials behave in a similar way to those of curve I. The proportionality limit lies at low values, and the deformation with increasing load is also quite small. Then, suddenly, a large extension occurs, even though the load remains constant or becomes smaller. The material begins to flow and the stress–strain curve sometimes runs nearly parallel to the abscissa. The point at which the flow begins is called the

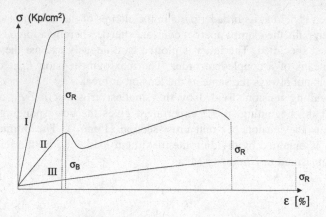

Fig. 2.22 Stress–strain diagram for various types of polymers (for explanation, see text): σ_B = yield strength; σ_R = tensile strength at break

upper yield point. The stress at this point is called the yield strength σ_B although the specimen has not broken. When all the macromolecules have been brought into a new (orientated) position by the flow process, the flow ceases and the stress increases again until finally the sample breaks. This phenomenon is used in the production of drawn fibers and films (see Sect. 2.4.2). The stress σ at this point is called the tensile strength at break σ_R, associated with the elongation at break ε_R. Many thermoplastics belong to this group, such as polyolefins, Nylon-6 and Nylon-6,6, and poly(ethylene terephthalate).

The third group comprises materials that show relatively large deformations even at low loads. In the stress–strain diagram (curve III in Fig. 2.22) there is no sudden drop in the stress, i.e., no flow limit. Furthermore, in the middle range the curve is not quite as flat as the curve for drawable materials (group II), i.e., the increase in strength resulting from reorientation of the macromolecules is a gradual one. A further increase in load leads eventually to failure. The stress at this point is called the tensile strength at break σ_R and the corresponding elongation at break ε_R. To this group belong all plasticized thermoplastics (e.g., soft PVC) as well as rubbers (elastomers).

Finally, the modulus of elasticity E (Young's modulus), which is a measure of the stiffness of the polymer, can be calculated from the stress–strain diagram. According to Hooke's law there is a linear relation between the stress σ and the strain ε:

$$\sigma = \varepsilon \cdot E \qquad (2.84)$$

so that the elastic modulus is given simply by the slope of the stress/strain curve in the linear region, i.e., below the proportionality limit. Thus,

$$E = \frac{stress}{relative\ change\ of\ length} \left[\frac{N/mm^2}{mm/mm}\right] = \left[\frac{N}{mm^2}\right] = [MPa] \qquad (2.85)$$

Hence, the elastic modulus corresponds in principle to the force per square millimeter that is necessary to extend a rod by its own length. Materials with low elastic modulus experience a large extension at quite low stress (e.g., rubber, $E \cong 1\ N/mm^2$). On the other hand, materials with high elastic modulus (e.g., polyoxymethylene, $E \cong 3500\ N/mm^2$) are only slightly deformed under stress. Different kinds of elastic modulus are distinguished according to the nature of the stress applied. For tension, compression, and bending, one speaks of the intrinsic elastic modulus (E modulus). For shear stress (torsion), a torsion modulus (G modulus) can be similarly defined, whose relationship to the E modulus is described in the literature.

In polymers the time dependence of an E modulus plays a more important role than in metals. If polymers are loaded with a constant stress they undergo a deformation ε, which increases with time. This process is named creep. Conversely, if a test specimen is elongated to a certain amount and kept under tension, the initial stress σ decreases with time. This decay is called stress relaxation. As a consequence of this time dependency of σ and ε it is important to state the conditions under which the elastic modulus was determined.

2.3.6.2 Dynamic-Mechanical Measurements

In dynamic-mechanical measurements the test specimen is not destroyed. The measurements are called dynamic because the mechanical properties are determined under oscillatory conditions. Of the numerous methods of measurement the oscillatory torsion experiment, using the so-called torsion pendulum, is one of the most widely used. In this test the specimen, having the form of a strip, is clamped firmly at the upper end, while an oscillating disc is fastened to the lower end. The turning motion of this disc can be followed optically and continuously recorded. A thermostatted chamber that can be heated or cooled surrounds the test specimen. After imparting an initial impulse to the torsion pendulum the decay of oscillation with time can be analyzed to yield two pieces of information. The vibration period allows the calculation of the torsion modulus (G modulus), which is a measure of the rigidity. From the decrease of amplitude with time one obtains a measure of the "internal mechanical absorption", called the mechanical loss factor δ (or logarithmic decrement). From such measurements it is possible to make valuable deductions about the molecular motions, not only about the motions of chain segments within a macromolecule (micro-Brownian motion), but also concerning the motions of the entire macromolecule relative to others (macro-Brownian motion).

Since these processes are strongly dependent on temperature, it is appropriate to carry out torsion-oscillation experiments over a wide range of temperature, and to plot the values for the modulus and d against temperature (see Fig. 2.23).

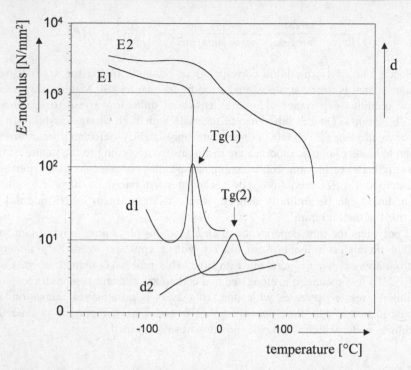

Fig. 2.23 Dependence of the elastic modulus E and the mechanical loss factor d on temperature for various polymers. Curves 1: elastomer (statistical copolymer of ethylene and propylene); curves 2: isotactic polypropylene (semicrystalline)

Besides temperature, frequency plays an important role, lying in the region from 1 to 1000 Hz; it is held constant during the measurement. An absorption maximum in the d curve corresponds to an inflection point in the modulus curve. Such diagrams are especially useful in determining the positions of softening point (glass transition temperature), melting point, melting range, and transition points, as well as indicating the influence of temperature on stiffness (G modulus or E modulus). If more absorption maxima occur, they are called α-, β-, γ-processes (the α-value being the absorption at the highest temperature).

The value of the modulus and the shape of the modulus curve allow deductions concerning not only the state of aggregation but also the structure of polymers. Thus, by means of torsion-oscillation measurements, one can determine the proportions of amorphous and crystalline regions, crosslinking and chemical non-uniformity, and can distinguish *random* copolymers from *block* copolymers. This procedure is also very suitable for the investigation of plasticized or filled polymers, as well as for the characterization of mixtures of different polymers (polymer blends).

Curve E1 in Fig. 2.22 corresponds to an elastomer (statistic copolymer from ethylene and propylene), characterized by a low value of the elastic modulus over a wide temperature range and a sudden increase at low temperature corresponding to the transition from the elastic to the brittle (glassy) state at the glass transition temperature $T_g(1)$ at $-50°C$.

In contrast, the curve E2 (isotactic polypropylene) is characteristic for partially crystalline polymers. The modulus is three decades higher than in an elastomer. At the glass transition temperature $[T_g(2) \sim 0°C]$ the decay of the E modulus is small; it does not drop to the lower level of the molten state before the melting point.

The corresponding curves for the mechanical loss factor d show the following characteristics: The transition to the glassy state for elastomers is seen in curve 1 as a characteristic "mechanical absorption". On the other hand, two absorption maxima are visible in the curve for the partially crystalline polymer d2. The first one at $10°C$ indicates the glass transition, the second one at about $145°C$ is coherent with the crystalline melting point.

2.3.6.3 Impact Strength and Notched Impact Strength

Besides the test methods in which the load is applied over a relatively long period and the deformation rate is small (as in stress–strain and hardness measurements), there are other methods of interest in which the material is placed suddenly under high stress. Such measurements include that of impact strength, which gives an idea of the brittleness and toughness of the material. For this purpose, the test specimen is broken by means of a weighted pendulum, the energy of fracture ($N \cdot mm = mJ$) being measured. The impact strength is given by the fracture energy divided by the area of cross-section and has the units $N \cdot mm/mm^2 = mJ/mm^2$. The test specimens may be either smooth or notched rectangular bars; accordingly one speaks of (normal) impact strength (a_n) and of notched impact strength (a_k). The notched impact method is simply a variant of the normal impact test, the specimen being notched in a V-shape before the test. On impact the shear is thereby concentrated at a particular point on the test bar.

There are two principle methods for measuring impact and notched impact strength, which in practice differ only in the way in which the test bar is held. In the *Charpy method* the test piece is suspended at both ends and is struck in the center by a weighted pendulum. In the *Izod method* the test piece is clamped at one end only and is struck at the free end by the pendulum.

2.3.6.4 Hardness

Hardness is defined as the resistance that one body offers against penetration by another. Hence, to judge the hardness of a material one measures the force that is required to obtain a certain depth of penetration. This force or depth of penetration is dependent not only on temperature but also on many factors not characteristic of the material, such as the form of the penetrating body (ball or needle) and the time factor. Unfortunately, there is no universal hardness test and numerous methods are in use, many of which do not cover the entire range of possible hardness and can, therefore, not be used on all polymers. These methods can be divided into two groups: in one group the depth of penetration is measured after removal of the load (e.g., Brinell test); in the other the depth of penetration is measured under full load. The latter methods are the most suitable for thermoplastics, thermosets, and elastomers. A very popular method is the ball hardness test: a steel ball (diameter 5 mm) is pressed into the sample (4-mm-thick plate) with constant force and the depth of penetration measured after 10 s and after 60 s loading. Since the area of deformation must be taken into account, ball hardness has the dimensions N/mm^2. The hardness values measured with different methods cannot be interconverted because each of these gives different results. For example, according to the Brinell method a rubber is very hard, because the permanent deformation is measured. In contrast, the ball pressure method points out the soft character of the polymer because the permanent and elastic deformation is simultaneously measured.

The hardness of a polymer can also be estimated from the modulus of elasticity E (high E modulus indicates high hardness). The advantage here is that every region of elasticity and every degree of hardness can be detected with a single kind of measurement (determination of stress–strain-behavior or torsional oscillation).

2.4   Correlations of Structure and Morphology with the Properties of Polymers

It is one of the advantages of polymers versus low-molecular-weight compounds that their properties do not only depend on the chemical composition alone. Macromolecules can differ from each other also in molecular weight, molecular-weight distribution, branching, crosslinking, stereoregularity, etc. Each of these parameters has an influence on the properties in solution and in bulk. Nearly all properties are dependent on the molecular weight. Some values increase more or less steadily with growing chain length, e.g., solution viscosity, others increase very fast from oligomers to polymers and then reach a plateau, in particular, many mechanical properties.

It is important to mention that the structure/properties relationships which will be discussed in the following section are valid for many polymer classes and not only for one specific macromolecule. In addition, the properties of polymers are influenced by the morphology of the liquid or solid state. For example, they can be amorphous or crystalline and the crystalline shape can be varied. Multiphase compositions like block copolymers and polymer blends exhibit very often unusual meso- and nano-morphologies. But in contrast to the synthesis of a special chemical structure, the controlled modification of the morphology is mostly much more difficult and results and rules found with one polymer are often not transferable to a second polymer.

2.4.1 Structure/Properties Relationships in Homopolymers

2.4.1.1 Correlations with Solution Properties

The solubility of polymers is determined by the interactions between macromolecules and the molecules of the solvent. But the prediction of the solubility of a macromolecule and hence the correlation to its chemical (and morphological) structure is much more complicated than for a low-molecular-weight compound. Nevertheless, some general rules do exist:

- A decisive and general role is played by the molecular weight: the higher the molecular weight, the lower is the solubility of the polymer and the higher is the solution viscosity at given conditions.
- Amorphous polymers dissolve much easier than crystalline ones. The latter are often soluble at elevated temperatures only, i.e., near the crystallite melting point T_m.
- Polar groups in a macromolecule mostly enhance solubility.
- Short chain branches increase solubility.
- Macromolecules with polar and stiff main chains (e.g., cellulose, polyaramides) are often only soluble via complexation.
- Crosslinking reduces the solubility dramatically. Medium or highly crosslinked polymers do not dissolve at all but form more or less swollen gels, depending on the degree of crosslinking.

In a similar way, some general rules can be derived for the viscosity of polymer solutions:

- The solution viscosity increases strongly with the molecular weight.
- Branched polymers show lower solution viscosities than linear ones with the same molecular weight.
- The degree of stereoregularity is in general only of very small influence on the solution viscosity for polymers with the same molecular weight.
- The stiffness of the main chain of a polymer is of great importance for the solution viscosity: the stiffer the chain is, the higher is the viscosity for polymers with the same molecular weight (see Sect. 2.3.3.3.1 for the dependency of K and a in the viscosity equation on the shape of macromolecules in solution).

2.4.1.2 Correlation with Bulk Properties

Thermal Properties

The thermal properties of polymers include their behavior during heating from the solid amorphous (glassy) or crystalline to the liquid (molten) state, but also their chemical and mechanical stability in the entire range of application.

In industrial practice temperature stability of a polymer means that it is able to maintain its mechanical properties up to a certain temperature and over a certain time period. Depending on the environmental conditions under which the thermal stability is measured one further differentiates between two cases: physical thermostability if the thermal treatment occurs in inert atmosphere and chemical thermostability if the thermal treatment is done, e.g., in the presence of air (thermooxidative stability).

The prediction of the chemical thermostability is based on the rules established for the thermal stability and the reactivity of chemical bonds for low-molecular-weight compounds. Instead, the physical thermostability depends on the transition points of the macromolecules, i.e., the glass transition temperature T_g in case of amorphous polymers, and additionally the crystalline melting point T_m in case of crystalline polymers.

In designing polymers with high physical stability valuable information is obtained from the Gibbs equation:

$$T_m = \frac{\Delta H}{\Delta S}$$

This equation shows that a high melting temperature T_m is obtained by raising the melt enthalpy ΔH and/or a lowering of the melt entropy ΔS. Since T_m and T_g are related via the Boyer/Beaman rule (see Sect. 2.3.4.3), this equation is also valid for the glass transition temperature.

The glass transition temperature T_g increases in the beginning rapidly with the molecular weight and reaches a plateau at high molecular weights. According to Fox and Flory the relation between T_g and the number average molecular weight M_n is given by:

$$T_g = T_{g,\infty} - \frac{K}{M_n}$$

where $T_{g,\infty}$ is the glass transition temperature at very (infinitely) high M_n and K is a constant. But it should be taken into consideration that this proportionality between T_g and M_n is only valid for limited regions of M_n. Crosslinking, stereoregularity (sometimes combined with crystallinity) and chain stiffness in general result in an increase of the glass transition temperature, whereas T_g decreases with branching. Similar rules also hold for the dependencies of the crystallite melting point on the molecular weight, stereoregularity, chain stiffness and branching. Crosslinked polymers in most cases are amorphous and do not crystallize.

The crystallinity and the corresponding density and melting temperature of the crystalline domains of (semi)crystalline polymers depend also on the molecular weight. First, with growing molecular weight the degree of crystallinity increases to a maximum, with further growing the degree of crystallinity drops a little due to changes in the kinetics of crystallization with increasing melt viscosity and the participation of some of the polymer chains in the formation of different crystallites, which hinders the growth of the crystallites. Therefore the rate of crystallization of very high-molecular-weight chains is slower than that of chains with lower molecular weight. This is also the reason why the rate of the growth of spherulites of isotactic polystyrene decreases with increasing molecular weight.

Table 2.12 Selected building blocks for thermally stable polymers

Aromatic units	Temperature-stable bond	Polymer class	Increase of Stiffness of main chain
	—S—	Polyarylene ether etc. $\Delta H \uparrow$	
	—O—		
	(C=O)		
	(ester O—C=O)	Aromatic polyesters and polyamides $\Delta S \downarrow$	
	(amide C=O, N–H)	Heterocyclic and ladder polymers $\Delta S \downarrow$	

To obtain polymers with high thermal and chemical stability and melting temperatures the following considerations are useful:

- A high melt enthalpy is obtained when the polymer is highly crystalline. The tendency to crystallize is improved when the molecules are linear with a regular chain structure and/or when high intermolecular forces exist. High intermolecular forces are caused through aromatic units, hydrogen bonds, and/or polar groups.
- A low melting entropy is experienced if the intramolecular motions of the polymer are hindered, for example, the free rotation of chain segments. This is preferentially be done by stiffening of the polymer chain. Table 2.12 shows a selection of building blocks for temperature-resistant polymers.

Another thermal property which is of great industrial importance is the melt viscosity, characterized, e.g., by the melt flow index, MFI, or the melt flow rate,

MFR (see Sect. 2.3.5.6). The melt viscosity is first of all influenced by the molecular weight according to:

$$MFI \approx M_w^{0.75},$$

and by the molecular weight distribution. But branching and crosslinking as well as chemical modifications (e.g., introduction of polar groups) are also factors of influence. A different behavior is experienced with thermotropic liquid crystalline polymers (see Sect. 1.3.4).

Mechanical Properties

The mechanical properties of polymer materials are characterized by the relation between applied mechanical stress and resulting deformation and can be described by a constitutive equation, e.g., in the case of very low deformations by Hooke's law. In its simplest form it gives the deformation of a sample under mechanical stress in form of a stress–strain diagram (see Sect. 2.3.6.1). For such investigations the load is applied over a relatively long time period and besides the tensile modulus of the test specimen the ultimate tensile strength and elongation at which the sample breaks (elongation at break) can be obtained. These values are influenced by many factors like processing conditions (they can cause, e.g., orientation of the macromolecules and inner strain) or content of additives (plasticizers, processing aids, stabilizers).

But for pure homopolymers some general rules can be derived: The tensile strength increases with molecular weight, with crosslinking, and with degree of steric order and stiffness of the main chain, whereas branching results either in increase or dropping of the mechanical strength, depending on the type of branches (number of branches per chain, short or long chain branches, distance of the branch points).

If a polymer material is placed suddenly under high stress, the impact strength can be measured. It provides information on the brittleness or the toughness of a sample. Similar as the tensile strength, the impact strength of homopolymers increases with molecular weight, stereoregularity and stiffness of the main chain, whereas branching shows the same influence as described above for the tensile strength. Remarkable is the effect of crosslinking: Weak crosslinking leads to some increase of impact strength, with stronger crosslinking a reverse tendency is observed and the polymer becomes more and more brittle.

A measure of the stiffness of a polymer is the modulus of elasticity (Young's modulus) E. It can be calculated from the stress–strain curve as the slope in the linear region of Hooke's law. It should be considered that due to the definition $E = \sigma/\varepsilon$ for rubberlike materials which show a rather large extension ε at quite low stress σ, the elastic modulus is lower than in the case of materials which are only slightly deformed under stress. For homopolymers, in general the modulus of elasticity increases with molecular weight, crosslinking, stereoregularity, and

Table 2.13 Correlations of structure with the properties of homopolymers

Property	M_w (increasing)	Branching	Crosslinking weak	Crosslinking strong	Stereo-regularity	Stiffness
Solubility	↓↓	↑	↓	↓↓	↓	↓
$\eta_{solution}$	↑	↓	↑	↑↑	→	↑
T_g	↗	↓	↑	↑	↑	↑↑
T_m	↗	↓	→	→	↑	↑
η_{melt}	↑↑	↓	↑	↑↑	→	↑
Tensile Strength	↑	↓↑	↑	↑↑	↑	↑↑
Impact Strength	↑	↓↑	↑	↓	↑	↑
Stiffness	↑	↓	↑	↑↑	↑	↑↑
Hardness	↑	↓	↑	↑↑	↑	↑

stiffness of the main chain, whereas like for tensile and impact strength no simple influence of branching can be predicted.

Also to the mechanical properties belongs the resistance of a sample against the penetration of another material, what often is described as hardness. The result of the measurement of this property is not only of practical interest and characteristic of the material but also dependant on the temperature, the shape of the penetrating body (e.g., ball or needle) and the rate of penetration. But the hardness of a polymer can also be estimated from the elastic behavior as a high modulus of elasticity indicates high hardness. Therefore, the general correlations of hardness with molecular weight, branching, crosslinking, and chain rigidity are similar to that of the stiffness (see above).

The general correlations of structure and properties of homopolymers are summarized in Table 2.13. Some experiments which demonstrate the influence of the molecular weight or the structure on selected properties of polymers are described in Examples 3.6 (degree of polymerization of polystyrene and solution viscosity), 3.15, 3.21, 3.31 (stereoregularity of polyisoprene resp. polystyrene), 4.7 and 5.11 (influence of crosslinking) or Sects. 4.1.1 and 4.1.2 (stiffness of the main chain of aliphatic and aromatic polyesters and polyamides).

2.4.2 Structure/Properties Relationships in Copolymers

Nearly all structure/properties relationships that were discussed for homopolymers are also valid for copolymers. Additional dependencies exist as a result of the composition and structure of the different types of copolymers.

2.4.2.1 Statistic Copolymers

In the case of statistic copolymers of two monomers (binary copolymers) the glass transition temperature steadily changes with the molar amounts of the two monomers. In many cases, a similar behavior is observed with some mechanical properties (tensile strength, impact strength, stiffness, and hardness) (see Chap. 1). Deviations can occur in copolymers, which contain only a few percent of one comonomer.

2.4.2.2 Alternating Copolymers

As discussed in Sect. 3.4 there are only few examples known of strongly alternating copolymers. They exhibit unusual properties: The radical polymerization of ethylene and tetrafluoroethylene leads to a crystalline material with a remarkable thermal, chemical, and light stability, although it contains 50 mol% ethylene units. Another example is the alternating copolymer of ethylene and norbornene obtained via metallocene-catalyzed polymerization. Whereas random copolymers of these monomers are amorphous thermoplasts, the alternating copolymer is crystalline. Moreover, the crystallites have dimensions below the wavelength of visible light, resulting in a crystalline but still highly transparent thermoplast. In none of these cases, however, general structure/properties relationships could be established, except one: The degree of alternation has to be 100%, because each diad of one monomer is a weak linkage in the polymer chain.

2.4.2.3 Block Copolymers

In general, block copolymers are heterogeneous (multiphase) polymer systems, because the different blocks from which they are built are incompatible with each other, as for example, in diene/styrene-block copolymers. This incompatibility, however, does not lead to a complete phase separation because the polystyrene segments can aggregate with each other to form hard domains that hold the polydiene segments together. As a result, block copolymers often combine the properties of the relevant homopolymers. This holds in particular for block copolymers of two monomers A and B.

The simplest dependency exists between composition and glass transition temperature: Independent from the ratio A/B one finds two values for T_g, one for the block from monomer A and one for the block of B. More complex are the dependencies with the mechanical properties. Here, parameters like the ratio A/B, number of blocks, block length, and alternation of the blocks play a decisive role. This is shown in Examples 3.47 and 3.48 with triblock copolymers of butadiene or isoprene with styrene. If the content of the diene blocks is around 20%, a stiff and elastic, transparent thermoplastic material is obtained. Instead, if the diene content is raised to about 70%, a highly elastic but still rather stiff thermoplastic elastomer is obtained. It has to be stressed that these properties can only be reached, when the polystyrene blocks are the terminal ones.

2.4.2.4 Graft Copolymers

Graft copolymers should in principle exhibit similar structure/properties relationships as block copolymers. The problem is that pure graft copolymers which are not accompanied by large amounts of homopolymers are – with few exceptions – very difficult to synthesize. This is the reason why reliable property data in relation to the structure of graft copolymers only in few cases are available, but these cannot be generalized.

2.4.3 Morphology/Properties Relationships

As already mentioned, the results and rules found in dependencies of properties with the morphology of polymeric materials are in most cases limited to special classes of polymers or even one polymer species only. Therefore, the intention of this subsection can only be to demonstrate the potential and the possibilities which exist to influence the properties of polymers by modification of their morphology. This will be done with relevant examples that are described in detail in Chaps. 3–6.

- By variation of the conditions of crystallization (see Sect. 1.3.3.3) polyethylene can be obtained either as folded lamellae, as extended chain crystals (high strength fiber), or as so-called shish kebabs (fibrils with a morphology similar to cellulose). All these variants differ in properties.
- Addition of small amounts of nucleating agents influence the spherulitic crystallization of polypropylene and improve transparency (Sect. 1.3.3.3 and Example 3.20).
- Under special conditions, stiff macromolecules (rigid rods) are able to self-organize and to form highly ordered domains in solution (polyaramides, Sect. 4.1.2 and Example 4.13) or in the melt (aromatic polyesters, Sect. 4.1.1 and Examples 4.5 and 4.6), which show liquid crystalline behavior. Hence, they can be further oriented in flow direction through external forces (see Sect. 1.3.4). By proper processing, materials with extraordinarily high moduli and tensile strength are obtained. In block copolymers and in polymer blends interesting meso- and nano-morphologies of the dispersed phase are sometimes observed. Variations in structure of the polymers or in phase distribution can lead to changes in properties (Sects. 3.4.2.1 and 5.6 and Example 5.23). The properties of polymers processed out of aqueous dispersions are influenced by the particle size and the particle size distribution (Example 3.47).

2.5 Processing of Polymers

Processing of a polymer can be performed with the polymer in various states of aggregation: in solution, in dispersion, and in the melt. The method chosen will depend on whether the polymer melts without decomposition and whether it is soluble. However, the nature of the application is also decisive. In practice a molded object can be prepared from a thermoplastic only via the melt, while for textile coatings the only feasible method is to process from solution or dispersion. But processing of polymers is not only of industrial interest. A previous processing step is also necessary for many physical and chemical investigations, which only can be done with a well-defined test specimen. Although suitable equipment is available for injection molding and extrusion on the laboratory scale which require amounts of substances in the range of 5–1000 g, the following simpler methods suffice for preliminary investigations.

2.5.1　Size Reduction of Polymer Particles

In many cases polymers are not obtained in a suitable finely divided form during preparation or recovery. It then becomes necessary to reduce the size of the particles before carrying out the processing test or the investigation. Conventional grinding in a mortar is usually ineffective because of the toughness of many polymers, unless the sample is first made brittle by cooling to low temperature with liquid nitrogen. During grinding, the polymer frequently acquires an electrical charge. Moistening the sample with a little ether can prevent this. Grinding can also be done in suitable mills. However, the generation of heat may cause low-melting polymers to begin to flow, leading to the formation of lumps. This can be avoided by the addition of small pieces of dry ice. Especially suitable are cooled analytical mills that have the additional advantage of a small milling volume.

In some cases, a combination of spray precipitation (see Sect. 2.2.5.6) and freeze-drying is recommended. For example, one can spray the polymer solution into a mortar, the bottom of which is covered with pieces of solid carbon dioxide the size of a hazel nut. The pieces are then ground more finely, the mortar placed in a desiccator and evacuated with an oil pump. The polymer solution can also be sprayed into a liquid cooled to low temperature, the liquid being immiscible with the solvent of the polymer, e.g., spraying an aqueous solution into cold ether. The polymer then precipitates in the form of a light flaky snow; decantation of the ether is followed by evacuation as described above.

2.5.2　Melt Processing of Polymers

The processing of polymers in the melt is the method most extensively used in technology. It is an essential requirement that the product does not undergo any significant degradation at the high processing temperature and under shear. The processing temperature must generally be considerably higher than the softening or melting temperature in order to reduce the high viscosity of the polymer melt. In this processing technique the polymer is heated above its melting or softening point, the viscous melt is brought into the desired form by mechanical forces and the formed object is finally cooled. This procedure is very widely applicable and allows the preparation of objects in practically any size or shape, for example, of molded bodies (by pressing, injection molding, or extrusion), of films and sheets (by extrusion and calendering), and of fibers (by extrusion or melt spinning).

Polymers that are difficult to process in the melt, either because the melt viscosity is too high or because they decompose, can often be processed at low temperatures through the addition of a plasticizer, leading to improved flow properties and lower processing temperature.

2.5.2.1 Preparation of Polymer Films from the Melt

Thin polymer films are very suitable for a number of physical investigations (microscopy, spectroscopy, mechanical measurements). They can be prepared in the laboratory as follows: A certain amount of finely powdered polymer is spread on a thin (0.1 mm) aluminum foil (15 × 15 cm), which is then covered with a second foil and the entirety placed between the plates of a hydraulic press heated to the melting point of the polymer. After pressing for 30–60 s the foil sandwich is removed from the press, and cooled with water or between two metal plates. Finally, the two aluminum foils are pulled carefully away from the thin polymer film. If a certain film thickness is required a suitable template can be laid between the aluminum foils. Optimal conditions, in respect to amount of polymer, press temperature, pressure, and time, must be determined empirically in every case. If the film is opaque, the temperature was probably too low. If it is too thin or contains gas bubbles (decomposition) the temperature was too high. Finally, the rate of cooling can also affect the properties of the film. Sometimes the aluminum foils can only be removed from the polymer film with difficulty; rapid cooling (with water or in a freezer) or prior coating of the aluminum foils with silicone oil or with an aqueous dispersion of poly(tetrafluoroethylene) usually prevents sticking.

If a suitable press is not available, one may improvise as follows: The heating plates of two electric irons are first bored to accept a thermocouple and then connected in parallel to the power supply through a variable transformer. A calibration curve is determined for the temperature attained at different voltages. For the preparation of a film the finely powdered polymer is placed, as described above, between the two hot-plates, which are then pressed together horizontally in a suitable vice. In this arrangement the pressure cannot be measured, but with a little skill the optimum conditions can be found.

2.5.2.2 Preparation of Fibers by Melt-Spinning
The simplest method of melt-spinning is to melt the polymer in a test tube, dip a glass rod in the melt and pull it out slowly. The threads obtained in this manner are short and irregular so that they are not well suited for subsequent investigations.

Much better for melt spinning is the extrusion of a polymer melt through a defined die from which the filament can be drawn to a fiber by stretching, e.g. by using rollers with different velocities. In the simplest case, a very small die can be used and a fiber can be extruded or drawn from a small scale compounder or an extruder. Laboratory spinning devices consisting of a heatable piston with a stamp pressing out a filament through a small die adapted with a drawing unit just requires only a few grams of polymer.

2.5.3 Processing of Polymers from Solution

If suitable equipment for processing the polymer melt is not available, or if the polymer will not melt, or melts only with decomposition, the processing may be carried out from solution. This technique, however, is limited to the fabrication of films, membranes, fibers, coatings, and adhesives.

2.5.3.1 Preparation of Films from Solution
The simplest way of making films in the laboratory is to pour a highly viscous solution of the polymer onto a glass or metal plate and to allow the solvent to

evaporate slowly. The polymer film is then carefully peeled off. The following points have to be considered: The solvent should not evaporate too quickly, otherwise the resulting film can wrinkle and tear. If necessary the experiment is carried out in an atmosphere of solvent (use a desiccator). On the other hand, the boiling point should not be too high; otherwise the last residues of solvent are difficult to remove. The concentration of the polymer should be such that the solution can still be poured, but does not run off the glass plate. An approximately 20% solution is usually about right, but the best conditions must be established by experiment. In order to obtain a film of uniform thickness the glass plate must be as level as possible and the polymer solution evenly distributed. This may be achieved most easily with the aid of a glass rod wrapped at both ends with some tape or twine. For more precise work, metal devices with adjustable layer thickness are preferable. When ready the film is lifted at one edge with the aid of a razor blade or small knife and then slowly peeled away from the glass plate. Films can also be made from aqueous polymer dispersions at room temperature if the glass transition temperature of the polymer is not too high.

In an analogous way porous films like asymmetric membranes can be prepared. For this purpose the polymer solution – normally with DMF as solvent – is coated on a glass plate and then the solvent is partially evaporated (for example, in a drying chamber at 80°C for a few minutes). The coated glass plate is cautiously immersed in water, whereby after about 1–5 min an optically opaque film which shows an internal pore structure, detaches from the glass plate. After the exchange of the organic solvent by water, the film is dried and can be used as a separating membrane for, e.g., the deionization of sea water. Appropriate polymers for such membranes are soluble aromatic polyamides, polysulfones, or cellulose derivatives. For an optically clear film, which contains no micropores, the procedure must be repeated with a smaller evaporation time.

2.5.3.2 Preparation of Fibers by Solution Spinning

There are two methods of spinning fibers from solution: dry spinning and wet spinning. In dry spinning the viscous polymer solution is forced through a jet (spinnerette) into a chamber filled with hot air (or hot nitrogen). The solvent is thereby evaporated leaving behind a ready-made thread. This spinning process cannot be conducted in a simple manner and is not readily carried out in the laboratory. On the other hand, wet spinning, in which the polymer solution is injected into a suitable precipitant, by which the polymer coagulates in the form of a thread, is readily conducted on a small scale. The quality of the thread depends very much on the precipitation conditions (precipitant, bath temperature) that must be determined experimentally in each case. The precipitant must be chosen so that complete coagulation does not occur too quickly. If necessary, a solvent/precipitant mixture can be used (e.g., polyacrylonitrile dissolved in DMF, precipitated in DMF/water mixture). For injecting the polymer solution into the precipitant bath, hypodermic syringes are very useful. They allow control of both the thread diameter and rate of spinning in a simple manner (see Example 3.11).

2.5.4 Processing of Aqueous Polymer Dispersions

Aqueous dispersions of polymers with a glass transition temperature near room temperature possess the property of irreversible film formation. If a poly(vinyl acetate) dispersion (solid content about 50 wt.%, see Example 3.4) is spread out in the form of a thin layer on a glass plate and the water is allowed to evaporate at room temperature, then the colloid particles agglomerate irreversibly and build a homogeneous film on the substrate, which cannot be redispersed in water. After removal from the glass plate the self-supporting film has good mechanical properties. The film formation and the properties of the film depend on the chemical composition (glass transition temperature) of the polymer, and also on the size of the colloid particles (see Example 3.48).

The technique of film formation from aqueous polymer dispersions is widely used in the field of paints ("dispersion paints", "latex-paints"), coatings, and adhesives. The equipment is simple: brush, scraper, rake, or roller.

Bibliography

Agassant JF, Avenas P, Sergent J-Ph, Carreau PJ (1991) Polymer processing: principles and modelling. Hanser, Munich

Allen G, Bevington J (eds) (1989) Comprehensive polymer science, vol 1. Pergamon, Oxford

Cherdron H, Brekner M-J, Osan F (1994) Cycloolefin copolymers. Angew Makromol Chem 223:121

Corish PJ (ed) (1991) Concise encyclopedia of polymer processing and application. Pergamon, Oxford

Eisele U (1990) Introduction to polymer physics. Springer, New York

Francuskiewics F (1994) Polymer fractionation. Springer, Berlin/Heidelberg/New York

Garton A (1992) Infrared spectroscopy of polymer blends, composites and surfaces. Hanser, Munich

Grellmann W, Seidler S (2007) Polymer testing. Hanser, Munich

Han CD (1976) Rheology in polymer processing. Academic, New York

Houben-Weyl (1987) Methoden der organischen Chemie, vol E20. Makromolekulare Stoffe, Thieme, Stuttgart/New York

Hummel DO, Scholl F (1986) Atlas of polymer and plastics analysis. 2nd edn, Hanser, Munich; Wiley-VCH, Weinheim/New York

Kämpf G (1986) Characterization of plastics by physical methods. Hanser, Munich

Klöpffer W (1984) Introduction to polymer spectroscopy. Springer, Berlin/Heidelberg/New York

van Krevelen DW (1990) Properties of polymers, their estimation and correlation with chemical structure, 3rd edn. Elsevier, Amsterdam

Mori S, Barth HG (1999) Size exclusion chromatography. Springer, Berlin/Heidelberg/New York

Morto-Jones DH (1989) Polymer processing. Chapman and Hall, New York

Osswald TA, Menges G (2003) Materials science of polymers for engineers. Hanser Gardner, 2nd edn. Cincinati, USA

Pasch H (1997) Analysis of complex polymers by interaction chromatography. Adv Pol Sci 128:1

Pethrick RA, Dawkins JV (1999) Modern techniques for polymer characterization. Wiley-VCH, Heidelberg

Schmidt-Rohr K, Spiess HW (1994) Multidimensional solid-state NMR and polymers. Academic, London

Synthesis of Macromolecules by Chain Growth Polymerization

3

Polymerization reactions can proceed by various mechanisms, as mentioned earlier, and can be catalyzed by initiators of different kinds. For chain growth (addition) polymerization of single compounds, initiation of chains may occur via radical, cationic, anionic, or so-called coordinative-acting initiators, but some monomers will not polymerize by more than one mechanism. Both thermodynamic and kinetic factors can be important, depending on the structure of the monomer and its electronic and steric situation. The initial step generates active centers that generally cause the reaction to propagate very rapidly via macroradicals or macroions; chain termination yields inactive macromolecules. It is important to note that in classical uncontrolled chain growth mechanism the molar mass of the formed polymers increases fast in the first reaction period but reaches a plateau value even at relatively low monomer conversion which leads to the fact that monomer as well as terminated final polymer chains are present in the reaction system. The most important initiators are summarized in Table 3.1.

3.1 Radical Homopolymerization

Radical polymerization is induced by an initiation step in which radicals are formed. Radicals can be generated thermally from the monomer, although such a mechanism of initiation has only been completely verified in the case of styrene (see Example 3.1). Radicals are usually generated by decomposition of an initiator (frequently also called the catalyst). The radicals so formed then react successively with the monomer molecules in the propagation step leading to growing radicals (macroradicals) which are finally deactivated by chain transfer or termination. Termination usually occurs by combination or disproportionation of two macroradicals; the radicals generated by the decomposition of the initiator are incorporated into the macromolecules as end groups, giving two such end groups per macromolecule for termination by combination, but only one for termination by disproportionation.

D. Braun et al., *Polymer Synthesis: Theory and Practice*,
DOI 10.1007/978-3-642-28980-4_3, © Springer-Verlag Berlin Heidelberg 2013

Table 3.1 Initiators for chain growth (addition) polymerizations

Radical	Cationic	Anionic	Coordinative
Inorganic and organic peroxo compounds, e.g., peroxodisulfates, peroxides, hydroperoxides, peresters	Protic acids	Proton acceptors	Organometallic compounds
Aliphatic azo compounds	Lewis acids with or without co-initiators	Lewis bases	Mixed catalysts (Ziegler-Natta catalysts)
Substituted ethanes, e.g., benzpinacol	Carbonium ions	Organometallic compounds	p-Complexes with transition metals, e.g., metallocenes
Redox systems with inorganic and organic components	Iodonium ions	Electron-transfer agents, e.g., alkali metals, alkali-aromatic complexes, alkali metal ketyls	Activated transition metal oxides
Controlled radical polymerization: stable nitroxy radicals, R–X/Cu(I)/ligand, thiocarbamates (transferters)			
Heat	Ionizing radiation		
UV radiation, high energy radiation			

In transfer reactions the growth of a chain is ended, for example, by transfer of a hydrogen atom from the molecule ZH, but at the same time a new polymer chain is started by the radical Z˙ that is formed simultaneously. Thus, several macromolecules result from one primary radical; therefore, the kinetic chain length, i.e., the total number of monomer molecules induced to be polymerized by one primary radical, is much larger than the degree of polymerization of the macromolecules formed. A general scheme is as follows:

Start:

$$R\text{-}R \longrightarrow 2\,R\cdot$$

Propagation:

Chain transfer:

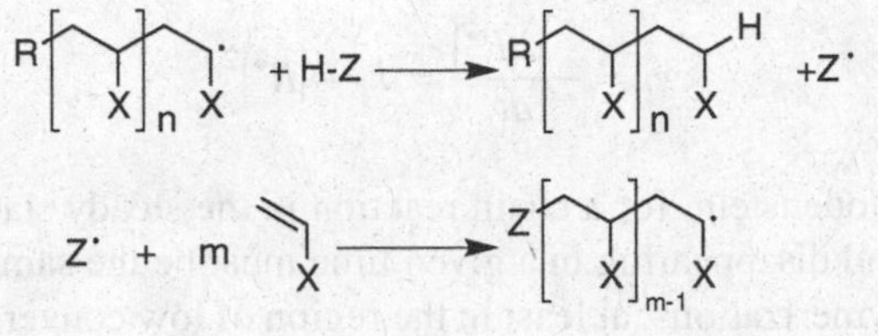

Chain termination by combination:

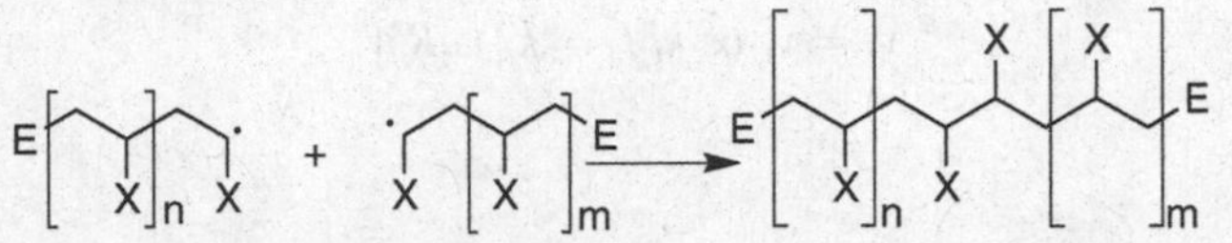

Chain termination by disproportionation:

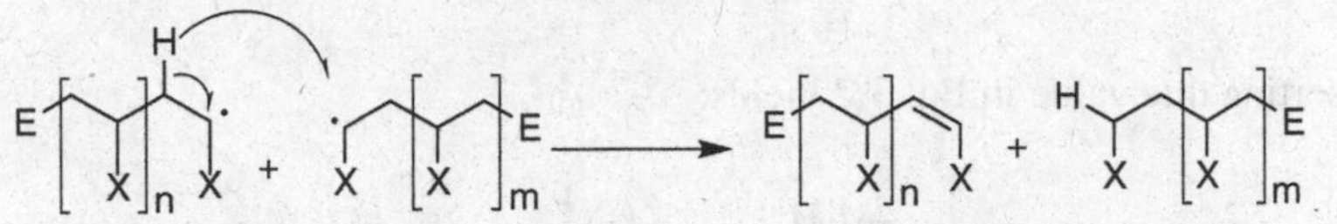

Let us for the moment disregard chain transfer reactions. Radical polymerization then consists of three component reactions: initiation, propagation of the polymer chains, and termination of chain growth. The rate of primary radical formation, v_i by decomposition of the initiator I, may be written:

$$v_i = k_i \cdot [I] \tag{3.1}$$

The rate constant k_i contains a factor that allows for the efficiency of initiation; not all the radicals generated by the initiator are capable of starting polymer chains, some are lost by combination or other reactions. The initiator efficiency is defined as the ratio of the number of initiator molecules that start polymer chains to the number of initiator molecules decomposed under the given conditions of the polymerization. With most radical initiators the efficiency lies between 0.6 and 0.9; it also depends on the nature of the monomer.

The rate of propagation v_p is given by:

$$v_p = \frac{-d[M]}{dt} = k_p \cdot [M] \cdot [R^\bullet] \tag{3.2}$$

Here it is assumed that k_p is independent of the number of monomer molecules already added; [R] denotes the concentration of radicals in the system.

The rate of the termination reaction v_t is given by:

$$v_t = \frac{-d[R^\bullet]}{dt} = k_t \cdot [R^\bullet]^2 \tag{3.3}$$

According to Bodenstein, for a chain reaction in the steady state, the number of radicals formed and disappearing in a given time must be the same. This applies to most addition polymerizations, at least in the region of low conversion. Under these conditions v_i and v_t may be equated:

$$v_i = v_t \text{ or } k_i[I] = k_t \cdot [R^\bullet]^2 \tag{3.4}$$

Therefore:

$$[R^\bullet] = \left(\frac{k_i}{k_t}\right)^{\frac{1}{2}} \cdot [I]^{\frac{1}{2}} \tag{3.5}$$

Inserting this value in Eq. 3.2 yields:

$$v_p = \frac{-d[M]}{dt} = k_p \left(\frac{k_i}{k_t}\right)^{\frac{1}{2}} \cdot [M] \cdot [I]^{\frac{1}{2}} \tag{3.6}$$

where v_p is identical with the overall rate of polymerization R_p, since at sufficiently large chain length it determines the consumption of the monomer M almost completely. Hence the rate of polymerization is proportional to the monomer concentration and the square root of the initiator concentration. At high initiator concentrations or with stable radicals formed, termination of this primary radicals can occur and the reaction kinetics change.

For initiation of polymerizations by light or high energy radiation, the initiator concentration [I] is replaced by the radiation intensity in the above kinetic equations.

Raising the temperature of a radical chain reaction causes an increase in the overall rate of polymerization since the main effect is an increase in the rate of decomposition of the initiator and hence the number of primary radicals generated per unit time. At the same time the degree of polymerization falls since, according to Eq. 3.3, the rate of the termination reaction depends on the concentration of radicals (see Example 3.2). Higher temperatures also favor side reactions such as chain transfer and branching, and in the polymerization of dienes the reaction temperature can affect the relative proportions of the different types of CRUs in the chains.

Although the above derivations involve certain simplifications, they nevertheless represent correctly the kinetics of many addition polymerization reactions. However, the behavior is different when the polymerization is conducted under heterogeneous conditions, e.g., in suspension or in emulsion (see literature cited in Sect. 2.2.4).

For radical polymerizations of some monomers in bulk a specific effect can appear when the conversion exceeds a certain value. In these cases the viscosity of the reaction mixture increases to such an extent as a result of the formation of macromolecules that the mobility of the growing macroradicals becomes severely restricted. Bimolecular termination, but not the propagation reaction, is then hindered, so that both the degree of polymerization and the reaction rate increase; the system is no longer in a steady state and the radical concentration rises continuously. The increasing reaction rate, coupled with the more difficult heat exchange in the very viscous medium, leads to a rise in temperature as a consequence of the heat evolution, and hence to an auto-acceleration of the reaction, which can become explosive in nature. This phenomenon is called the gel effect or Trommsdorff-Norrish effect, but does not occur to the same extent in all monomers. It is especially noticeable with methyl methacrylate (see Example 3.8). Basically, the gel effect is based on restrained diffusion since it can be also observed if the polymerization is run strictly isothermal. Somewhat similar behavior is observed in the polymerization of some monomers in poor solvents in which the resulting macromolecules are more tightly coiled than in good solvents, where the polymer chains are more strongly solvated and more mobile. The interplay of radical formation, propagation, and termination of the growing chains determines the overall rate and degree of polymerization, provided there are no chain transfer reactions. When a growing polymer chain undergoes chain transfer its growth is terminated, but at the same time a new polymer chain is started; the kinetic chain is therefore uninterrupted.

The kinetic chain length is given by the number of monomer molecules consumed per initiation step. Since the efficiency of most initiators is not known quantitatively it is necessary to compare the rate of the propagation reaction with either the rate of initiation or the rate of termination. If there is no chain transfer, the kinetic chain length v for termination by disproportionation is equal to the number-average degree of polymerization:

$$P_n = v = \frac{v_p}{v_i} = \frac{v_p}{v_t} \tag{3.7}$$

On the other hand, for termination by combination the degree of polymerization is equal to twice the kinetic chain length:

$$P_n = 2 \cdot v = 2 \cdot \frac{v_p}{v_i} = 2 \cdot \frac{v_p}{v_t} \tag{3.8}$$

If chain transfer does take place the rate v_f of the chain transfer reaction $(R^{\bullet} + ZH \rightarrow RH + Z^{\bullet})$ must be taken into account:

$$P_n = \frac{v_p}{v_i + v_f} \tag{3.9}$$

and for termination by combination:

$$P_n = \frac{v_p}{\frac{1}{2}v_i + v_f} \tag{3.10}$$

Taking the reciprocal of P_n and inserting the expressions for v_p, v_i, and v_f from Eqs. 3.4, 3.5 and 3.6, one obtains:

$$\frac{1}{P_n} = \frac{v_i}{\alpha \cdot v_p} + \frac{v_f}{v_p} = \frac{k_i \cdot [I]}{\alpha \cdot k_p \cdot \left(\frac{k_i}{k_t}\right)^{\frac{1}{2}} \cdot [I]^{\frac{1}{2}} \cdot [M]} + \frac{k_f \cdot [R^\bullet] \cdot [ZH]}{k_p \cdot [R^\bullet] \cdot [M]} \tag{3.11}$$

where $\alpha = 1$ for termination by disproportionation and $\alpha = 2$ for termination by combination.

Since the chain carrier ZH may be the monomer, the initiator, the solvent (S), an added transfer agent (regulator), or the polymer already formed, a more general form of Eq. 3.11 is:

$$\frac{1}{P_{n,0}} = K \cdot \frac{[I]^{\frac{1}{2}}}{[M]} + C_M + C_I \cdot \frac{[I]}{[M]} + C_S \cdot \frac{[S]}{[M]} + C_{ZH} \cdot \frac{[ZH]}{[M]} + \ldots \tag{3.12}$$

where C_x denotes the chain transfer constant k_f/k_p, appropriate to the chain transfer agent X (see Example 3.8). Since the chain transfer constant C_I for most initiators is approximately zero, Eq. 3.12 shows that at moderate conversion the reciprocal of the degree of polymerization is a linear function of the square root of the initiator concentration. Since in turn $[I]^{1/2}$ is proportional to the overall rate of polymerization R_p, Eq. 3.6, the degree of polymerization is lower, the faster the polymerization occurs.

Transfer reactions with solvent and with those compounds termed regulators are especially important because of their marked effect on the molecular weight of the polymer being formed. While the transfer constants for most solvents are not very big (e.g., for benzene reacting with the growing polystyrene radical at 60°C, C_S is of the order of 10^{-5}), there are some with relatively high transfer constants, so that the polymer formed has a relatively short chain length. A particularly well-investigated case is that of the polymerization of styrene in carbon tetrachloride, where the transfer constant is about 10^{-2}. The resulting polystyrene is of very low molecular weight and consists of a mixture of oligomers.

$$P^\bullet + CCl_4 \longrightarrow P–Cl + Cl_3C^\bullet$$

$$Cl_3C^\bullet + M \longrightarrow Cl_3C–M^\bullet$$

Such a process is called telomerization. The polymer thus contains chloro and trichloromethyl end groups. More important, however, is the ability to control the reduction of molecular weight by the use of regulators. The molecular weight of a polymer can only be controlled to a limited extent by adjustment of the monomer concentration, initiator concentration, and temperature. Hence, in industry it is common practice to add transfer agents. For this purpose one may use various thiols which, because of their high transfer constants (e.g., for 1-dodecanethiol in the polymerization of styrene, $C_{ZH} = 19$) need only to be added in very small amounts (about 0.1% with respect to monomer). The simplest way of determining the transfer constant of such a regulator, using Eq. 3.12, is by polymerization experiments at constant initiator and monomer concentrations and varying concentrations of the transfer agent ZH.

In the absence of regulators one may write:

$$\frac{1}{P_{n,0}} = K \cdot \frac{[I]^{\frac{1}{2}}}{[M]} + C_M + C_I \cdot \frac{[I]}{[M]} + C_S \cdot \frac{[S]}{[M]} \qquad (3.13)$$

Then with addition of regulator ZH:

$$\frac{1}{P_n} = \frac{1}{P_{n,0}} + C_{ZH} \cdot \frac{[ZH]}{[M]} \qquad (3.14)$$

Plotting the reciprocal of the number-average degree of polymerization for polymers obtained at different regulator concentrations against $[ZH]/[M]$, a straight line is obtained which intersects the ordinate at $1/P_n$ and has a slope equal to the transfer constant of the regulator C_{ZH} (also see Sect. 2.2.5.5 and Example 3.8b).

Finally, many substances can inhibit polymerization reactions. Amongst these are molecular oxygen, Cu(I) ions, nitric oxide, phenols such as hydroquinone and 4-*tert*-butylpyrocatechol, quinones, certain aromatic amines such as *N*-phenyl-*β*-naphthylamine, nitro compounds, and some sulfur compounds. The mechanism of action of most inhibitors is not yet fully clarified; the inhibitors react either with the primary radicals or with the growing chains to yield products that are no longer active in propagation. Stable free radicals such as *N,N*-diphenyl-*N'*-picrylhydrazyl are also effective inhibitors. Since this radical is strongly colored its consumption during the inhibition period can be followed photometrically:

Inhibitors are frequently used to stop a polymerization quickly, for example, in kinetics investigations. Another important application is the stabilization of monomers against undesired polymerization during storage. Autoxidation of unsaturated monomers by the action of atmospheric oxygen frequently results in the formation of peroxidic compounds that can generate radicals at relatively low temperatures, thus initiating polymerization; inhibitors are added as stabilizers to suppress this undesired and uncontrolled polymerization (see Sect. 2.2.5.5). These must of course be removed before using the monomer for polymerization reactions (see Example 3.1). Since the inhibitors are consumed by growing polymer chains, the time during which they prevent polymerization (incubation time, induction period) depends on their concentration in the stabilized monomer.

Molecular oxygen plays a special role in radical polymerizations. It is known to react very rapidly with hydrocarbon radicals with the formation of peroxy radicals:

$$R^{\bullet} + {}^{\bullet}O - O^{\bullet} \rightarrow R - O - O^{\bullet}$$

It is thus clear why atmospheric oxygen must be carefully excluded during radical polymerizations. Peroxy radicals are much less reactive than most alkyl (or aryl) radicals, but they can add a further monomer molecule, regenerating an alkyl radical, which can react again with oxygen. The rate of consumption of monomer, relative to that in the absence of oxygen, is substantially reduced. An alternating addition of unsaturated monomer and oxygen is observed, resulting in the formation of a polymeric peroxide (copolymerization with molecular oxygen):

Normal polymerization commences only after complete consumption of the oxygen; this is then accelerated by the formation of additional initiating radicals through the thermal decomposition of the polymeric peroxide. Thus, molecular oxygen at first inhibits the polymerization, but after its consumption there is an accelerating action.

Unlike ionic polymerizations, radical chain polymerizations have so far been found to occur only with unsaturated compounds. In some cases they can be induced purely thermally, or by means of light or high-energy radiation; generally, however, radical initiators such as peroxo compounds, azo compounds, and redox systems are used.

The initiation of a radical polymerization of a monomer can be achieved with practically every peroxo or azo compound. This means that in these cases the type of initiator influences only the rate and degree of polymerization, the nature of the end groups and branching but not the polymerizability of the monomer as such. This is not the case with redox systems as radical initiators. As a consequence, the determination of whether a new substance polymerizes radically or not is rather simple: to the purified compound (as a 30–50% solution) is added under nitrogen 1% of dibenzoyl peroxide and this mixture is heated for several hours to 60–120°C. The occurrence of turbidity or an increase of viscosity (if the polymer is soluble in the reaction mixture) are first indications that a polymerization has taken place. Final proof is the analysis of the reaction mixture after separation of the polymer

Table 3.2 Some parameters influencing radical polymerization

Parameter		Rate of polymerization/ conversion	Molecular weight	Example
Initiator concentration	↑	Increase	Decrease	3–6, 3–11b
Temperature	↑	Increase	Decrease	3–1
Monomer concentration	↑	Increase	Increase	3–7
Polymerization time	↑	Decrease	Decrease[a]	3–6
Chain transfer agent		Constant	Decrease	3–8b

[a]Exception: "Living polymerization"

either by filtration or by precipitation in a solvent that is miscible with the monomer and does not keep the polymer in solution.

A short summary of the parameters which influence the radical polymerization are given in Table 3.2

3.1.1 Polymerization with Peroxo Compounds as Initiators

Organic and inorganic peroxo compounds are especially important as initiators of radical polymerizations. Hydroperoxides, dialkyl peroxides, diacyl peroxides, and peresters are typical organic peroxo compounds. Since they dissolve not only in organic solvents but also in most monomers, they are suitable for solution polymerization as well as bulk or bead polymerization. Their decomposition into radicals can be brought about either thermally, or by irradiation with light, or by redox reactions (see Sect. 3.1.3). The rate of decomposition of organic peroxo compounds depends on their structure and on the temperature. For initiation by thermal decomposition of peroxo compounds an acceptable rate of polymerization is generally attained only above 50°C. Some peresters, for example, diethyl peroxydicarbonate, are exceptional and decompose rapidly at room temperature, thus, because of the danger of explosion, they should be added only in dilute solution.

A peroxo compound that is frequently used (concentration 0.1–1 wt% with respect to monomer) is dibenzoyl peroxide (see Example 3.4). It decomposes in solution at temperatures of about 50–80°C, mainly into benzoyloxy radicals; at higher temperatures phenyl radicals are formed to an increasing extent by elimination of carbon dioxide, so that the end groups of the resulting polymer are either hydrolyzable benzoic ester groups or non-hydrolyzable phenyl groups:

In practice *tert*-butyl peroxobenzoate is often used as a thermally decomposing radical initiator. It decomposes into benzoyloxy and methyl radicals, and acetone:

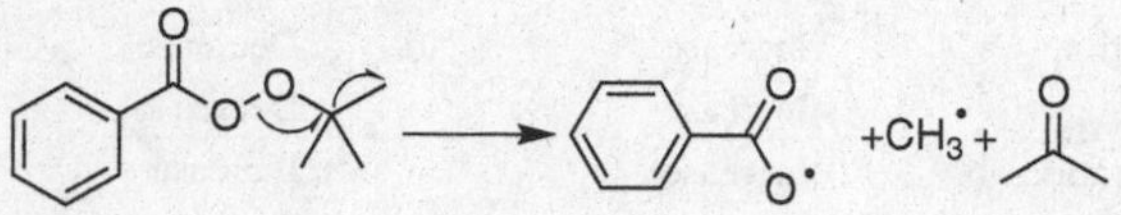

The kinetic relationship, according to which the rate of polymerization increases and the average degree of polymerization decreases with increasing initiator concentration, is satisfied by most monomers when either unsubstituted or substituted dibenzoyl peroxides are used as initiators.

Thermally activated organic peroxo compounds are generally used for polymerization in bulk or in organic solvents, as well as for bead polymerization; instead, inorganic peroxo compounds are the most suitable for initiating polymerization in aqueous solution or emulsion. Hydrogen peroxide is mainly used as a component of a redox initiator (see Example 3.9); in contrast, potassium or ammonium peroxodisulfate (concentration 0.1–1 wt% with respect to monomer) are very frequently used without a reducing agent, since even at 30°C they decompose thermally into radicals that can initiate polymerization.

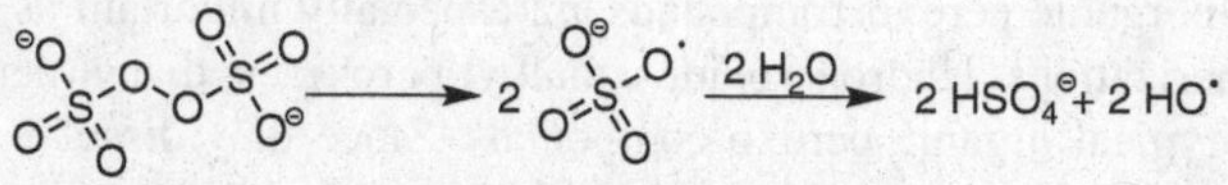

Ammonium peroxodisulfate is more soluble in water than the potassium salt; furthermore, it dissolves in more polar organic solvents (e.g., DMF), so that it is sometimes also used for initiating polymerizations in organic media. In polymerizations initiated by peroxodisulfates the reaction medium is liable to become acidic, so that buffering is generally necessary (see Example 3.2).

A list of some peroxo compounds that generate free radicals is given in Table 3.3, extensive information can be found in the literature. The initiators are selected according to their thermal half-lives to ensure that at the polymerization temperature they provide a source of free radicals. The rate equation for the thermal half-life is given by: $t_{1/2} = 0.693 \cdot k_d$, where k_d is the rate constant for the thermal decomposition. In technical applications one often uses the temperature at which within a certain time interval one half of the initiator is decomposed (e.g., quoted as 10 h half-life temperature).

Example 3.1 Thermal Polymerization of Styrene in Bulk (Effect of Temperature)

Safety precautions: Before this experiment is carried out, Sect. 2.2.5 must be read as well as the material safety data sheets (MSDS) for all chemicals and products used.

Table 3.3 Peroxo compounds for initiation of radical polymerization

Peroxo compound	Formula	Suitable temperature for polymerizations [°C]
Hydrogen peroxide		30–80
Potassium peroxo-disulfate		40–100
Dibenzoyl peroxide		40–100
Cumyl hydroperoxide		50–120
Di-tert-butyl peroxide		80–150

Monomeric styrene is freed from phenolic inhibitors by shaking twice with 10% sodium hydroxide solution, washing three times with distilled water, drying over calcium chloride or silica gel and distilling into a receiver (see Sect. 2.2.5.3) under reduced pressure of nitrogen (bp 82°C/100 Torr, 46°C/20 Torr). It is stored in a refrigerator until required.

4 g (38.4 mmol) of destabilized styrene is weighed into each of five thick-walled Pyrex Schlenk tubes (content 15–20 ml). The tubes, equipped with a suitable adapter (see Sect. 2.2.5.4), are now cooled in a methylene chloride/dry ice cold bath, thereby freezing the styrene (mp −30.6°C); after evacuation with an oil pump and thawing, the tubes are filled with nitrogen. This sequence is repeated twice more. Finally the tubes are sealed off under nitrogen. The samples are polymerized at 80°C, 100°C, 110°C, 120°C, and 130°C, respectively, by placing them in an appropriately adjusted thermostat or vapor bath (Caution: the tubes may explode; place them behind shield, or cover them with cloth!). After exactly 6 h the sealed tubes are rapidly cooled by immersion in cold water (wear safety goggles) and then opened. The contents are each dissolved in 20–30 ml toluene and the solution run slowly from a dropping funnel into 200–300 ml methanol with stirring, thereby precipitating the polystyrene. The polymers are filtered off using sintered glass crucibles (porosity 2) and dried to constant weight in vacuum at 50°C. The observed yield (in %) is plotted as a function of polymerization temperature. Using an Ostwald viscometer (capillary diameter 0.3 mm) the limiting viscosity numbers of all samples are determined in toluene at 20°C, and the average degrees

of polymerization derived (see Sect. 2.3.3.3). These values are plotted as a function of temperature.

Example 3.2 Polymerization of Styrene with Potassium Peroxodisulfate in Emulsion

Safety precautions: Before this experiment is carried out, Sect. 2.2.5 must be read as well as the material safety data sheets (MSDS) for all chemicals and products used.

A 250-ml standard apparatus (round bottom flask, several inlets for stirrer, reflux condenser, nitrogen flux or vaccum, thermometer, heating bath) is evacuated and filled with nitrogen three times. The following are then added under nitrogen: 122 mg (0.45 mmol) of potassium peroxodisulfate, 50 mg of NaH_2PO_4, 1.0 g of sodium oleate or sodium dodecyl sulfate, and 100 ml of water that has been boiled under nitrogen (sodium dihydrogen phosphate is added because styrene polymerizes best in weakly alkaline medium). When everything has dissolved, 50 ml of destabilized styrene are added with constant stirring; the resulting oil-in-water emulsion is heated at 60°C for 6 h with steady stirring under a slow stream of nitrogen. After cooling the polystyrene latex, 30 ml are pipetted into a beaker; the polymer is precipitated by addition of an equal volume of a concentrated solution of aluminum sulfate, if necessary by boiling; a further 30-ml sample is precipitated by dropping into 300 ml of methanol. Finally the latex remaining in the flask is coagulated by addition of concentrated hydrochloric acid. The samples are washed with water and methanol, filtered, and dried in vacuum at 50°C. The total yield and the limiting viscosity number (degree of polymerization) of one sample are determined. The values are compared with those obtained under similar conditions for polymerization conducted in bulk (Example 3.1) and in solution (Example 3.7).

Example 3.3 Polymerization of Vinyl Acetate with Ammonium Peroxodisulfate in Emulsion

Safety precautions: Before this experiment is carried out, Sect. 2.2.5 must be read as well as the material safety data sheets (MSDS) for all chemicals and products used.

A 500-ml standard apparatus (round bottom flask, several inlets for stirrer, reflux condenser, nitrogen flux or vaccum, thermometer, heating bath) is evacuated and filled with nitrogen. 5 g of poly(vinyl alcohol) are then placed in the flask (see Example 5.1) and dissolved in 100 ml of distilled water by stirring at 60°C; next are added 2.2 g of oxethylated nonylphenol and 0.4 g (1.8 mmol) of ammonium peroxodisulfate buffered by the addition of 0.46 g sodium acetate in order to prevent the hydrolysis of the monomeric vinyl acetate.

The solution is heated to 72°C and 25 g (0.29 mol) of vinyl acetate (freshly distilled under nitrogen) are added dropwise. The temperature of the water bath is then raised to 80°C. As soon as the internal temperature reaches 75°C, a further 75 g (0.87 mol) of vinyl acetate are added dropwise at such a rate that the internal temperature is maintained between 79°C and 83°C at moderate reflux (total time about 20 min); finally a further 0.1 g (0.44 mmol) of ammonium peroxodisulfate in 1 ml of distilled water is added. Refluxing soon abates and the internal temperature rises to about 86°C. The reaction mixture is allowed to polymerize for another 30 min on the water bath at 80°C. On cooling, a creamy dispersion is obtained that contains less than 1% monomer (corresponding to a solid content of about 50%). The polymer can now be precipitated by addition of a threefold excess of saturated sodium chloride solution. The poly(vinyl acetate) dispersion can also be spread out as a thin layer on a glass plate; on drying in air the polymer particles coalesce and form a homogeneous, very cohesive film that is resistant to water. These kinds of dispersions are very stable and insensitive to the addition of pigments or electrolytes, as well as to temperature variations (within certain limits) and are therefore extensively used as paints for wood or plaster surfaces, as well as for cementing wood and for impregnation of leather, paper, and textiles.

Example 3.4 Polymerization of Vinyl Acetate in Suspension (Bead Polymerization)

Safety precautions: Before this experiment is carried out, Sect. 2.2.5 must be read as well as the material safety data sheets (MSDS) for all chemicals and products used.

0.15 g of a hydrolyzed copolymer of styrene and maleic anhydride (see Example 5.3) are dissolved in 150 ml of hot distilled water to give a 1% solution, which is then neutralized with a few drops of ammonia solution. The solution of the ammonium salt is placed in a standard apparatus (round bottom flask, several inlets for stirrer, reflux condenser, nitrogen flux or vaccum, thermometer, heating bath) with dropping funnel (mounted on the reflux condenser), and heated to about 70°C by means of a water bath set at 80°C.

0.6 g (2.5 mmol) of dibenzoyl peroxide are dissolved in 100 g (1.16 mol) of freshly distilled vinyl acetate and the clear solution allowed to run in through the reflux condenser over a period of about 30 min with vigorous stirring. The water bath temperature is held steady at 80°C and the rate of addition of the vinyl acetate is so regulated that moderate refluxing is maintained. After the addition of monomer is complete the internal temperature rises to about 80°C, the reflux having ceased a few minutes previously.

In order to remove the small amount of unconverted vinyl acetate, steam is blown through the suspension for about 30 min, the flask being fitted with a condenser for distillation. The suspension is finally cooled externally to room temperature and diluted with cold water to about 500 ml. Only now is the stirrer switched off and after settlement of the bead polymer the aqueous layer is drawn off. The product is washed by repeated slurrying with cold water and subsequent decantation until the wash water no longer foams and is therefore free of suspending agent. The moist bead polymer is dried as a thin layer in vacuum at room temperature. The limiting viscosity number is determined in acetone at 30°C and the average molecular weight derived (see Sect. 2.3.3.3).

Example 3.5 Polymerization of Methacrylic Acid with Potassium Peroxodisulfate in Aqueous Solution

Safety precautions: Before this experiment is carried out, Sect. 2.2.5 must be read as well as the material safety data sheets (MSDS) for all chemicals and products used.

11 ml of distilled water are heated to 80°C in a 50-ml standard apparatus (round-bottom flask, several inlets for stirrer, reflux condenser, nitrogen flux or vaccum, thermometer, heating bath) with two dropping funnels. At this temperature 6 g (0.07 mol) of methacrylic acid, purified by vacuum distillation under nitrogen, and a solution of 0.18 g (0.66 mmol) of potassium peroxodisulfate in 4 ml of water are slowly introduced dropwise into the flask over a period of 10–15 min, while stirring with a magnetic stirrer. The methacrylic acid polymerizes immediately, as may be seen from the increased viscosity of the solution. After the additions are complete, the temperature is held for another hour at 80°C.

After polymerization the rather viscous solution is added dropwise to 50 ml of 0.1 N HCl, whereupon the polymer precipitates. The polymer is filtered, if necessary broken up, extracted in a Soxhlet apparatus with petroleum ether for 5 h, and finally dried to constant weight in vacuum at 50°C. The yield is quantitative.

Poly(methacrylic acid) is soluble in water, methanol, 1,4-dioxane, and DMF. The solution viscosity of the polymer is measured in water at 20°C, using concentrations of 0.5, 0.7, 1.0, 1.5, 2.0, 2.5, 3.0, 3.5, and 4.0 g/l (Ostwald viscometer, capillary diameter 0.3 mm). A plot of h_{sp}/c against c gives a curve that is typical of polyelectrolytes (see Sects. 1.3.1.2 and 2.3.3.3.1).

If, however, the viscosity is measured at the same concentrations in 1 N NaCl solution the behavior is identical with that for non-electrolyte polymers. It is best to proceed as follows. 30 g of NaCl are dissolved in water and made up to 100 ml in a graduated flask; this solution is 5.1 normal. To prepare the solutions for measurement at the aforementioned concentrations, the required amounts of poly (methacrylic acid) are weighed into 10-ml graduated flasks, dissolved in about 5 ml of water, and 2 ml of the 5.1 N NaCl solution added. The solutions are finally made up to the mark with water to give a solution of 1 N with respect to NaCl.

3.1.2   Polymerization with Azo Compounds as Initiator

Azo compounds that are especially suitable as initiators for radical polymerization are those in which the azo group is bonded on both sides to tertiary carbon atoms that carry nitrile or ester groups in addition to alkyl groups (Table 3.4). They are stable at room temperature, but decompose thermally above 40°C, or photochemically below 40°C, giving substituted alkyl radicals and liberating nitrogen. This nitrogen can be a nuisance at high initiator concentrations, both, in dilatometric measurements (gas bubbles in the measuring capillary) and in bulk polymerizations (the solid polymer is then frequently permeated with minute gas bubbles). Such azo compounds decompose in a manner that is essentially independent of solvent and strictly according to a first order rate law so that they are

Table 3.4 Azo-compounds for the initiation of radical polymerization

Azo-initiator	Formula	Suitable temperature for polymerization in °C
2,2′-Azobisisobutyronitrile (AIBN)		50–70
2,2′-Azobis (2-amidino-propane). 2HCl		40–60
2,2′-Azobis (4-cyano-pentanoic acid)		40–70

especially suited for kinetic investigations (see Sect. 3.1 and Examples 3.6 and 3.8). The most important azo compound in this connection is 2,2′-azobisisobutyronitrile (AIBN).

The yield of initiating radicals is, however, generally smaller than would be expected. In the case of AIBN this is because a certain amount of tetramethyl-succinic acid dinitrile is formed by combination of the primary radicals, while some methacrylonitrile and *iso*-butyronitrile are formed by disproportionation of the primary radicals. Azo compounds are especially suited as initiators for polymerization in bulk or in organic solvents.

Example 3.6 Bulk Polymerization of Styrene with 2,2′-Azobisisobutyronitrile in a Dilatometer

Safety precautions: Before this experiment is carried out, Sect. 2.2.5 must be read as well as the material safety data sheets (MSDS) for all chemicals and products used.

A polymerization reaction can be followed very conveniently and with great accuracy by observing the resulting contraction of volume in a dilatometer. This contraction results from the considerable difference in density between monomer and polymer. Knowing the initial volume V and contraction ΔV during polymerization, the percentage conversion U in the absence of a diluent is given by the following Eq. 3.15.

$$U \cdot K = 100 \cdot \frac{\Delta V}{V} \qquad (3.15)$$

Here, K is a constant that can be calculated from the specific volumes of the monomer and polymer at the appropriate temperature. One can, however, also determine the relationship between U and ΔU by direct experiment.

Dilatometric Measurements

The following amounts of 2,2′-azobisisobutyronitrile (AIBN) are weighed into 4 graduated 25-ml flasks: 35, 110, 180, and 250 mg (0. 21, 0.67, 1.10, and 1.52 mmol). The flasks are filled to the mark with destabilized styrene (at 20°C) and the amount of styrene divided by 25 ml gives the density at 20°C (neglecting the partial volume of AIBN).

Four dilatometers are filled with the above solutions of AIBN in styrene, and placed in a large water thermostat at 60°C (± 0.05°C) so that the filled parts of the capillaries are completely immersed. The thermostat can easily be constructed from a large glass tank, a powerful stirrer, an immersion heater, a contact thermometer and a relay. The dilatometers are put into a metal test tube rack fitted with suitable mountings to hold them rigidly in the thermostat. After inserting a filled dilatometer in the thermostat the meniscus rises in the capillary until thermal equilibrium is reached (if necessary some of the styrene solution may be withdrawn from the capillaries by means of a syringe or thin glass capillary). The level remains steady for a short induction period because of the presence of dissolved oxygen, and then falls as polymerization commences. The meniscus level is read every minute after insertion of the dilatometer in the thermostat and plotted against time. When the reaction slows down, it is sufficient to take readings every 5 min. Zero reaction time is taken as the intersection of the horizontal line and the initial slope. When the volume has fallen by 0.1–0.2 ml the meniscus level is quickly noted and the reaction immediately quenched by immersion of the dilatometer in ice/water.

The dilatometers are emptied as follows. The dilatometer, cooled in ice to below 10°C, is inclined carefully over a small beaker, the capillary is lifted from the dilatometer, the polymer solution poured into a beaker and the capillary and dilatometer bulb washed out several times with small amounts of toluene. The polymer solutions are each added dropwise to an 8- to 10-fold excess of methanol. The amounts of polymers precipitated are determined gravimetrically.

Evaluation of the Dilatometric Measurements

The change in volume ΔV is determined from the initial and final dilatometer readings in each experiment, as given by the plot of meniscus level against time. The constant K can now be calculated by using Eq. 3.15 for each dilatometric measurement and the results can be averaged. The statistical error is estimated from the scatter of the data (without applying the Gaussian formula).

The rate of polymerization (in % conversion per hour) is plotted against the square root of the initiator concentration (in mol%), according to Eq. 3.6. The limiting viscosity numbers [η] and hence the degrees of polymerization are also determined and plotted against the reciprocal square root of the initiator concentration.

The statistical error determined for K is only a limited measure of the accuracy of the dilatometric measurements. Since the main errors will be similar for each measurement, the accuracy of the method is best determined by estimating the limits of error of each individual measurement. The main source of error and its approximate magnitude should be indicated.

Example 3.7 Polymerization of Styrene with 2,2′-Azobisisobutyronitrile in Solution (Effect of Monomer Concentration)

Safety precautions: Before this experiment is carried out, Sect. 2.2.5 must be read as well as the material safety data sheets (MSDS) for all chemicals and products used.

23 mg (0.14 mmol) of AIBN are weighed into each of seven tubes or Erlenmeyer flasks with ground joints. Using adapters attached with springs, the tubes are evacuated and filled with nitrogen three times. Under a flow of nitrogen 0.5, 1.0, 1.5, 2.0, 2.5, and 3.0 ml (4.36, 8.72, 13.07, 17.42, 21.78, and 26.13 mmol) of destabilized styrene are pipetted into the tubes and each is diluted to 15 ml with pure toluene (distilled under nitrogen); the seventh tube is charged with 15 ml of styrene only (bulk polymerization). Using a slight positive pressure of nitrogen the adaptors are removed from the tubes and immediately closed with ground glass stoppers secured with springs. The tubes are placed in a boiling water bath and cooled after 6 h. After dilution with toluene the polymer solutions are run from a dropping funnel into 300 ml of stirred methanol. The polymer flakes are filtered off and dried at 50°C to constant weight.

The yield [Y] and degree of polymerization [P] are plotted against the monomer concentration. The results are compared with those for the sample polymerized in bulk.

Example 3.8 Polymerization of Methyl Methacrylate with 2,2′-Azobisisobutyronitrile in Bulk

Safety precautions: Before this experiment is carried out, Sect. 2.2.5 must be read as well as the material safety data sheets (MSDS) for all chemicals and products used.

(a) Observation of the Trommsdorff Effect (Gel Effect)

100 ml of methyl methacrylate are distilled under nitrogen into a graduated receiver or dropping funnel with pressure-equalizing tube, into which 100 mg (0.61 mmol) of 2,2′-azobisisobutyronitrile (AIBN) have previously been weighed. Ten tubes with ground joints and suitable adapters are evacuated, filled with nitrogen, and 10 ml (93.6 mmol) of methyl methacrylate with AIBN introduced to each tube. The adapters are removed from the tubes under slight positive pressure of nitrogen, the tubes are immediately closed with glass stoppers secured with springs, and stored until needed in a refrigerator.

To start the experiment all the tubes are placed in a rack at the same time and allowed to warm to room temperature; finally they are placed in a thermostat at 50°C. The tubes are removed at intervals of 1 h and immediately cooled in an acetone/dry ice bath. The samples that are still fluid are diluted with approximately 50 ml of chloroform and dropped into about 500 ml of stirred heptane or petroleum ether. For the very viscous or solid samples 1–2 g are dissolved in 50–100 ml of chloroform and the solution is added dropwise to 500–1,000 ml of heptane or petroleum ether with stirring. The polymers are filtered off and dried to constant weight in vacuum at 50°C. The yield, the limiting viscosity number (measured in chloroform at 20°C) and the degree of polymerization are plotted against reaction time.

The conversion can also be followed refractometrically, since the change of refractive index during polymerization is directly proportional to the conversion. The measurements can be made with an Abbé refractometer, by placing a drop of the still-liquid sample on the prism using a glass rod in the usual way. However, the determination of the refractive index of the highly viscous or solid samples can only be done with special equipment. The conversion corresponding to the measured refractive index is derived from a calibration line connecting the value of the pure monomer ($n_{20} = 1.4140$) and that of the pure polymer ($n_{20} = 1.4915$).

(b) Control of the Molecular Weight by Chain Transfer

Five tubes with ground joints are filled, as described in (a), with 10 ml (93.6 mmol) of methyl methacrylate (containing 0.1 wt% of AIBN). 0.1, 0.5, 1.0, and 2.0 mol% 1-dodecanethiol are added as regulator to four of the tubes, while the fifth serves as reference. The tubes are stoppered and stored in a refrigerator until needed. To begin the experiment the tubes are warmed up to room temperature at the same time and placed in a thermostat at 50°C. After 2 h the tubes are taken out, the contents each dissolved in 30 ml of chloroform and the solutions added dropwise to 300 ml of heptane or petroleum ether under stirring. The polymers are reprecipitated from chloroform into heptane or petroleum ether, filtered and dried in vacuum at 50°C. The limiting viscosity number of all samples is determined in chloroform at 20°C and the average degree of polymerization derived (see Sect. 2.3.3.3). The value for the transfer constant cannot be determined very accurately from these values since the chain transfer Eq. 3.14 is strictly valid only for the number-average degree of polymerization. To determine the transfer constant $1/Pn$ is plotted against the mole ratio of thiol to monomer, $[ZH]/[M]$. A straight line is obtained, intersecting the ordinate at $1/P_0$ (experiment without thiol); the slope gives the transfer constant C_{ZH}. On a second graph the yield is plotted against $[ZH]/[M]$. It can be seen that the rate of polymerization is unaffected by the occurrence of chain transfer.

3.1.3 Polymerization with Redox Systems as Initiators

Numerous redox reactions generate radicals that can initiate polymerization. Chief amongst oxidizing agents are organic and inorganic peroxides; reducing agents include either low valency metal ions or non-metallic compounds that are readily oxidized, for example, certain sulfur compounds. There are also redox systems that consist of a mixture of a peroxo compound with metal ions (e.g., Fe^{2+}) and a second reducing agent such as a hydrogen sulfite. In this case, the iron(III) ion produced by the redox reaction between the peroxo compound and an iron(II) compound is

Table 3.5 Some common oxidizing and reducing agents that are suitable for initiating radical polymerization by redox reactions

Oxidizing agents	Reducing agents
Hydrogen peroxide	Ag^+, Fe^{2+}, Co^{2+}, Ti^{3+}
Peroxodisulfates	Hydrogen sulfites, sulfites, thiosulfates, mercaptans, sulfinic acids
Diacyl peroxides	Amines (e.g. N,N-dimethylaniline), certain sugars
Hydroperoxides	Benzoin/Fe^{2+}
Peracid ester	Hydrogen sulfite/Fe^{2+}

reduced again to the iron(II) state by the hydrogen sulfite, so that only a very small amount of iron(II) ions is required.

Table 3.5 lists some suitable oxidizing and reducing agents.

In a redox system consisting of a peroxo compound and an iron(II) salt, the initiating radicals are formed by electron transfer from Fe^{2+} to the peroxo compound, causing the peroxy link to be cleaved, with simultaneous formation of a radical and an anion. In a second step, the oxidized metal reacts with another hydroperoxide to form a peroxy radical and a proton:

$$R_{\diagdown O}{\diagup}^{O}{\diagdown}_H + Fe^{2+} \longrightarrow R\text{-}O^{\cdot} + OH^{\ominus} + Fe^{3+}$$

$$R_{\diagdown O}{\diagup}^{O}{\diagdown}_H + Fe^{3+} \longrightarrow R\text{-}O\text{-}O^{\cdot} + H^{\oplus} + Fe^{2+}$$

It must be emphasized that, in contrast to the initiation of polymerization with peroxo compounds or azo compounds, not every redox system is suitable for initiating polymerization of every unsaturated monomer. Before attempting to polymerize a new compound with a redox system it is, therefore, advisable first to test its radical polymerizability with dibenzoyl peroxide.

Furthermore, the effectiveness of a redox system is influenced by a number of factors, and the redox components must be carefully balanced in order to attain optimum polymerization conditions. The most favorable conditions do not always correspond to a stoichiometric ratio of oxidizing and reducing agents. However, at constant molar ratio of oxidizing agent to reducing agent it is generally the rule that the rate of polymerization increases, and the mean degree of polymerization decreases, with increasing initiator concentration (see Example 3.11). The order of addition of the components can also be important; while it is normal to add the reducing agent first (in order to remove any oxygen which may be present), with subsequent dropwise addition of the oxidizing agent, there are cases where the reverse order must be applied. In aqueous medium, the pH value is also important; if it is necessary to work in alkaline medium, iron salts can only be used in combination with complexing agents such as sodium pyrophosphate ($Na_4P_2O_7$).

Redox polymerizations are usually carried out in aqueous solution, suspension, or emulsion; rarely in organic solvents. Their special importance lies in the fact that they proceed at relatively low temperatures with high rates and with the formation of high molecular weight polymers. Furthermore, transfer and branching reactions are relatively unimportant. The first large-scale commercial application of redox polymerization was the production of synthetic rubber from butadiene and styrene (SBR1500) at temperatures below 5°C (see Example 3.44).

Example 3.9 Polymerization of Acrylamide with a Redox System in Aqueous Solution

Safety precautions: Before this experiment is carried out, Sect. 2.2.5 must be read as well as the material safety data sheets (MSDS) for all chemicals and products used.

5 g of pure acrylamide are dissolved in 500 ml of water (that has been boiled and distilled under nitrogen) in a 1 l standard apparatus (round bottom flask, several inlets for stirrer, reflux condenser, nitrogen flux or vaccum, thermometer, heating bath). To the solution are added 25 ml of a 0.1 M aqueous solution of iron(II) ammonium sulfate and 25 ml of a 0.1 M aqueous solution of hydrogen peroxide. The solution is gently stirred and nitrogen passed through the flask for 5 min in order to displace the atmospheric oxygen. The polymerization is conducted at room temperature (ca. 20°C). After 30 min the viscous solution is run dropwise with vigorous stirring into 4 l of methanol to which a few drops of concentrated hydrochloric acid have been added. The precipitate, colored brown by iron(III) hydroxide, is filtered off and dissolved in 50 ml of water. To this solution ammonia solution is added, the precipitated iron(III) hydroxide is filtered off, and the polymer solution is added dropwise to 500 ml of methanol. After filtration, the poly(acryl-amide) is dried to constant weight in vacuum at 20°C; yield about 50%. Poly (acrylamide) is soluble in water but is not fusible. The limiting viscosity number is determined in water at 25°C (capillary diameter 0.35 mm).

Example 3.10 Fractionation of Polyacrylamide by Gel Permeation Chromatography in Water

Safety precautions: Before this experiment is carried out, Sect. 2.2.5 must be read as well as the material safety data sheets (MSDS) for all chemicals and products used.

10 g of Sephadex G100 are placed in a beaker and swollen with 400 ml water for 2 days. In a 2 m chromatography column (diameter 1.5–2 cm) a layer of glass wool and some glass beads are placed above the outlet tap. The tube is then partially filled with water and the swollen Sephadex is carefully run in to avoid trapping air bubbles. The gel particles gradually settle and the supernatant water is drawn off. The gel is now washed with fresh water until the runnings no longer show turbidity when added to methanol. (Fresh Sephadex often contains a small amount of water-soluble material that may have an adverse effect on the fractionation.) Finally the water is run off until the liquid meniscus is a few mm above the gel particles; the liquid level must never be allowed to fall below the level of the gel.

A solution of 250 mg of poly(acrylamide) from Example 3.9 in 25 ml of distilled water is introduced at the top of the column. At the same time the collection of the eluate is commenced, taking 10 ml portions (measuring cylinder) for each fraction. The flow time under these conditions is about 10 ml/h. When the meniscus of the poly(acrylamide) solution reaches the gel layer, distilled water is added to the column as eluting agent.

The fractions collected are each run dropwise into 100 ml of stirred methanol to which two drops of concentrated hydrochloric acid have been added. Turbidity or precipitation will be observed from about the 6th fraction to about the 20th fraction. The precipitated fractions are filtered off, washed with methanol and dried to constant weight in vacuum at 20°C. For each fraction the viscosity is measured in water at 25°C using a capillary viscometer (capillary diameter 0.35 mm) and at as high a concentration as possible (10 g/l) in order to minimize errors. The limiting viscosity number, and hence the molecular weight, is estimated (see Sect. 2.3.3.3). Adjacent fractions for which there may be insufficient material for a viscosity measurement, can be combined where necessary.

After the fractionation, the amount of the individual fractions in mg is plotted against the elution volume. One can test whether there has been any loss of polymer by comparing the total mass of all the fractions with the initial amount; the loss should not be more than 3%.

The integral and differential molecular weight distribution curves are finally determined as described in Sect. 2.3.3.4. For summing the percentage amounts of the fractions, one proceeds from the last fraction having the lowest molecular weight to the first fraction having the highest molecular weight.

Example 3.11 Polymerization of Acrylonitrile with a Redox System in Aqueous Solution (Precipitation Polymerization)

Safety precautions: Before this experiment is carried out, Sect. 2.2.5 must be read as well as the material safety data sheets (MSDS) for all chemicals and products used.

(a) Effect of the Ratio of Oxidizing Agent to Reducing Agent

The following solutions are prepared in distilled water:

5% solution of sodium disulfite ($Na_2S_2O_5$),

5% solution of potassium peroxodisulfate ($K_2S_2O_8$),

$$0.010 \text{ g FeSO}_4 \cdot 7H_2O \text{ in } 100 \text{ ml water} + 2 \text{ ml conc. } H_2SO_4.$$

Four 250-ml round-bottomed flasks are evacuated and filled with nitrogen using a suitable adapter (see Sect. 2.2.5.2). The reagents listed in Table 3.6 are then introduced; the potassium peroxodisulfate solution is added last to all four samples at about the same time. The flasks are shaken briefly and allowed to stand under nitrogen at 20°C. For each sample the time of appearance of the first turbidity is noted. After 20 min the precipitated polymer is filtered off by suction from the four samples, washed with water, then with methanol, and dried overnight at 50°C in vacuum. The yield, the rate of polymerization and the limiting viscosity number (measured in DMF at 25°C) are plotted against the mole ratio of oxidizing agent to reducing agent.

Table 3.6 Test series for Example 3.11a

Sample	H_2O^a ml	Acrylonitrile[b] ml	$Na_2S_2O_5$ ml sol.	mmol	$FeSO_4$ ml sol.	mmol	$K_2S_2O_8$ ml sol.	mmol
1	175	15	0.5	0.13	2.5	$9 \cdot 10^{-4}$	2.5	0.46
2	173	15	2.5	0.66	2.5	$9 \cdot 10^{-4}$	2.5	0.46
3	170	15	5.0	1.32	2.5	$9 \cdot 10^{-4}$	2.5	0.46
4	165	15	10.0	2.63	2.5	$9 \cdot 10^{-4}$	2.5	0.46

[a]The water is previously boiled under nitrogen
[b]Destabilized by distillation under nitrogen

Table 3.7 Test series for Example 3.11b

Sample	H_2O^a ml	Acrylonitrile[b] ml	$Na_2S_2O_5$ ml sol.	mmol	$FeSO_4$ ml sol.	mmol	$K_2S_2O_8$ ml sol.	mmol
1	181	15	0.5	0.13	0.5	$1.8 \cdot 10^{-4}$	0.5	0.09
2	175	15	2.5	0.66	2.5	$9.0 \cdot 10^{-4}$	2.5	0.46
3	167	15	5.0	1.32	5.0	$1.8 \cdot 10^{-3}$	5.0	0.93
4	152	15	10.0	2.63	10.0	$3.6 \cdot 10^{-3}$	10.0	1.85

[a]The water is previously boiled under nitrogen
[b]Destabilized by distillation under nitrogen

(b) Effect of Initiator Concentration at Constant Ratio of Oxidizing Agent to Reducing Agent

Four 250-ml flasks are filled with the components listed in Table 3.7 as described above under (a).

The samples are allowed to react for 20 min at 20°C and then worked up as described above under (a). The rate of polymerization (in % conversion per minute) is plotted against the square root of the initiator concentration c_i (in mol $K_2S_2O_8$ per liter); the limiting viscosity number and the degree of polymerization are plotted against $c_i^{-1/2}$.

(c) Inhibition of Polymerization

5, 20, 100, and 200 mg of hydroquinone are weighed into four 250-ml round-bottomed flasks that are then evacuated and filled with nitrogen. To each are then added 15 ml of acrylonitrile (destabilized by distillation under nitrogen), 165 ml of degassed water, 10 ml of $Na_2S_2O_5$ solution, and 2.5 ml of $FeSO_4$ solution; finally 2.5 ml of $K_2S_2O_8$ solution are added to all four samples (20°C) at about the same time. The time required for the appearance of the first turbidity (incubation time, induction period) is noted. The induction periods are compared with each other and with that for sample 4 of experiment (a).

(d) Solution-Spinning of Poly(acrylonitrile)

3.5 g of one of the poly(acrylonitrile)s obtained above are dissolved in 25 ml of DMF. The viscous solution is poured into a 1-cm-wide glass tube which is drawn out to a jet at the lower end and dips into a dish of cold water. The polymer solution flows continuously out of the jet under its own weight, the poly(acrylonitrile) being precipitated in the form of an endless filament. This is guided through the water bath and wound on to a rotating drum driven slowly by a motor. It is also possible to use a hypodermic syringe in place of the drawn-out glass tube; the filament thickness and rate of spinning can then be varied easily.

Example 3.12 Polymerization of Isoprene with a Redox System in Emulsion

Safety precautions: Before this experiment is carried out, Sect. 2.2.5 must be read as well as the material safety data sheets (MSDS) for all chemicals and products used.

Most emulsion polymerizations are performed with water-soluble initiators; however, the following experiment describes a redox polymerization where one component (dibenzoyl peroxide) is water-insoluble, while the other is water-soluble.

A 100-ml standard apparatus (round-bottomed flask, several inlets for stirrer, reflux condenser, nitrogen flux or vaccum, thermometer, heating bath) is evacuated and filled with nitrogen three times. The following solutions are prepared: (a) 500 mg of sodium oleate (or sodium dodecyl sulfate) in 16 ml of degassed water; (b) 125 mg (0.32 mmol) of iron(II) ammonium sulfate and 125 mg of sodium pyrophosphate (as buffer) in 4 ml of degassed water. The latter solution is maintained at 60–70°C for about 15 min, with occasional shaking, and is then poured into the flask together with solution (a).

The mixture is cooled to room temperature and 20 ml (0.2 mol) of isoprene (distilled under nitrogen) containing 50 mg (0.21 mmol) of dissolved dibenzoyl peroxide are added. Vigorous stirring produces a stable emulsion that becomes more viscous as polymerizations proceeds. After 6 h at room temperature the isoprene is almost completely polymerized. The polymer is precipitated from the latex in the form of large flakes by adding it dropwise to 500 ml of methanol containing 500 mg of *N*-phenyl-β-naphthylamine as stabilizer for the polyisoprene; flocculation can be improved by the addition of a few drops of concentrated hydrochloric acid. The solid elastic product is filtered off, washed with methanol, and dried in vacuum at 50°C. The solubility is tested in different solvents, the limiting viscosity number determined (toluene, 25°C), also the proportion of 1,2- and 1,4-repeating units, and the *cis/trans* ratio. These values are compared with those obtained in the polymerization of isoprene with butyllithium (Example 3.21). The main application of polyisopren is in tires.

3.1.4   Polymerization Using Photolabile Compounds as Initiators

Radical polymerization can in several cases be initiated by subjecting the unsaturated monomer to light or high energy radiation. A very versatile variant of photopolymerization is the use of photolabile compounds as initiators which upon irradiation are fragmented to form radicals. This generation of radicals can occur in two ways. In the first case the photoinitiator itself absorbs light and decomposes. In the second case the photoinitiator is not able to absorb the light directly. Instead, a second compound (sensitizer) is needed to absorb the light and to transfer the energy to the photoinitiator which than decomposes immediately. Some typical UV-initiators are benzoin, benzoin ether, benzil, and benzil ketals. Especially useful for the initiation of a radical polymerization in the visible spectral range of light is the yellow-colored campherquinone.

Photoinitiation is limited to thin layers due to the low penetration of light, nevertheless it possesses several advantages compared to the common techniques. Thus, it is possible to control the rate and degree of polymerization and also the number and length of crosslinks, by the intensity of light.

These advantages are commercially used in the so-called photolithography, a technique that allows the production of very tiny and accurate nanometer scale structures on the surface of semiconductors (e.g., silicon wafers).

The principles of photolithography are as follows: Onto the surface of a semi-conductor is spread a thin film of a photosensitive composition that consists of several chemicals. This composition is called photoresist and it is characterized by the effect that it changes its solubility upon irradiation. Thus, when the thin film is irradiated through a mask, the structure of this mask is projected onto the film which changes its solubility at the exposed segments. If the photoresist contains also bi- or higher functional monomers (Example 3.13), then crosslinked and insoluble polymers are formed. By treatment with a solvent the nonirradiated parts of the photoresist-film are removed while the crosslinked polymer remains on the surface of the substrate as an image of the structured mask, i.e., the structure of the integrated circuit. Without photolithography the modern microelectronics would not be possible.

The same principle is also applied in the manufacturing of printing plates for modern printing processes. Moreover, photopolymerization is used for coating of metals or wood and it finds also application in dentistry.

Example 3.13 Photopolymerization of Hexamethylene Bisacrylate

Safety precautions: Before this experiment is carried out, Sect. 2.2.5 must be read as well as the material safety data sheets (MSDS) for all chemicals and products used.

In a 50-ml round-bottomed flask 5 g hexamethylene bisacrylate and 0.2 g of 2-hydroxy-2-methyl-1-phenylpropane-1-one as a photoinitiator are stirred until a homogeneous solution is obtained. The flask is evacuated for a short time to remove bubbles from the solution. An IR spectrum is taken of the homogeneous solution.

Part of this solution is now spread on a glass plate with a spatula and irradiated with UV light (Hg medium-pressure lamp), until a hard crosslinked and insoluble film is obtained (5 min). The distance between the source of radiation and the substrate should be about 20 cm. This is an example of a photo-cured coating.

In a second experiment a mask, e.g., a coin, is placed between the substrate and the source of radiation by means of a holder. After 5 min the unexposed part of the film is removed by treating with an organic solvent. This is an example of surface structuring by photolithography.

The irradiated film shows an absorption band in IR spectrum (KBr pellet) at $1,635$ cm^{-1}. The increase in intensity of this absorption with time is a measure of the conversion rate. Parallel to this the solubility/swelling of the film in THF or toluene should be qualitatively determined.

3.1.5 Polymerization of Cyclodextrin Host-Guest Complexes in Water

Polymerization reactions in aqueous medium can be carried out in homogeneous solution if the monomers and the polymers are soluble in water as in the case of acrylamide or methacrylic acid (see Examples 3.5, 3.9, and 3.35). Since most of the monomers are only sparingly soluble in water, suspension or emulsion techniques have to be applied in these cases.

Very recently a new method was developed that opens the possibility to polymerize even hydrophobic monomers in aqueous solution. This method is based on the finding that hydrophobic monomers can be made water-soluble by incorporation in the cavities of cyclodextrins. It has to be mentioned that no covalent bonds are formed by the interaction of the cyclodextrin host and the water-insoluble guest molecule. Obviously only hydrogen bonds or hydrophobic interactions are responsible for the spontaneous formation and the stability of these host-guest complexes. X-ray diffraction pattern support this hypothesis. Radical polymerization then occurs via these host-guest complexes using water-soluble initiators. Only after a few percent conversion the homogeneous solution becomes turbid and the polymer precipitates.

While the polymer precipitates, the cyclodextrin stays in solution and can be reused again as host-molecule.

Variations of this method are possible in several ways. First of all, cyclodextrin which is available on a large scale by enzymatically catalyzed modification of starch can be tailored by chemical reactions. Furthermore, copolymerizations between different host-guest complexes are possible whereby in some cases the reactivity ratios differ from those reported in literature.

Cyclodextrin can also be used in order to stabilize monomers that would otherwise oxidize upon contact with air. The cyclodextrin host shields the monomer from oxygen in the air. Monomers such as pyrrole can be stored as a cyclodextrin complex for months without any noticeable degradation. The complex is a colorless powder that does not change color over time unlike pure pyrrole, which would oxidize and therefore turn black via yellow. Another advantage of the method is the fact, that the complex is odorless whereas pyrrole itself has an unpleasant smell.

In general, polypyrrole can be prepared via electrochemical or chemical oxidative polymerization of pyrrole involving different highly reactive intermediates. Also, the pyrrole/cyclodextrin complex can be polymerized in aqueous solution under oxidative conditions by adding potassium peroxodisulfate as an oxidizing agent. Polypyrrole is insoluble and infusible but achieved high interest since it can be used as conductive polymer. Its conjugated double bonds are able to conduct electricity if the polymer is doped. In the undoped state the polymer behaves as a semiconductor.

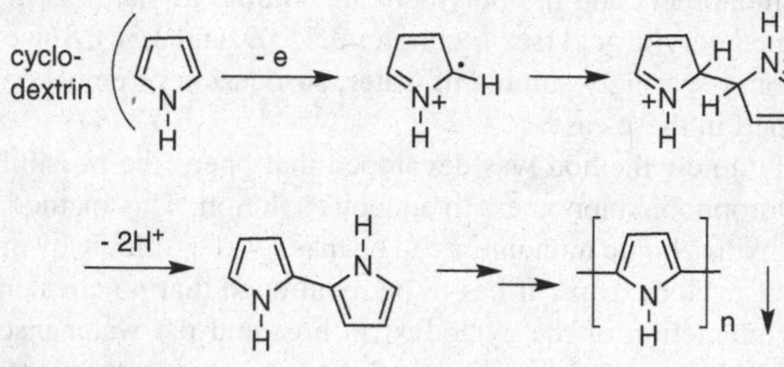

Example 3.14a Free Radical Polymerization of Cyclodextrin Host-Guest Complexes of Butyl Acrylate from Homogeneous Aqueous Solution (Precipitation Polymerization)

Safety precautions: Before this experiment is carried out, Sect. 2.2.5 must be read as well as the material safety data sheets (MSDS) for all chemicals and products used.

27.1 g (20.35 mmol) of 1.8-fold methylated ß-cyclodextrin (m-ß-CD) are dissolved in 70 ml of deionized water in a 250-ml standard apparatus (round bottom flask, several inlets for stirrer, reflux condenser, nitrogen flux or vaccum, thermometer, heating bath) fitted with magnetic stirrer. The resulting solution is flushed with nitrogen for 15 min. To this solution 2.0 g (15.61 mmol) of *n*-butyl acrylate are added. The shaken yellowish dispersion is sonicated for 20 min yielding a clear solution of the complexed monomer *n*-butyl acrylate/m-ß-CD. After complexation, 0.21 g (0.78 mmol) of potassium peroxodisulfate and 0.08 g (0.78 mmol) of sodium hydrogen sulfite are added to the solution. The polymerization is conducted at room temperature (20–25°C). After some minutes the clear monomer solution changes to a colorless cloudy polymer dispersion. The polymerization can be stopped after 3 h by cooling the reaction mixture in an ice bath and adding 100 ml of water. The precipitate is filtered off. After dissolving the crude polymer in 15 ml of THF the solution is poured dropwise under vigorous stirring into 200 ml of water, yielding colorless polymeric products. After filtration the material is free of monomers and m-β-CD which can be proved by TLC using iodine for visualizing. Conversion 80%; $M_n = 20,000$ (GPC in THF at 25°C).

Example 3.14b Oxidative Polymerization of a Cyclodextrin Host-Guest Complex of Pyrrole from Homogeneous Aqueous Solution (Conducting Polymer)

4 g (4.11 mmol) α-cyclodextrin are dissolved in 40 ml of water. 0.24 ml (0.23 g = 3.43 mmol) pyrrole are added to the solution. The mixture is stirred for 4 h at room temperature (20–25°C). The complex precipitates as colorless crystals. The complex is filtered off and washed with 5 ml of cold water. The complex may be allowed to dry at room temperature on the filter paper or it can be used as it is.

1 g of the complex is completely dissolved in 25 ml of water at 60°C. 0.65 g potassium peroxodisulfate are added as an oxidizing agent. As the polymerization progresses a black precipitate forms. The polymer is filtered off after approximately 30 min reaction time, washed with hot water and dried in an oven at 50°C.

After drying, the filter paper displays a resistance of 40–100 kΩ, measured with a multimeter by pressing the electrodes into the polypyrrole powder at a distance of 1 cm. The resistance may vary depending on the location of measurement, since the polypyrrole thickness and distribution may vary on the filter paper. Without the polypyrrole, the filter paper has a resistance of at least 10^{10} Ω, which in most cases lies beyond the measuring range of the multimeter. Thus it can be shown that the polypyrrole reduces the resistance significantly.

3.1.6 Controlled Radical Polymerization

Recently there has been progress in achieving higher control in free radical chain addition polymerization by suppressing chain termination reactions and reducing the content of free radicals in the system. All basic concepts involve a reversible chain termination reaction leading to a "dormant" chain which "sleeps" most of the time and is active only for a very short time to allow monomer "insertion" into the lable bond. These systems are called "quasiliving" or controlled radical polymerizations and show features of the "living" ionic chain addition reaction. In a well controlled radical system the monomer conversion is first order, molar mass increases linearly with monomer conversion and the molar mass distribution M_w/M_n is below 1.5. In addition, chain-end functionalionization as well as subsequent monomer addition allow the preparation of well-controlled polymer architectures, e.g., block copolymers and star polymers by a radical mechanism which had been up to now reserved for ionic chain growth polymerization techniques.

Three major systems are distinguished so far:

(a) Atom transfer radical polymerization (ATRP)
(b) Stable free radical polymerization (SFRP) or nitroxy-mediated radical polymerization (NMP)
(c) Reversible addition-fragmentation termination (RAFT)

The mechanism of the ATRP is based on the reversible activation of alkyl halides by redox reaction of a complexed metal with the halogen terminal group of the initiator or the growing chain end. Thus, the initiating step is the homolytic cleavage of the carbon-halogen bond in the organic halide R-X by oxidation of the metal M_t in the metal complex M_t^n-Y/ligand. Consequently. the initiating radical species R$^\bullet$ and the oxidized metal complex are formed. At this point, R$^\bullet$ can add monomer units to enable chain propagation, or else it can react with the halogen on the oxidized metal to regenerate the dormant species R-X. The activation constant K_{act} leading to a free radical at the growing chain end is low compared to the deactivation constant K_{deact} and therefore the equilibrium is strongly on the side of the dormant species reducing the amount of free radicals. Secondary or tertiary chloride or bromide compounds can be used as initiator in combination with copper (1)chloride and, e.g., (substituted) bypyridine or tetradendates (e.g., MeTREN) are applied as ligand. The ratio initiator/copper/ligand as well as the reaction temperature has to be optimized for each monomer system, however, a broad variety of monomers including styrenics as well as acrylates and methacrylates, was polymerized under controlled conditions so far.

ATRP is a very potent method for preparing block copolymers by sequential monomer addition as well as star polymers using multifunctional initators. Furthermore, it can be applied also in heterogenous polymerization systems, e.g., emulsion or dispersion polymerization. In Example 3.15a the ATRP of MMA in miniemulsion (see also Sect. 2.2.4.2) is described.

The SFRP or NMP has been studied mainly using the stable free radical TEMPO (2,2,6,6-tetramethyl-1-piperidinyloxy) or its adducts with, e.g., styrene derivatives. It is based on the formation of a labile bond between the growing radical chain end or monomeric radical and the nitroxy radical. Monomer is inserted into this bond when it opens thermally. The free radical necessary to start the reaction can be created by adding a conventional radical initiator in combination with, e.g., TEMPO or by starting the reaction with a preformed adduct of the monomer with the nitroxy radical using so-called unimolecular initiators (Hawker adducts).

The thermal lability of the R–C–O–N bond system controls the reversibility of the chain termination and limits also the use of NMP. SFRP of styrene at about 130°C is studied intensively. In this case, high control and high-molar-mass products could be achieved. It was found that the thermal autopolymerization of the styrene monomer plays an important role in the mechanism of the reaction. Therefore, first experiments using different monomers in the presence of TEMPO and a radical initiator failed with regard of the control. However, new nitroxy adducts with a different R–O–N bond stability have been developed, e.g., by Hawker which work also for styrene derivatives as well as for acrylates.

Hawker adduct

End-group functionalization in NMP can be achieved by using a functional radical initiator in combination with TEMPO.

In the RAFT mechanism, the chain equilibrium process is based on a transfer reaction, no radicals are formed or destroyed, and, when the RAFT agents behave ideally, the kinetics can be compared to the one of a conventional free radical polymerization. The release of initiating radicals through chain transfer (b) at the beginning and the addition-fragmentation step (d), necessary to minimize the irreversible termination events, are the basis on which RAFT mechanism relies. During the reaction, chains are alternatively converted from propagating radicals to polymeric transfer agents and vice versa, generating an equilibrium. This enables the incremental growths of the chains with conversion, giving living character to the process. The choice of RAFT agents is very important for the achievement of well-defined products: they should have a high transfer constant regarding the monomers being polymerized, which means a high rate of addition, and suitable leaving groups for the propagating radical. As a result, dithioesters, trithiocarbonates, and certain dithiocarbamates can be successfully employed to obtain narrow polydispersity for styrenes and (meth)acrylates in batch polymerization. RAFT can be performed at a broad temperature range and has a high tolerance to a large variety of functionalities (e.g., OH, COOH, NR_2). Very similar is the MADIX (macromolecuelar design via interchange of xanthates) approach which employs xanthates.

Example 3.15a Controlled Radical Polymerization (ARTP) of Methyl Methacrylate in Miniemulsion

Safety precautions: Before this experiment is carried out, Sect. 2.2.5 must be read as well as the material safety data sheets (MSDS) for all chemicals and products used.

The reaction described in this example is carried out in miniemulsion. Miniemulsions are dispersions of critically stabilized oil droplets with a size between 50 and 500 nm prepared by shearing a system containing oil, water, a surfactant and a hydrophobe. In contrast to the classical emulsion polymerization (see Sect. 2.2.4.2), here the polymerization starts and proceeds directly within the preformed micellar "nanoreactors" (= monomer droplets). This means that the droplets have to become the primary locus of the nucleation of the polymer reaction. With the concept of "nanoreactors" one can take advantage of a potential thermodynamic control for the design of nanoparticles. Polymerizations in such miniemulsions, when carefully prepared, result in latex particles which have about the same size as the initial droplets. The polymerization of miniemulsions extends the possibilities of the widely applied emulsion polymerization and provides advantages with respect to copolymerization reactions of monomers with different polarity, incorporation of hydrophobic materials, or with respect to the stability of the formed latexes.

Experimental Conditions

All ingredients are carefully degassed and brought under an atmosphere of nitrogen prior to use. All material transfers are carried out using degassed syringes (see also Example 3.19). All reactions are carried out under an atmosphere of nitrogen in 10-ml round-bottom flasks closed by a rubber septum which can be passed by syringe needles. A solution of n-decane (0.422 g, hydrophobic) and poly(methyl methacrylate) (0.075 g; co-hydrophobic) in the monomeric methyl methacrylate (3 g, 29.96 mmol) is transferred into the 10-ml reaction flask containing a mixture of 4,4'-dinonyl-2,2'-bipyridine (0.081 g, 0.198 mmol) and CuCl (0.009 g, 0.099 mmol). Stirring at room temperature for several minutes leads to an intensively brown solution of the formed copper(I) complex. Stirring is continued until all copper salt has disappeared.

In a second 10-ml reaction flask, water (18 g) and Brij 78 [poly(oxyethylene stearyl ether); 0.45 g, 15 wt%] are mixed under vigorous stirring. The above solution of copper complex, hydrophob and co-hydrophob in MMA is then added to the resulting clear solution of the emulsifier, and the resulting mixture is vigorously stirred at room temperature using a magnetic stirring bar until a fine primary emulsion has been formed. Subsequently, the emulsion is sonicated for approx. 20 min in a conventional ultrasonic bath. Here, it is possible to follow the decrease of droplet size – and thus the increase of internal surface – via surface tension measurements. The resulting miniemulsion is heated in an oil bath to 70°C and then the initiator 2-bromo-2-methylpropionic acid methyl ester (0.018 g, 0.099 mmol) is added. Heating and stirring of the reaction mixture is continued for further 3 h (Table 3.8).

At regular intervals, samples are taken of the emulsion for GC (0.3 ml each) and SEC (0.5 ml each) analysis using degassed syringes. The following reaction times proved to be appropriate for taking samples for GC and SEC: 1: immediately after addition of the initiator; 2: after 15 min; 3: after 30 min; 4: after 60 min; 5: after 90 min; 6: after 120 min; 7: after 150 min; 8: after 160 min; 9: after 180 min; 10: after full reaction time.

For GC analysis, the emulsion samples are diluted in THF or acetone (1.5 ml). For SEC samples, the emulsions are dissolved in THF (3–5 ml, containing 0.06% toluene as an internal SEC standard). The solution SEC is filtered over aluminum oxide (to remove the copper residues) and then through a syringe filter prior to the injection into the SEC.

For the evaluation of the obtained GC data, the decrease of MMA concentration is plotted vs. time. From the plot the apparent rate of conversion can be determined.

Table 3.8 Recipe for Example 3.15 (T = 333 K/P_n ~ 300)

	n:n	n [mmol]	Weight [g]	Volume [ml]
MMA	300	29.96	3	3
MbiB	1	0.099	0.018	
CuCl	1	0.099	0.009	
dNbPy	2	0.198	0.081	
		m [g]	wt%[a]	v:v
Water		18		6
MMA		3		1
Brij 78		0.45	15	
D/PMMA		0.422/0.075		

MMA	Methyl methacrylate	Monomer
MbiB	2-Bromo-2-methylpropionic acid methyl ester	Initiator
dNbPy	4,4′-Dinonyl-2,2′-bipyridine	Ligand
D	Decane	Hydrophob/GC-Standard
PMMA	Poly(methyl methacrylate) (7H, Röhm)	Co-hydrophob
Brij 78	Poly(oxyethylene stearyl ether)	Emulsifier

[a]wt% of emulsifier with respect to monomer

Also, the degree of conversion can be calculated for each data point. The resulting plot shows until what time the process occurs in a controlled manner and where the uncontrolled free radical polymerization sets in.

For the evaluation of the SEC measurements, a second plot can be created where the obtained values of M_n (left coordinate, should increase linearly with conversion) and the polydispersity indices (PDI, right coordinate, should be lower than 1.3) are plotted vs. the conversion. This plot again can be used to discuss the degree of control, and the time period where control was achieved during the chain growth process. This time is in general approx. 2 h.

Example 3.15b Controlled Radical Polymerization (RAFT) of Trimethylsilylpropargyl Methacrylate and Subsequent Polymer Analogous Click Reaction

Safety precautions: Before this experiment is carried out, Sect. 2.2.5 must be read as well as the material safety data sheets (MSDS) for all chemicals and products used.

(a) Synthesis of the monomer trimethylsilyl propargylmethacrylate

In a 500 ml inert gas flask, 3-(trimethylsilyl)prop-2-in-1-ol (3 g , 23.4 mmol) and triethyl-amine (12 ml, 86.3 mmol) are dissolved in THF (200 ml). The clear, colorless solution is cooled down by using an ice bath and in an inert gas stream methacryloyl chloride (2.4 ml, 24.8 mmol) is added slowly via a syringe. After completed addition, the ice bath is removed and the solution is stirred at room temperature further 2 h. Subsequently, remains of methacryloyl chloride are neutralized by careful addition of water (2 ml). The white precipitation is filtered off, all solvents are removed using a rotary evaporator and the residue is dissolved again in a little amount of chloroform. Next, the product is purified by column chromatography (solvent: chloroform) to obtain the monomer as a clear, colorless liquid. Yield: 3.83 g (19.5 mmol; 83%).

^{1}H-NMR: δ_H (500,13 MHz, CDCl$_3$) 6.17 (1H), 5.61 (1H), 4.76 (2H), 1.96 (3H), 0.18 (9H) ppm.

(b) RAFT polymerization of trimethylsilyl propargylmethacrylate

CTA   **P1**

A 20 ml Schlenk tube is equipped with the CTA 2-cyanoprop-2-yl-dithiobenzoate (see Bibliography) (11.6 mg, 0.0524 mmol), the monomer (1 g, 5.09 mmol) and the initiator 2,2′-azobis(isobutyronitrile) (AIBN) (0.8 mg, 0.00487 mmol) in anisole (0.8 ml). The clear, pink colored solution is degassed by using three freeze-pump-thaw cycles. Next, the tube is immerged into a 70°C heated oil bath for 22 h. The polymerization is stopped by immersion of the reaction tube into liquid nitrogen for about 5 min. Then, the product mixture is diluted by roughly the same volume of deuterated chloroform (CDCl$_3$) to take a sample for NMR analysis (determination of conversion). The product P1 is obtained via precipitation in a methanol/water (1:1) mixture followed by filtration over a G4 frit and drying in vacuum. The polymer P1 is isolated in 50–60% yield and unde the given condition as a monomer conversion of about 80% is obtained as verified by NMR. The product is characterized by ^{1}H NMR and SEC. The obtained molar masses are in good agreement as the theoretical ones.

^{1}H-NMR: δ_H (500.13 MHz, CDCl$_3$) 4.6 (2H), 2.15–1.4 (2H), 1.35–0.7 (3H), 0.2 (9H) ppm. SEC (RI-detector): M_n: 16,900 g/mol, M_w: 20,000 g/mol, PDI: 1.18.

In the following step, the trimethylsilyl propargyl units are deprotected. In a 50 ml round flask, the protected polymer P1 (0.401 g, 2.045 mmol) is dissolved in THF (20 ml) and the solution is cooled down with an ice bath. Next, 2.7 ml (2.7 mmol) of a 1 M THF solution of tetrabutylammonium fluoride (TBAF) are added dropwise via a syringe. After 10 min, the ice bath is removed and the solution is stirred further 1.5 h. Subsequently, the product is precipitated in a methanol/water

(1:1) mixture, filtered (G4 frit) and dried in vacuum. The completely deprotected polymer P2 is obtained as a white solid in 60–70% yield and with a molar mass somewhat lower than the parent polymer (M_n: 14,000 g/mol, M_w: 16,000 g/mol). The success of the deprotection can be verified by NMR by the loss of the signals of the trimethylsilylgroup (0.2 ppm) and the appearance of the CH signal at 2.49 ppm (when measured in $CDCl_3$)

(c) 1,3-Dipolar cycloaddition of polymer P2 with benzyl azide

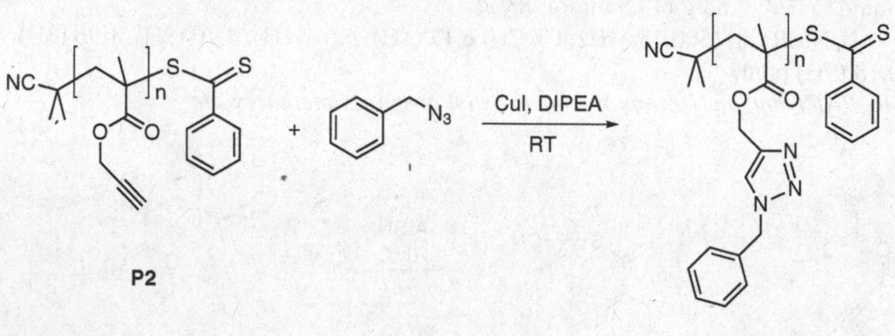

P2

P3

In a 100 ml inert gas flask, 110 mg (0.89 mmol) of the deprotected polymer P2 are dissolved in 20 ml DMF. Afterwards, benzyl azide (Alfa Aesar, 94%) (200 mg; 1.41 mmol), a minimal amount of copper iodide and 200 mg (1.55 mmol) of diisopropylethylamine (DIEA) are added. The addition of the substances is done in an inert gas stream. Subsequently, the reaction vessel is sealed and the mixture is stirred at room temperature over night (15 h). Next, the solution is added dropwise to a methanol/water mixture (1:1) leading to precipitation of the polymer. The product (P3) is filtered (G4 frit), washed with water and ethanol and dried in vacuum. A yield of 80–90% is achieved and complete reaction of the propargyl groups with the azide through formation of the triozole can be verified by ^{1}H NMR. SEC analysis shows some increase in molar mass.

^{1}H-NMR: δ_H (500, 13 MHz, $CDCl_3$): 8.1–7.6 (1H), 7.6–6.9 (5H), 5.8–5.3 (2H), 5.3–4.6 (2H), 2.15–1.15 (2H), 1.15–0.2 (3H) ppm. SEC (RI-detector): M_n: 19.000 g/mol, M_w: 24.000 g/mol, PDI: 1.2–1.3.3.2

3.2   Ionic Homopolymerization

Like radical polymerizations, ionic polymerizations also occur by a chain mechanism. In contrast to radical polymerizations the chain carriers are macroions:

carbonium ions in the case of cationic polymerizations and carbanions in the case of anionic polymerization of $C = C$ compounds:

Cationic Polymerization

$$R^{\oplus} + n \underset{X}{\overset{}{=}} \longrightarrow R\left[\begin{array}{c} \\ \\ X \end{array}\right]_{n-1} X^{\oplus}$$

Anionic Polymerization

$$R^{\ominus} + n \underset{Y}{\overset{}{=}} \longrightarrow R\left[\begin{array}{c} \\ \\ Y \end{array}\right]_{n-1} Y^{\ominus}$$

Again, in contrast to radical polymerization, there is no chain termination by combination, since the growing chains (macroions) repel each other electrostatically because of their like charges. Chain termination occurs only by reaction of the growing chain ends with substances such as water, alcohols, acids, and amines. The ions produced by reaction of these substances can sometimes initiate new chains (chain transfer). Under certain conditions the ionic propagation species retain their ability to grow over extended periods of time, even after complete consumption of monomer ("living polymers", see Sect. 3.2.1).

Radical initiators have so far been employed successfully only for the polymerization of compounds containing C = C bonds. The number of ionically polymerizable monomers is considerably larger, and includes also compounds containing a C = O or a C = N group, and a series of heterocyclic compounds. In some cases migration of an atom or group occurs during polymerization (isomerization polymerization). A particular characteristic of some ionic polymerizations is that they proceed stereo-specifically, leading to tactic polymers.

In contrast to radical polymerizations, ionic polymerizations proceed at high rates even at low temperatures, since the initiation and propagation reactions have only small activation energies. For example, isobutylene is polymerized commercially with boron trifluoride in liquid propane at $-100°C$ (see Example 3.16). The polymerization temperature often has a considerable influence on the structure of the resulting polymer.

3.2.1 Ionic Polymerization Via C = C Bonds

The tendency of unsaturated compounds to undergo cationic or anionic polymerization differs greatly according to the type of substituent at the double bond. Some monomers can polymerize only cationically or only anionically but there are also some that can polymerize by both mechanisms (see Table 2.2). Electron-repelling groups (alkyl, phenyl, alkoxy) cause polarization, such that the unsubstituted carbon atom of the double bond carries a partial negative charge; hence, protons or other suitable cations can attach themselves to this unsubstituted carbon atom with the positive charge being transferred to the substituted carbon atom which can likewise add monomer by the propagating reaction. In cationic polymerization the propagating chains can be terminated only by addition of reactive anions, since the combination of two macrocations is not possible; b-elimination of a proton from the chain end may also take place in a transfer reaction.

Initiation:

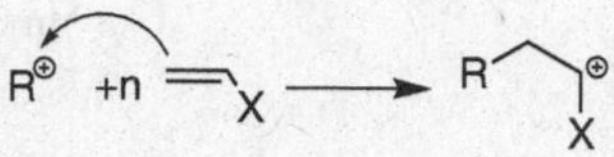

Propagation:

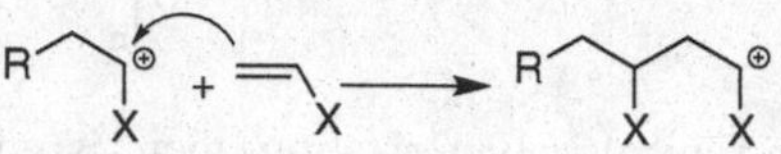

Termination:

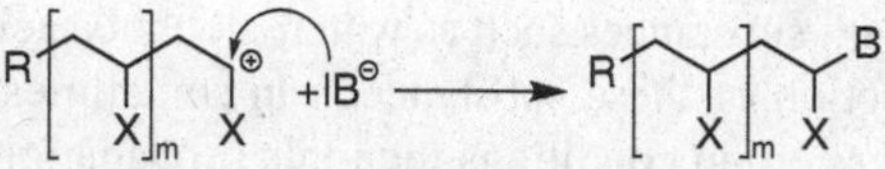

Transfer:

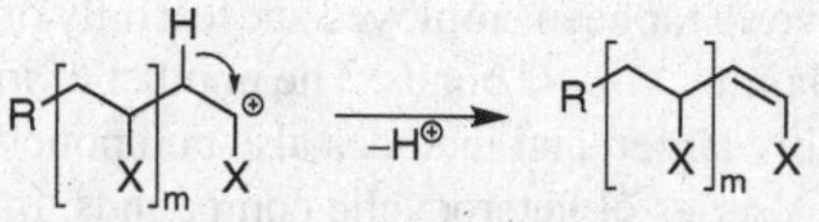

Nucleophiles such as water, alcohols, esters, acetals, and ethers can also act as transfer agents by reacting with the macrocations, at the same time forming a new cation that initiates the growth of a new chain; the kinetic chain thus remains unbroken.

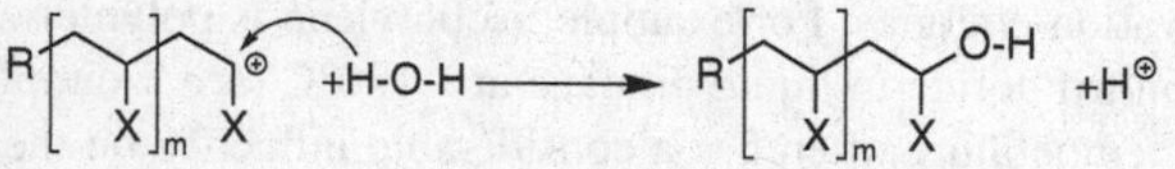

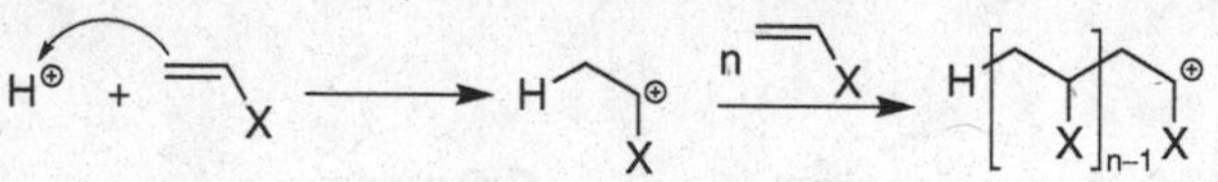

Cationic polymerizations can be initiated with protic acids (e.g., sulfuric, perchloric, trifluoroacetic acid), with Lewis acids (see Sect. 3.2.1.1), and with compounds that form suitable cations (e.g., iodine, acetyl perchlorate). Some monomers are also polymerized by high-energy radiation according to a cationic mechanism.

By choice of appropriate conditions it is sometimes possible to stop the reaction at the dimer stage; for example, in the case of styrene this leads to an unsaturated linear dimer, or, by ring closure, to a cyclic dimer:

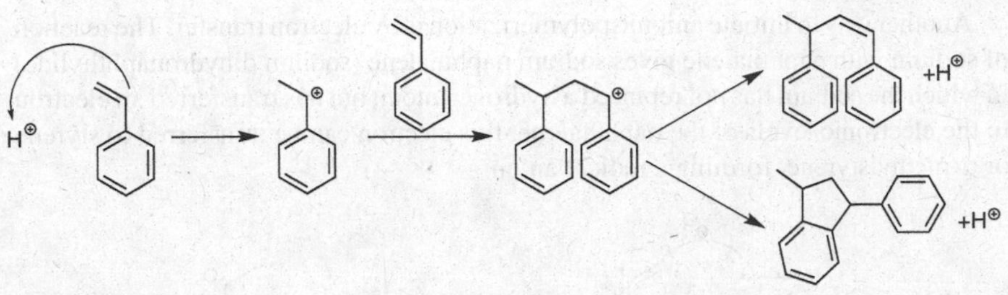

According to this reaction, the dimerization of divinylbenzene leads to polyindane.

For unsaturated compounds with electron-withdrawing substituents (carboxyalkyl, nitro, nitrile, vinyl), polymerization can be initiated anionically (e.g., by OH⁻, NH₂⁻, or carbanions).

The addition of the anion takes place at the unsubstituted carbon atom, which, in this case, carries a partial positive charge. Since the growing chain end is a genuine anion, chain termination can occur by addition of a reactive cation. As in cationic polymerization, combination of two growing ends is not possible. Chain transfer with electrophiles can also occur.

Initiation:

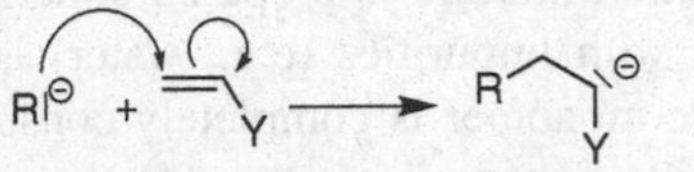

Propagation:

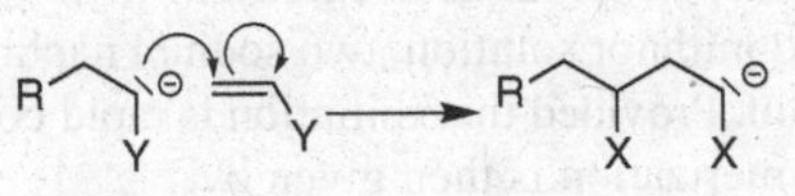

Chain transfer:

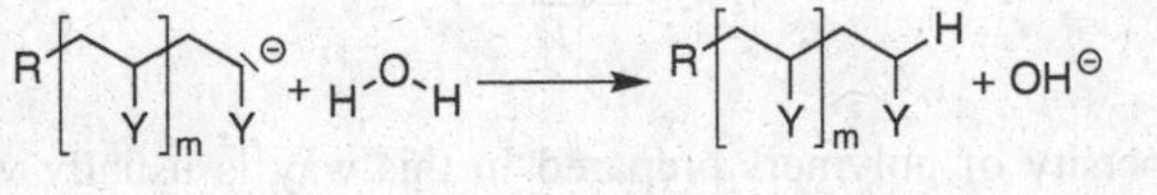

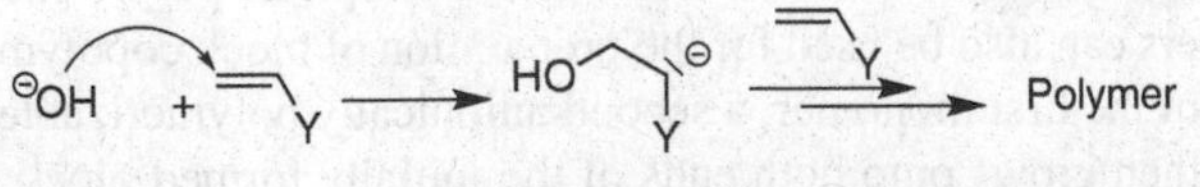

For some monomers (e.g., nitroethylene and 2-cyano-2,4-hexadienoic acid ester, $CH_3-CH = CH-CH = C(CN)-COOR$), anionic polymerization can be conducted in aqueous alkaline solution. Other anionic initiators are Lewis bases, e.g., tertiary amines or phosphines, and organometallic compounds (see Sect. 3.2.1.2). Since the polarizability of unsaturated compounds depends very much on the substituents and on the solvent used, there are considerable differences in the effectiveness of the initiators mentioned.

Another way to initiate anionic polymerization is by electron transfer. The reaction of sodium with naphthalene gives sodium naphthalene (sodium dihydronaphthylide) in which the sodium has not replaced a hydrogen atom, but has transferred an electron to the electronic levels of the naphthalene; this electron can be transferred to styrene or α-methylstyrene, forming a radical anion:

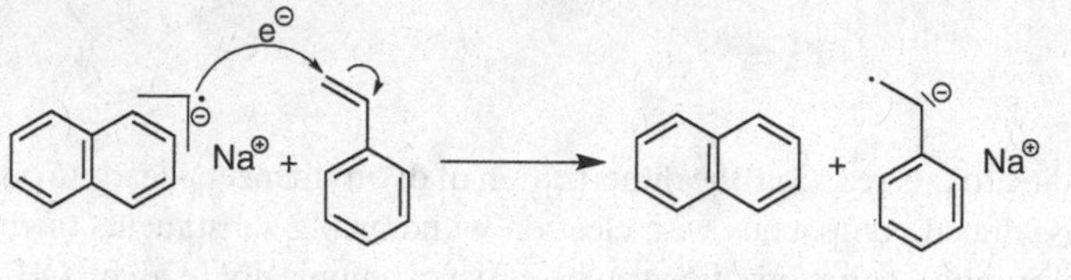

Such radical anions combine very quickly forming a dianion which can then add styrene at both ends:

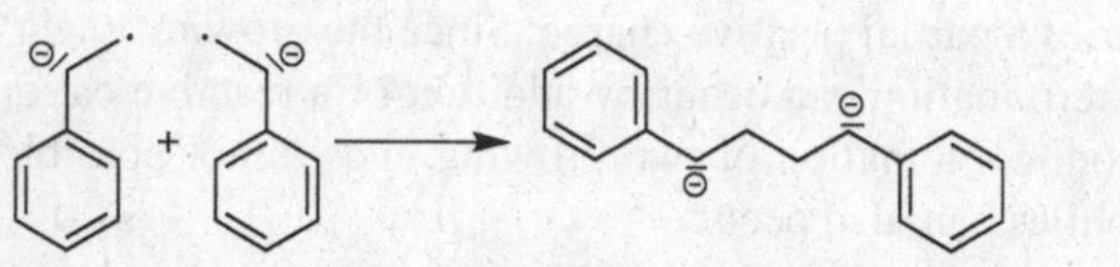

Provided that the reaction mixture is prepared under stringent conditions, that reaction of the dianions with impurities (e.g., water) is prevented, the polymer chains can grow until the monomer is completely consumed. If another batch of styrene is added, the "living polymer" can grow further. Finally a "dead" polymer results, if a chain breaker is added (e.g., proton donor).

Under ideal conditions, i.e., complete exclusion of impurities and very rapid mixing of monomer and initiator solution, two sodium naphthalene molecules give rise to one polymer chain. Provided that initiation is rapid compared with propagation, the degree of polymerization is then given by:

$$P = \frac{2 \cdot [M]}{[I]} \tag{3.16}$$

The polydispersity of polymers prepared in this way is usually very low; for example, a value M_w/M_n of 1.05 was found for a sample of poly(α-methylstyrene). Living polymers can also be used for the preparation of block copolymers; after the consumption of the first monomer, a second anionically polymerizable monomer is added which then grows onto both ends of the initially formed block. By termination of the living polymer with electrophilic compounds the polymer chains can be provided with specific end groups; for example, living polystyrene reacts with carbon dioxide to give polystyrene with carboxylic end groups.

In the anionic polymerization of α-methylstyrene with sodium naphthalene the reaction proceeds to an equilibrium and it is possible to observe the temperature dependence of the equilibrium between monomer and polymer. After addition of the monomer, the deep green color of the initiator solution is transformed into the red color of the carbanions. At low temperatures (−70 to −40°C) the living polymer is formed and the solution becomes viscous. After warming, the macroanions depolymerize again, but reform reversibly on cooling. The temperature at which the equilibrium is established is called the ceiling temperature for the given monomer concentration; for α-methylstyrene it lies at about 60°C, but for most vinyl monomers it occurs above 250°C. For some monomers with polymerizable C = O bonds, or for cyclic monomers, the ceiling temperature is relatively low, for example, formaldehyde or trioxane (126°C) and THF (85°C). In spite of its thermo-dynamic instability, poly(α-methylstyrene) can be isolated if the living anionic chain ends are capped, e.g., by reaction with water or carbon dioxide. The thermal degradation of poly(α-methylstyrene) with stable end groups to monomer proceeds with measurable speed only above 200°C where chain scission begins to occur (see Example 5.12).

In certain cases of ionic polymerization the chain growth is stereoregular. It was first shown to be possible to make tactic polymers by stereospecific catalysis using organometallic catalysts (see Sect. 3.2.1.2), but other initiators have since been found that are suitable for stereoregular polymerization. For example, under certain conditions styrene can be polymerized with many organo-alkali-metal compounds to give tactic polystyrene. A certain degree of stereoregulation during chain growth has also been observed in cationic and even in radical polymerizations of some monomers.

Stereospecific polymerization has particular significance for the preparation of stereoregular polymeric dienes. In the radical polymerization of butadiene or isoprene the molecular chains always consist of varying proportions of adjacent cis- and trans-1,4-units as well as 1,2- and 3,4- linked units, depending on the polymerization conditions; but it is now possible, using particular ionic initiation systems to make a "synthetic natural rubber" that contains more than 90% cis-1,4-isoprene repeating units (see Example 3.21).

It can be seen that both the solvent and the catalyst affect the structure of the polymer produced. For example, the structure of the polyisoprene differs strongly with the alkali metal, even when used in the same solvent medium. Experiments with a typical organometallic complex catalyst, consisting of trialkylaluminum and titanium tetrachloride, show that the same initiator can lead to quite different structures in the products of polymerization of isoprene and of butadiene.

3.2.1.1 Cationic Polymerization with Lewis Acids as Initiators

The following Lewis acids are suitable for initiating cationic polymerization: BF_3, $AlCl_3$, $TiCl_4$, $SnCl_2$, $SnCl_4$, and $FeCl_3$. So-called co-initiators (cocatalysts) play an essential part in the initiation mechanism. Among these are proton-donating compounds (protic acids, water, alcohols) and compounds such as alkyl halides that form ionic complexes with Lewis acids which can then dissociate to give cations capable of initiating polymerization. Thus, in the case of water or alcohol as cocatalyst the real initiator is a proton:

$$BF_3 + ROH \rightleftharpoons R-\underset{H}{\overset{\ominus}{\underset{|}{O}}}{\overset{BF_3}{\oplus}} \rightleftharpoons R-\overset{BF_3}{\underset{\ominus}{O}} + H^{\oplus}$$

and, when alkyl halides are used, an alkyl cation:

$$SnCl_4 + RCl \rightleftharpoons [RClSnCl_4] \rightleftharpoons R^{\oplus} + {}^{\ominus}SnCl_5$$

The initiator concentrations required for cationic polymerizations are smaller than those for radical polymerizations; frequently 10^{-3}–10^{-5} mol of initiator per mol monomer is sufficient to achieve a high rate of reaction. The effect of initiator concentration on the rate and average degree of polymerization depends on the monomer and a variety of other factors and does not follow a consistent pattern.

The type and amount of cocatalyst required for optimum polymerization conditions must be determined for each case; generally the amount of cocatalyst required (in mol) is much less than that of the initiator.

In polymerizations of unsaturated compounds with Lewis acids the required reaction temperatures are below room temperature, down to $-100°C$ or even lower (see Example 3.16). On the other hand, cyclic monomers (see Sect. 3.2.3) frequently require higher temperatures.

Cationic polymerizations of unsaturated compounds are practically always carried out in solution. In radical polymerization, dilution with a solvent, under otherwise similar conditions, always results in a decrease of rate and molecular weight; but in cationic polymerization, addition of one to four parts by volume of a suitable solvent (with respect to monomer) often causes a significant increase of the molecular weight and sometimes also of the rate. The main reasons for this behavior reside in the effect of solvent polarity and the cocatalytic action of the solvent. The choice of solvent is thus extremely important. The rate of polymerization and the molecular weight generally increase with increasing polarity and relative permittivity (dielectric constant) of the solvent; some solvents can form complexes with Lewis acids that then initiate polymerization by a carbonium ion mechanism. The solvent can also interfere with the course of a cationic polymerization through chain transfer reactions. Taking into account the above-mentioned limitations, the following solvents are suitable for cationic polymerizations: toluene, cyclohexane, methylene chloride, carbon tetrachloride, dichloroethylene, trichloroethylene, chlorotoluene, nitrotoluene, and liquid sulfur dioxide.

In solution polymerizations catalyzed by Lewis acids, the polymerization frequently does not begin immediately after addition of the initiator, and there is an induction period that cannot be completely eliminated even by careful purification of the starting materials. In contrast, some cationic polymerizations proceed so quickly (flash polymerization), even after dilution, that conversion is already complete after a few minutes (e.g., isobutylene, Example 3.16).

Finally, it should be noted that cationic polymerizations are very sensitive to impurities. These can act as cocatalysts, accelerating the polymerization, or as inhibitors (e.g., tertiary amines); they can also give rise to chain transfer or chain termination and so cause a lowering of the degree of polymerization. Since these effects can be caused by very small amounts of impurities (10^{-3} mol% or less), careful purification and drying of all materials and equipment is imperative.

Example 3.16 Cationic Polymerization of Isobutylene with Gaseous BF₃ at Low Temperatures in Bulk

Safety precautions: Before this experiment is carried out, Sect. 2.2.5 must be read as well as the material safety data sheets (MSDS) for all chemicals and products used.

Monomeric isobutylene is passed from a cylinder through a drying column filled with solid potassium hydroxide pellets or *IA* zeolith and condensed in a dry cold trap. 10 ml of the condensate are collected in a dropping funnel installed directly below the trap.

A dry 100-ml standard apparatus (round-bottom flask, several inlets for stirrer, reflux condenser, nitrogen flux or vaccum, thermometer, heating bath) is cooled to $-80°C$ in a methanol/dry ice bath; 10 ml isobutylene and 5 g of dry ice are then added while stirring (the dry ice should be taken from the middle of a larger piece in order to be as free from water as possible). A dry 10-ml syringe pipette is filled with gaseous boron trifluoride (see Example 3.24) from a cylinder via an empty wash bottle; this is then injected into the liquid monomer. Polymerization begins immediately and gives a rubber-elastic product. The cold bath is removed after 45 min so that excess monomer slowly evaporates on warming to room temperature. The polyisobutylene obtained in this way is soluble in aliphatic, cycloaliphatic, and chlorinated hydrocarbons. The limiting viscosity number is determined in cyclohexane at $24°C$ and the molecular weight is derived (see Sect. 2.3.3.3). Depending on the molar mass polyisobutylene finds use as rubber and additive.

Example 3.17 Cationic Polymerization of Isobutyl Vinyl Ether with BF₃-Etherate at Low Temperatures

Safety precautions: Before this experiment is carried out, Sect. 2.2.5 must be read as well as the material safety data sheets (MSDS) for all chemicals and products used.

35 ml of dry isobutyl vinyl ether are washed five times with distilled water, dried over $MgSO_4$ and distilled (bp $80°C$) under nitrogen over solid NaOH into a dry receiver (see Sect. 2.2.5.3). About 80 ml of propane, taken from a cylinder, are passed through a drying column filled with solid KOH pellets and condensed in a dry cold trap.

A dry 500-ml standard apparatus (round bottom flask, several inlets for stirrer, reflux condenser, nitrogen flux or vaccum, thermometer, heating bath) is heated under vacuum with a heat gun in order to remove traces of water. It is then filled with nitrogen and cooled to $-70°C$ in an acetone/dry ice bath, the gas outlet being protected against back diffusion (see Sect. 2.2.5.2). The propane in the cold trap is transferred to the flask and 20 ml (0.15 mol) of isobutyl vinyl ether are added from the receiver. 0.30 ml (0.39 mmol) of freshly distilled BF₃-etherate are added dropwise with vigorous stirring and the mixture is held at $-70°C$ for 30 min; another 0.3 ml of BF₃-etherate are then dropped in and stirring is continued at this temperature for a further 90 min. During polymerization, which takes place at the surface of the initiator droplets, a very slow stream of nitrogen is passed through the apparatus. Finally the initiator is destroyed by addition of an excess of cyclohexylamine and the mixture is slowly warmed to room temperature; the propane evaporates, leaving the polymer behind in the form of small lumps. It is dried in vacuum at 50°C. Before drying, a small sample is dissolved in toluene and reprecipitated in petroleum ether; this is used for the determination of the limiting viscosity number in toluene at 20°C and also for the determination of the melting point (90°C) with the product being crystalline (see Sects. 2.3.3.2 and 2.3.5.4). Polyisobutylvinylether finds use as an adhesive.

Example 3.18 Cationic Polymerization of α-Methylstyrene in Solution

Safety precautions: Before this experiment is carried out, Sect. 2.2.5 must be read as well as the material safety data sheets (MSDS) for all chemicals and products used.

Monomeric α-methylstyrene is freed from phenolic inhibitors by shaking twice with 10% sodium hydroxide. It is then washed three times with distilled water, dried over calcium hydride, and distilled before use over calcium hydride under reduced pressure of nitrogen (bp 54°C/12 Torr).

Methylene chloride is refluxed over P_2O_5 (best deposited on a solid carrier) for 1 h and then distilled into a dry receiver.

A carefully dried 100-ml three-necked flask, fitted with stirrer, thermometer, and nitrogen inlet, is evacuated and filled with nitrogen; 5.9 g (0.05 mol) of α-methylstyrene and 50 ml of methylene chloride are pipetted in and cooled to −78°C in a methanol/dry ice bath. 0.040 ml (0.075 mmol) of concentrated sulfuric acid are now added, causing polymerization to begin immediately, as shown by the slight temperature rise. After 3 h the viscous solution is added dropwise to about 500 ml of methanol in order to precipitate the polymer that is filtered off after settling and washed with methanol; a portion of the polymer is reprecipitated from a 2% toluene solution into methanol. Both samples are dried to constant weight in vacuum at 50°C; yield: about 70%. The reprecipitated sample is used to determine the softening range (150–200°C) and the limiting viscosity number in toluene at 25°C. This experiment can also be carried out with styrene; boron trifluoride or tin tetrachloride may also be used as initiators.

3.2.1.2 Anionic Polymerization with Organometallic Compounds as Initiators

There are numerous organometallic compounds that are capable of initiating the polymerization of unsaturated monomers. The following ones are of general importance: organic compounds of the alkali metals (e.g., butyllithium), organic compounds of zinc and cadmium (e.g., diethylzinc, diisobutylzinc), and organomagnesium compounds. Polymerizations with organometallic compounds as initiators are generally carried out in solution. The following solvents can be used: aliphatic and aromatic hydrocarbons (hexane, heptane, decalin, benzene, toluene), and cyclic ethers (THF, 1,4-dioxane). Polarity and solvating power of the solvent have a major effect on the course of the polymerization and on the structure of the resulting polymer when stereospecific initiators are used (see Example 3.20).

For the polymerization of unsaturated monomers with organometallic compounds, the initiator concentration must generally be between 10^{-1} and 10^{-4}/mol of monomer; cocatalysts are usually unnecessary. Polymerization frequently occurs at temperatures below 20°C. Raising the temperature increases the rate of polymerization, but usually decreases the tacticity or tactic content when stereospecific initiators are used. An induction period is rarely observed.

All the above-mentioned initiators are very sensitive towards substances with active hydrogen. Care must therefore be taken to exclude acids, water, thiols, amines, and acetylene derivatives. Oxygen, carbon dioxide, carbon monoxide, carbonyl compounds, and alkyl halides which can react with the initiator, also interfere with the reaction. Careful purification and drying of the starting materials and apparatus is, therefore, absolutely essential, especially when dealing with "living polymers" (see Example 3.19).

Example 3.19 Anionic Polymerization of α-Methylstyrene with Sodium Naphthalene in Solution ("Living Polymerization")

Safety precautions: Before this experiment is carried out, Sect. 2.2.5 must be read as well as the material safety data sheets (MSDS) for all chemicals and products used.

All operations must be carried out under especially careful exclusion of air and moisture, otherwise the experiment will fail.

Purification of the Monomer and Solvent

Monomeric α-methylstyrene is destabilized and dried as described in Example 3.18. Pure THF (previously distilled through a column) is refluxed over potassium under nitrogen for at least a day. A small amount of pure benzophenone is then added; if the THF is sufficiently pure a blue color appears, due to the formation of metal ketyl. If the blue color fails to appear, the purification with potassium must be continued. The THF is distilled from this ketyl solution under nitrogen before use.

Preparation of the Initiator Solution

First of all the parts of the apparatus (see Fig. 3.1) are carefully cleaned and assembled, evacuated under a flame, and filled with dry nitrogen three times.

Very reactive, finely divided sodium, that is easy to handle and relatively safe, can be prepared by mixing 5 g of sodium and 50 g of neutral aluminum oxide powder under nitrogen with vigorous stirring at 150–170°C. A fine powder is obtained that can easily be dosed under nitrogen using the burette shown in Fig. 3.1.

To prepare the initiator solution, 1.28 g (10 mmol) of very pure naphthalene are introduced into tube (1) through the joint (8), while passing a slow stream of nitrogen through stopcock (6). After inserting a magnetic stirrer, 100 ml of purified

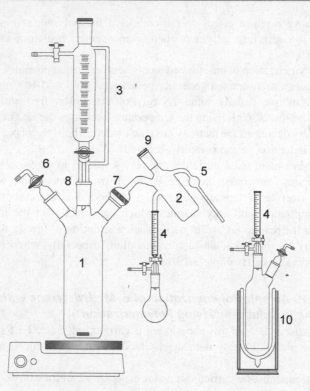

Fig. 3.1 Apparatus for the preparation of the initiator solution (explanation see text)

THF are run in through joint (8) with exclusion of air. The 10-ml burette (3) (bore of the stopcock barrel of this powder burette is 4 mm) is filled with the sodium/aluminum oxide powder and mounted on joint (8). The apparatus is now completely closed to the external atmosphere. 5 ml of the sodium powder are then run into the stirred solution in tube (1); the burette can now be removed and replaced with a ground-glass stopper. The sodium is allowed to react with the naphthalene for 30 min, while stirring is continued; the solution in tube (1) is then transferred to tube (2) by tilting the apparatus and applying a pressure of nitrogen. The aluminum oxide powder is held back by means of the sintered disc (7) (porosity 1). Capillary (5) can now be filled and sealed off after cooling to −78°C; this solution can be used to obtain the electron spin resonance spectrum and thus to demonstrate the radical character of sodium naphthalene. 10 ml of the initiator solution are run from burette (4) into a conical flask containing some distilled water.

The content of sodium naphthalene is determined by titration of the hydrolysis product with 0.1 M hydrochloric acid. The solution made by this procedure should contain about 100 mmol of initiator per liter.

Polymerization Procedure

A Schlenk tube is dried by heating under vacuum and is then charged with 50 ml of purified THF under nitrogen. The tube is attached through the joint to the burette (4) under a gentle stream of nitrogen. To remove traces of impurities that might still be present, a few drops of initiator solution are run into the vessel. As soon as the green color persists 1 mmol of initiator (about 10 ml depending of the result of the titration) is run into the reaction vessel from burette (4). The vessel is removed from the burette under nitrogen, closed with a stopper and cooled to −78°C in a Dewar vessel (10). 5.90 g (0.05 mol ≈ 6.5 ml) of α-methylstyrene are now injected in one dose from a hypodermic syringe, the solution being vigorously shaken. The solution immediately turns red as a result of the formation of the carbanions.

After 1 h at −78°C an aliquot of the solution is removed from the reaction vessel under a nitrogen stream using a syringe filled with nitrogen; this is dropped into a tenfold excess of methanol in order to precipitate the poly(α-methylstyrene). Another 5.9 g of α-methylstyrene are now added to the solution remaining in the reaction vessel, the red color of which should not be affected by the removal of the sample aliquot. After a further hour another sample is taken and a fresh batch of monomer added. The experiment is finally brought to an end after another hour, the remaining solution being dropped into methanol to precipitate the polymer. The polymers obtained during the experiment are reprecipitated from THF into methanol, filtered and dried in vacuum at 50°C.

This experiment can be carried out under identical conditions also with styrene.

Determination of the Molecular Weight of the Poly(α-Methylstyrene)s

If the polymerization of α-methylstyrene with sodium naphthalene proceeds without termination according to the above mechanism, the degree of polymerization can be represented by the following simple relation:

$$P = 2 \cdot \frac{\text{amount of monomer consumed (in mol)}}{\text{amount of initiator (in mol)}} \tag{3.17}$$

Thus, for 0.05 mol of α-methylstyrene and 0.001 mol of initiator the polymer should have a degree of polymerization of 100, corresponding to a molecular weight of 11,800 g/mol.

The limiting viscosity numbers of the samples are determined in toluene at 25°C and the molecular weights are derived. The observed and calculated values are tabulated and compared.

Example 3.20 Preparation of Isotactic and Syndiotactic Poly(Methyl Methacrylate) with Butyllithium in Solution

Safety precautions: Before this experiment is carried out, Sect. 2.2.5 must be read as well as the material safety data sheets (MSDS) for all chemicals and products used.

Preparation of the Initiator Solution

0.5 g (72 mmol) of finely cut lithium and 4.64 g (50.1 mmol) of freshly distilled butyl chloride are added to 50 ml of pure toluene under nitrogen in a well-dried, nitrogen-filled, 250 ml three-necked flask, fitted with stirrer and nitrogen inlet. A reflux condenser is then attached and the reaction started by stirring and slowly heating to about 80°C. The mixture is refluxed for a further 4 h and then allowed to stand. The butyllithium solution so obtained is about 1 M. For the exact determination of the butyllithium content, 2 ml of this solution are withdrawn under nitrogen with the aid of a hypodermic syringe, added to 20 ml of pure methanol and titrated with 0.1 M hydrochloric acid using methyl red as indicator.

Polymerization Procedure

Methyl methacrylate, toluene, and 1,2-dimethoxyethane are fractionated using a column, and finally distilled over calcium hydride into a receiver under dry nitrogen (see Sect. 2.2.5.3).

Two 250 ml three-necked flasks are baked out under vacuum, fitted with stirrer, nitrogen inlet, and self-sealing closure (see Sect. 2.2.5.2), evacuated and filled several times with dry nitrogen. 100 ml of toluene are introduced to one of the flasks and 100 ml of 1,2-dimethoxyethane to the other; to both are added 0.006 mol of butyllithium (about 6 ml of 1 M solution) using a hypodermic syringe, and the solutions cooled to −78°C; 10 ml (0.1 mol) of methyl methacrylate are then injected into each. After 30 min the polymerizations are terminated by addition of 10 ml of methanol, and the polymers precipitated by dropping each solution into 1.5 l of low-boiling petroleum ether. The polymers are filtered off, the damp polymer then being dissolved in toluene and centrifuged for about 30 min at 4,000 rpm in order to separate insoluble residues (inorganic hydrolysis products and some crosslinked polymers). The polymer is reprecipitated in a 15-fold amount of petroleum ether, filtered, and dried in vacuum at 40°C. Yield in toluene as solvent: 60–70% (isotactic polymer); yield in 1,2-dimethoxyethane as solvent: 20–30% (syndiotactic polymer). The limiting viscosity numbers of the two samples are measured in acetone at 25°C (see Sect. 2.3.3.3); also the IR spectra in potassium bromide discs (see Sect. 2.3.2.2) are recorded. From the latter, the isotactic and syndiotactic content can be determined both qualitatively and quantitatively.

Example 3.21 Stereospecific Polymerization of Isoprene with Butyllithium in Solution

Safety precautions: Before this experiment is carried out, Sect. 2.2.5 must be read as well as the material safety data sheets (MSDS) for all chemicals and products used.

(a) Preparation of 3,4-Polyisoprene

Monomeric isoprene is purified as described for styrene in Example 3.1; before use it is run under nitrogen through a 30-cm column packed with neutral aluminum oxide; cyclohexane and 1,2-dimethoxyethane are refluxed over sodium for 6 h and distilled off under nitrogen.

A carefully dried 250 ml standard apparatus (round bottom flask, several inlets for stirrer, reflux condenser, nitrogen flux or vaccum, thermometer, heating bath) fitted with a glass-sealed magnetic stirrer, is evacuated and filled with nitrogen (dried over phosphorus pentoxide) three times. 20 ml of 1,2-dimethoxyethane, 100 ml of cyclohexane, and 6.8 g (0.1 mol) of isoprene are pipetted in under a stream of nitrogen. The magnetic stirrer is switched on and 0.1 mmol of butyllithium (0.5 ml of a 0.2 M solution in toluene) is added; the solution turns lemon yellow and gradually warms up, a sign that polymerization has been triggered. The reaction may fail to commence on account of impurities present in the system, in which case more initiator solution is pipetted in until the yellow color of the reaction mixture persists. Polymerization ceases after 3 h, and the polymer is precipitated by dropping the solution into ethanol. It is filtered off and the polymer is dried in vacuum at 50°C. Yield: 90–95%. Before characterizing the polymer as described under (c), about 1 g is reprecipitated from 2% toluene solution into a tenfold excess of ethanol.

(b) Structural Investigations of Polymeric Dienes by IR Spectroscopy

The type of arrangement of the monomeric units in polymeric dienes can be determined qualitatively and quantitatively by IR spectroscopy. For this purpose thin films are prepared by dropping an approximately 2% solution in carbon disulfide (spectroscopically pure) on to suitable rock salt plates and allowing the solvent to evaporate at room temperature. The plates are placed in the spectrometer beam and the IR spectrum is recorded. The different types of chemical linkages are associated with characteristic IR bands as summarized in Table 3.9.

The spectra may first be evaluated qualitatively. The polyisoprene prepared in solution shows a pronounced band at 888 cm^{-1}, which indicates a high proportion of 3,4-linkages. For the product of bulk polymerization this band is much reduced in favor of absorptions at 1,127 and 1,315 cm^{-1}, indicating predominantly *cis*-1,4-linkage of the monomeric units in this case. The polymer made by radical polymerization in emulsion (see Example 3.12) shows the presence of all possible structural units, although the proportion of *cis*-1,4-linkages is low.

A distinction between 1,2- and 1,4-arrangement is possible through selective epoxidation of the olefinic components using *m*-chloroperbenzoic acid. Double bonds in the polymer main chain with higher electron density are more nucleophilic and therefore more reactive than pendant double bonds. In addition, a good distinction can be drawn by NMR spectroscopy.

3.2.2 Ionic Polymerization Via C = O Bonds

Polymerization by opening of C = O bonds was investigated many years ago in the case of formaldehyde. As may be expected from its polar mesomeric structure formaldehyde can polymerize both anionically and cationically:

Table 3.9 Characteristic IR absorption bands for the different possible repeat units in polymeric dienes

Polybutadiene	Structure	Wavenumber in cm^{-1}
1,2	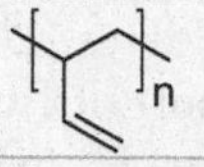	909
trans-1,4	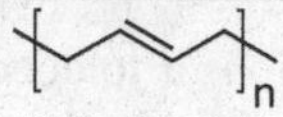	1355/971
cis-1,4		1311/741–725
Polyisoprene		
1,2	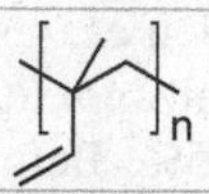	909
3,4	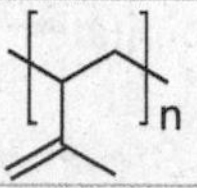	888
trans-1,4	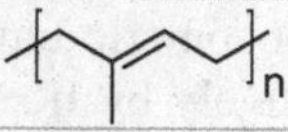	1325/1148
cis-1,4	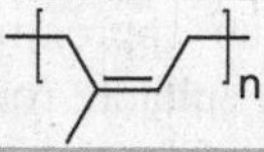	1315/1127

Anionic polymerization can be initiated with tertiary phosphines or amines, with organometallic compounds or with alcoholates. With all of these, initiation occurs by nucleophilic attack on the positive carbonyl carbon atom:

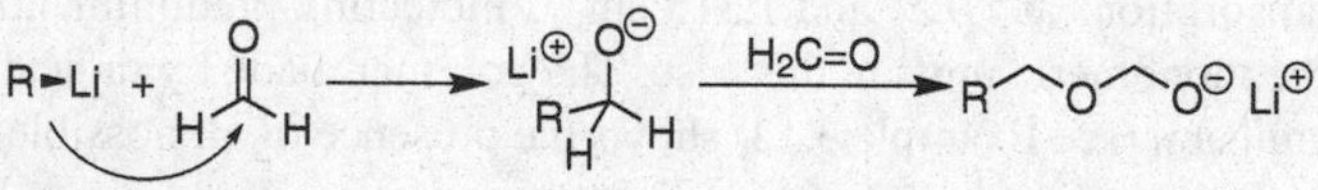

Cationic polymerization of formaldehyde (which should be carried out under the driest possible conditions to avoid transfer reactions) can be initiated with protic acids, Lewis acids (see Sect. 3.2.1.1), or other compounds that yield cations such as acetyl perchlorate or iodine:

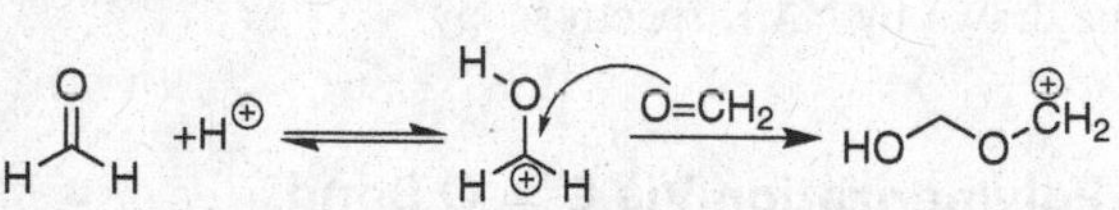

Polymers of formaldehyde with semiacetal end groups are thermally unstable; they decompose at temperatures as low as 150°C, splitting off monomeric formaldehyde. Upon acetylation of the hydroxy end groups, thermal stability up to 220°C is achieved; alkylations also provides stability against alkali, but not against acids since these are capable of splitting the acetal bonds in the polymer chains (see Example 5.13).

Other carbonyl compounds, such as acetaldehyde or propionaldehyde can also be polymerized to high-molecular-weight products; however, their stability is lower than that of polyoxymethylenes with protected end groups.

The industrial synthesis of polyformaldehyde [poly(oxymethylene)] occurs by anionic polymerization of formaldehyde in suspension. For this the purification and handling of monomeric formaldehyde is of special importance since it tends to form solid paraformaldehyde. After the polymerization the semiacetal end groups have to be protected in order to avoid thermal depolymerization (Example 5.13). This is achieved by esterfication with acetic anhydride (see Example 5.7). As in the case of trioxane copolymers (see Sect. 3.2.3.2) the homopolymers of formaldehyde find application as engineering plastics.

Example 3.22 Anionic Polymerization of Formaldehyde in Solution (Precipitation Polymerization)

Safety precautions: Before this experiment is carried out, Sect. 2.2.5 must be read as well as the material safety data sheets (MSDS) for all chemicals and products used.

Caution: All operations with formaldehyde have to be carried out in a closed hood.

Preparation of a Solution of Monomeric Formaldehyde

The apparatus for the preparation of monomeric formaldehyde consists of a 1 l two-necked flask (A), one neck of which is closed with a loosely fitting stopper so that it can be removed quickly in the event of development of excess pressure. To the second neck is fitted a 2 m long, angled tube (diameter at least 2 cm), to the end of which is attached another 500 ml two-necked flask (B) filled with Raschig rings. The second neck of flask B serves as gas outlet and is connected to a 1 l cold trap. The outlet of the trap is fitted with a suitable device (see Sect. 2.2.5.2) to prevent the ingress of atmospheric moisture and CO_2, the latter being a powerful inhibitor of polymerization. In the first flask (A), which is not yet connected to the rest of the apparatus, are mixed 100 g of polyoxymethylene (or alternatively paraformaldehyde) and 100 g of dry paraffin oil. This is then heated in an oil bath to 130°C until about 10–15% of the polyoxymethylene has decomposed; the resulting gaseous formaldehyde is not yet passed into the apparatus, but into a washing bottle with dry ethylene glycol or glycerin; this procedure serves to remove water from the polyoxymethylene to be decomposed.

In the meantime the rest of the apparatus is flamed out under vacuum and filled with dry nitrogen; the cold trap is filled with 500 ml of sodium-dried ether and cooled to −78°C. Flask A is now attached via the 2 m glass tube. The gaseous formaldehyde generated by further pyrolysis polymerizes partially in the glass tube and in flask B to low-molecular-weight polyoxymethylene, which contains as end groups most of the water formed during depolymerization (thus resulting in purification by prepolymerization). Care must be taken that the polymer deposited does not block the glass tube; if necessary the pyrolysis must be interrupted so that the tube can be cleaned. Sticking of the ground joints can be prevented by generous greasing. The pyrolysis should be carried out rather quickly (within about 1 h), since otherwise too much monomer is lost by prepolymerization; an oil bath temperature up to 200°C is required. The resulting ether solution contains about 70 g of formaldehyde (about 4 M). Another method of preparing very pure monomeric formaldehyde is from 1,3,5-trioxane, which can be decomposed in the gas phase at 220°C on a supported phosphoric acid catalyst.

Polymerization Procedure

150 ml of the ethereal formaldehyde solution are placed in a previously flamed 250 ml standard apparatus (round bottom flask, several inlets for stirrer, reflux condenser, nitrogen flux or vaccum, thermometer, heating bath) cooled to −78°C, taking care to exclude atmospheric moisture. (For larger quantities care must be taken to make adequate provision for the removal of the heat of polymerization.) The flask is fitted with a self-sealing closure and a pressure equalizer protected with a soda-lime tube. 1 mg of pyridine, dissolved in 5 ml of dry ether, is used as initiator; this is carefully injected into the formaldehyde solution (cooled to −78°C) over a period of 15 min, with vigorous stirring. After 1 h at −78°C the conversion reaches more than 90%. Should the rate of polymerization be lower because of impurities, the amount of initiator can be raised. The precipitated polyoxymethylene is filtered off, washed with ether and dried in vacuum at room temperature; it melts between 176°C and 178°C. The limiting viscosity number is determined on a 1% solution in DMF at 140°C ($\eta_{sp}/c \approx 0.08$ l/g, corresponding to an average molecular weight of 80,000). The thermal stability can be tested before and after blocking the hydroxy end groups (see Examples 5.7 and 5.13).

Polyoxymethylene, obtained by the polymerization either of formaldehyde or of 1,3,5-trioxane, is a highly crystalline product which is insoluble in all solvents at room temperature with the exception of hexafluoroacetone hydrate; at higher temperatures it dissolves in some polar solvents (e.g., at 130°C in DMF or DMSO). If the unstable semi-acetal end groups are blocked (see Example 5.7) polyoxymethylene can be processed without decomposition as a thermoplastic at elevated temperatures in the presence of stabilizers.

3.2.3   Ring-Opening Polymerization

Many heterocyclic compounds can be polymerized by ring opening under certain conditions with ionic initiators, to produce linear macromolecules. Amongst these are cyclic ethers, cyclic sulfides, cyclic acetals, cyclic esters (lactones), cyclic amides (lactams), and cyclic amines. Ring-opening polymerizations are carried out under similar conditions, and frequently with similar initiators to those used for ionic polymerizations of unsaturated monomers (see Sect. 3.2.1); they are likewise sensitive to impurities. In some cases one succeeds in the ring-opening polymerization of cyclic olefins with formation of straight-chain unsaturated polymers (metathesis polymerization, see Example 3.33).

3.2.3.1 Ring-Opening Polymerization of Cyclic Ethers

The ring-opening polymerization of cyclic ethers having 3-, 4-, and 5-membered rings (e.g., epoxides, oxetanes, THF) yields polymeric ethers. Six-membered rings (1,4-dioxane) are not capable of polymerization.

Epoxides such as epoxyethane (ethylene oxide) can be polymerized cationically (e.g., with Lewis acids) and anionically (e.g., with alcoholates or organometallic compounds).

Polymers with extremely high molecular weights result from the polymerization of ethylene oxide initiated by the carbonates of the alkaline earth metals, e.g., strontium carbonate, which must, however, be very pure. Poly(ethylene oxides) having molecular weights up to about 600 are viscous liquids; above that they are wax-like or solid, crystalline products that are readily soluble not only in water but also in organic solvents such as benzene or chloroform. Polymers of propylene oxide and generally substituted ethylene oxides can be produced in both atactic amorphous and isotactic crystalline forms. Optically active poly(propylene oxide)s can be obtained from propylene oxide.

Polymerization of four-membered cyclic ethers (oxetanes) is also brought about by cationic initiators (e.g., Lewis acids) and by anionic initiators (e.g., organometallic compounds). The polymer of 3,3-bis(chloromethyl)oxetane is distinguished by its very high softening point and by its unusual chemical stability.

THF can be polymerized only with cationic initiators, for example, boron trifluoride or antimony pentachloride. The initial step consists of the formation of a cyclic oxonium ion; one of two activated methylene groups in the a-position to the oxonium ion is then attacked by a monomer molecule in an S_N2-reaction, resulting in the opening of the ring. Further chain growth proceeds again via tertiary oxonium ions and not, as formerly assumed, via free carbonium ions:

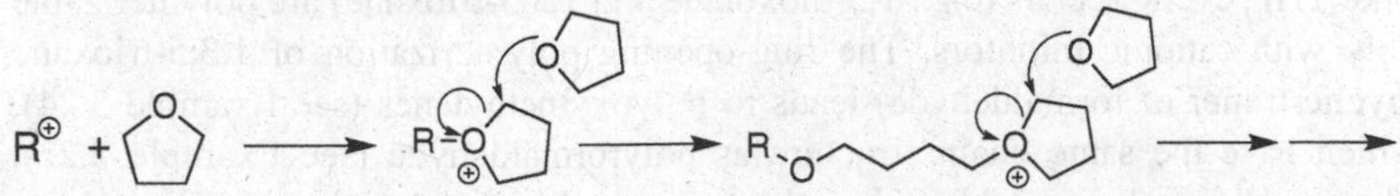

Deviations from this mechanism are observed when so-called superacids (e.g., fluorosulfonic acid, FSO_3H) are used as initiators (macroion/macroester equilibrium).

Poly(oxytetramethylene), poly(tetrahydrofuran), may assume the state of a viscous oil, a wax, or a crystalline solid (melting range around 55°C), depending on the molecular weight. Poly(tetrahydrofuran) telechelics prepared with two OH end groups and in molar masses of 500–4,000 g/mol are used widely as soft block in segmented polyurethanes and polyesters (see Sect. 3.4.2.1).

Example 3.23 Polymerization of THF with Antimony Pentachloride in Bulk

Safety precautions: Before this experiment is carried out, Sect. 2.2.5 must be read as well as the material safety data sheets (MSDS) for all chemicals and products used.

THF is purified as described in Example 3.19 and distilled shortly before use into a dry receiver (see Sect. 2.2.5.3).

A dry 250 ml round-bottomed flask, fitted with an adapter (see Sect. 2.2.5.4) is flamed under vacuum and filled with nitrogen. 15 g (0.21 mol) of THF are then added by connecting the receiver to the adapter and applying a slight vacuum to the flask while admitting nitrogen to the receiver through a side arm and stopcock. The flask is cooled to −30°C and 0.6 g (2.0 mmol) of freshly distilled antimony pentachloride are added under nitrogen. The flask is now detached from the adapter under a slight positive pressure of nitrogen and immediately closed with a ground glass stopper, secured with springs. The mixture is swirled around and allowed to stand at room temperature for 24 h, during which time the mixture becomes viscous. It is then treated with 2 ml of water and 150 ml of THF, and refluxed for about 30 min until a homogeneous solution is produced. The still viscous solution is diluted with a further 100 ml of THF and filtered to remove the insoluble portion (hydrolysis product of the initiator). The polymer is precipitated by dropping the solution into 3 l of water with vigorous stirring; it is filtered off and dried in vacuum at 20°C. The polymer obtained in the above preparation is a somewhat sticky solid material with a melting range around 40°C. It is soluble in toluene, carbon tetrachloride, chlorobenzene, THF, 1,4-dioxane, and acetic acid; it is insoluble in water, methanol, and acetone. The limiting viscosity number is determined in toluene at 20°C (see Sect. 2.3.3.3).

3.2.3.2 Ring-Opening Polymerization of Cyclic Acetals

Like THF, cyclic acetals (e.g., 1,3-dioxolane and 1,3,5-trioxane) are polymerizable only with cationic initiators. The ring-opening polymerization of 1,3,5-trioxane (cyclic trimer of formaldehyde) leads to polyoxymethylenes (see Example 3.24), which have the same chain structure as polyformaldehyde (see Example 3.22). They are thermally unstable unless the semiacetal hydroxy end groups have been protected in a suitable way (see Example 5.7). Like the cyclic ethers, the polymerization of 1,3,5-trioxane proceeds via the addition of an initiator cation to a ring oxygen atom, with the formation of an oxonium ion which is transformed to a carbenium ion by ring opening. The chain propagates by the addition of further 1,3,5-trioxane molecules:

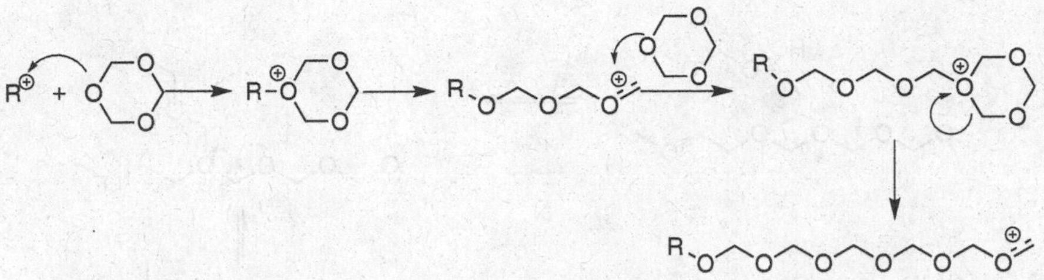

The solubility of polyoxymethylene is very poor so that the ring-opening polymerization of 1,3,5-trioxane proceeds heterogeneously both in bulk (melt) and in solution. 1,3,5-Trioxane can also be readily polymerized in the solid state; this polymerization can be initiated both by high-energy radiation and by cationic initiators (see Example 3.24).

Generally, the molecular weight and the molecular-weight distribution are determined by two side reactions. Moreover, the end groups and in case of copolymers, their sequence length distribution are determined by the following two side reactions:

Hydride transfer or hydride migration is initiated by the electrophilic attack of the poly(oxymethylene) cation from the methylene bridge of its own or of a neighboring macromolecule. A hydride ion is thus split off, and a methoxy end group is formed.

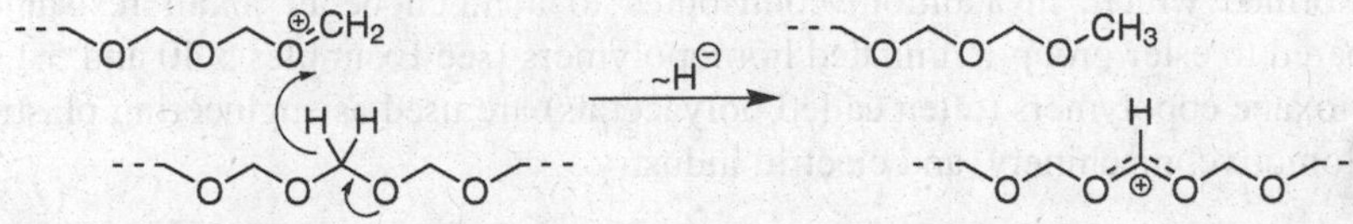

The newly created cation is stabilized through conjugation with the free electron pairs of the neighboring O atoms and is broken into a chain with the formate end group and another poly(oxymethylene) cation.

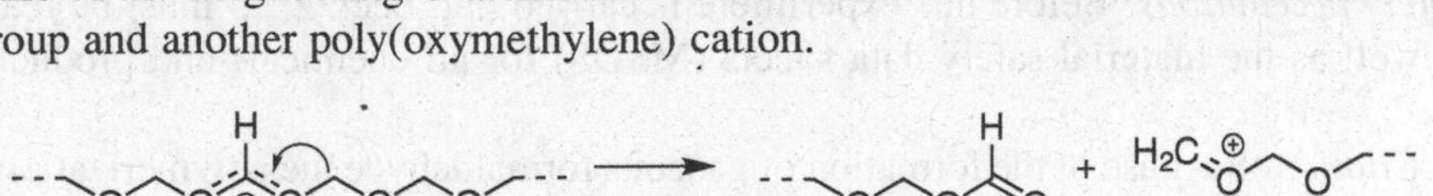

Both end groups can be determined quantitatively. A second side reaction is the transacetalization. Here a poly(oxymethylene) cation attacks an oxygen of a poly (oxymethylene) chain with formation of an oxonium ion that decomposes. Through continued cleavage and recombination of poly(oxymethylene) chains one obtains polymers which are chemically and molecularly largely homogeneous. For the case of a trioxane/ethylene oxide copolymer the following reaction scheme can be formulated:

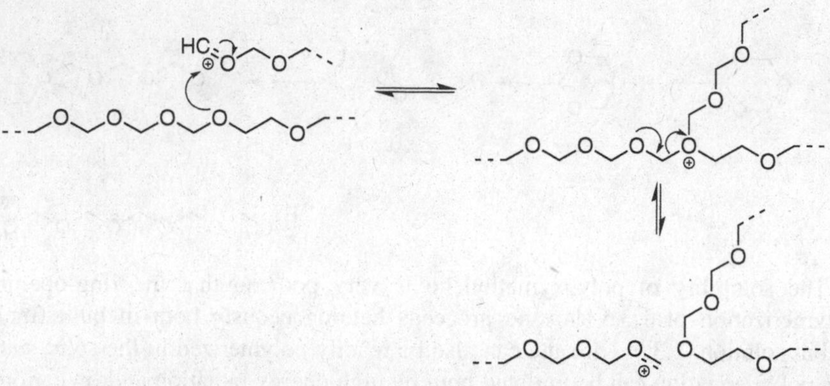

The transacetalization also proceeds intramolecularly. It then leads to the formation of cyclic acetals which participate as monomers in the propagation reaction.

The industrial production of copolymers of trioxane with ethylene oxide or dioxolane (1–5%) is conducted as a bulk polymerization in special equipment. The incorporation of small amounts of C–C bonds into the C–O chain has a remarkable effect on the thermal and chemical stability. In homopolymers the thermal decomposition starts at the semiacetal end groups ("unzipping") and leads to a complete destruction of the polymer chain, whereas this reaction stops in copolymers already at the first C–C bond. A thermally stable OH end group is thus formed which, in addition, contributes to a much better alkali resistance compared to ester group-terminated homopolymers (see Examples 3.40 and 5.13).

Trioxane copolymers (often called polyacetals) are used as engineering plastics in automotive, machinery, and electric industry.

Example 3.24 Polymerization of Trioxane with BF₃-Etherate as Initiator

Safety precautions: Before this experiment is carried out, Sect. 2.2.5 must be read as well as the material safety data sheets (MSDS) for all chemicals and products used.

Caution: Because of the formation of gaseous formaldehyde the polymerizations have to be carried out in a closed hood.

(a) Polymerization in the Melt

150 g of commercial 1,3,5-trioxane are refluxed (bp 115°C) with 9 g of sodium or potassium under nitrogen for 48 h, using an air condenser with suitable protection against ingress of atmospheric moisture. 20 g of the purified monomer are distilled into a 50 ml flask that has previously been flamed in vacuum. The flask contains a glass-sealed magnetic stirrer and is equipped with a self-sealing closure (see Sect. 2.2.5.2). The contents are heated to 70°C and 0.05 ml (0.4 mmol) of

BF_3-etherate ($d_{20} = 1.125$ g/ml) are injected as an approximately 10% solution in nitrobenzene. Special care must be taken that the molten 1,3,5-trioxane is mixed thoroughly with the initiator and a homogeneous mixture is obtained as quickly as possible. Immediately after the addition of initiator to the molten monomer the polyoxymethylene begins to precipitate; after only 10 s the whole reaction mixture has solidified. The polymerization is terminated with acetone and the polymer filtered off on a glass sinter after thorough mixing and, if necessary grinding. The polymer is boiled twice with acetone for 20 min, filtered, and dried in vacuum at room temperature. Yield: about 50%; melting range 177–180°C. The viscosity number is determined for a 1% solution in DMF at 140°C ($\eta_{sp}/c \approx 0.06$ l/g, corresponding to an average molecular weight of 60,000). The thermal degradation can also be studied before and after blocking the hydroxy end groups (see Examples 5.7 and 5.13).

(b) Polymerization in the Solid State

Gaseous boron trifluoride is required for this experiment. Suitable gas cylinders are available commercially. If necessary boron trifluoride can be prepared as follows: 15 g of powdered sodium tetrafluoroborate are mixed in a 100-ml flask with 2.5 g of boron trioxide and 15 ml of concentrated sulfuric acid. The flask is fitted with a gas delivery tube, to which is attached a piece of PVC tubing, pinched tight at the open end having a slit in the side to allow pressure equalization. The gaseous BF_3 is generated by heating to 160–170°C and can be withdrawn by means of a hypodermic syringe through the PVC tubing.

Commercial 1,3,5-trioxane is purified by sublimation under normal pressure at 50°C, followed by recrystallization of 60 g from 500 ml of dry cyclohexane (yield: about 36 g). Especially well-formed crystalline needles are obtained which, after filtering off the solvent, can be used without further drying.

20 g of 1,3,5-trioxane are placed in a 300 ml conical flask that is then flushed with nitrogen and sealed with a polyethylene film. 10 ml of BF_3 gas are injected through the film.

After a short time the trioxane crystals which initially have a glassy appearance, become cloudy. After 1 min the polymerization has advanced so far that a sample is nearly totally insoluble in methanol. The conical flask is kept for another hour at 80°C to allow the polymerization to die away. The product can be worked up and treated further as described under (a). Yield: >50%.

(c) Polymerization in Solution (Precipitation Polymerization)

90 g of 1,3,5-trioxane, purified as in (a), are distilled into a 500 ml flask (that has previously been flamed under vacuum) containing 200 ml of 1,2-dichloroethane that has been dried over P_2O_5; atmospheric moisture must be excluded. The flask is closed with a self-sealing closure (see Sect. 2.2.5.2). 0.06 ml (0.5 mmol) of BF_3-etherate dissolved in 7 ml of 1,2-dichloroethane are now injected while shaking the flask, which is then warmed to 45°C. After an induction period of about 1 min, solid poly (oxymethylene) begins to separate, until finally the contents of the flask are completely solidified. After 1 h the product is well ground with 200 ml of acetone,

filtered, boiled well with acetone as under (a) and dried. In order to remove occluded initiator residues the polymer is boiled with 1 l of ether containing 2 wt% of tributylamine; it is finally filtered and dried in vacuum at room temperature. Yield: 90–95%; melting range: 176–178°C. The limiting viscosity is determined in DMF at 140°C (molecular weight ~60,000). The thermal stability can also be measured; see under (a). The amount of initiator can be reduced dramatically if the trioxane is purified very carefully. For this, trioxane is refluxed over sodium/potassium alloy for 4 days under a nitrogen atmosphere and distilled into the polymerization flask.

3.2.3.3 Ring-Opening Polymerization of Cyclic Esters (Lactones)

Cyclic esters of ω-hydroxycarboxylic acids can be polymerized by ring-opening to give linear aliphatic polyesters. According to the type of initiator and monomer the polymerization occurs either by alkyl or by acyl cleavage:

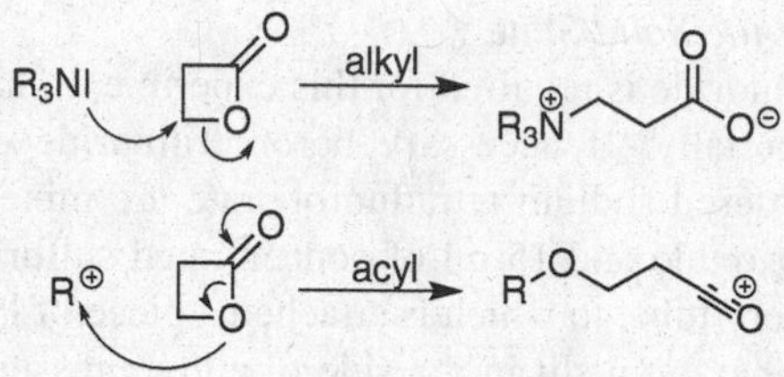

The polymerizability depends on the ring size and on the number, size, and position of the substituents.

The ring-opening polymerization of dilactide (dimeric cyclic ester of lactic acid) allows the preparation of high molecular weight, optically active polyesters of lactic acid. The configuration of the asymmetric carbon atoms of the monomer is retained when the polymerization is initiated with $SnCl_4$ or Et_2Zn, for example:

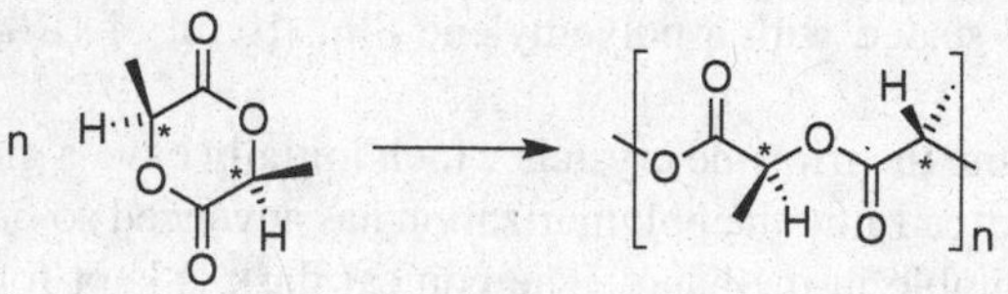

Example 3.25a Ring-Opening Polymerization of Dilactide with Cationic Initiators In Solution

Safety precautions: Before this experiment is carried out, Sect. 2.2.5 must be read as well as the material safety data sheets (MSDS) for all chemicals and products used.

NOTE: L-Lactide is hygroscopic and should be stored over P_4O_{10}.

5 g of L-lactide, 50 ml of pure toluene, and a magnetic stirrer are placed in a reaction vessel that has been flamed under vacuum and flushed with nitrogen. The vessel is closed with a pressure-tight rubber cap and heated to 110°C in an oil bath. 0.5 ml of initiator solution (3.32 g of $SnCl_4$/100 ml of toluene) is then injected

through the rubber cap with a hypodermic syringe. After about 3 h the viscous solution is cooled and added dropwise to 300 ml of vigorously stirred methanol. The filtered polymer is dissolved in 50 ml of 1,4-dioxane and reprecipitated in methanol. After drying in vacuum at about 70°C one obtains 4.2 g (yield 74%) of a white polymer with a crystalline melting point (DTA) of 153°C. It is soluble in 1,4-dioxane, chloroform, and acetonitrile, and insoluble in methanol, ether, and hexane. The solution viscosity (see Sect. 2.3.3.3) is determined in 1,4-dioxane at 25°C using an Ostwald viscometer (capillary diameter = 0.4 mm). The polymer can be characterized by IR spectroscopy and by ORD and CD measurements.

Enzyme Calaysed Ring-Opening Polymerization of ε-Caprolacton

Lipase-catalysed ring opening polymerization was first disolved in 1993 for ε-caprolacton (ε-CL) and δ-valerolactone (δ-VL) by two independent groups. Lipase (triacylglycerol acylhydrolase, EC Sect. 3.1.1.3) is an enzyme that catalyzes both the hydrolysis of a fatty acid lycerol ester in vivo with bond-cleaving and the polymerization reaction to give polyester in vitro with bond-forming, when the lipase catalyst an substrate monomer are appropooriately combined for the reaction.

Among the various lactones, ε-CL is the most extensively studied monomer. The ROP was induced by many lipases from different origin. *Candida antarctica* [lipase CA or CA lipase B (CALB) immobilized on an acrylic resin, produced commercially as Novozym 435] was found to be the most effective lipase for the ROP of ε-CL.

Monomer activation

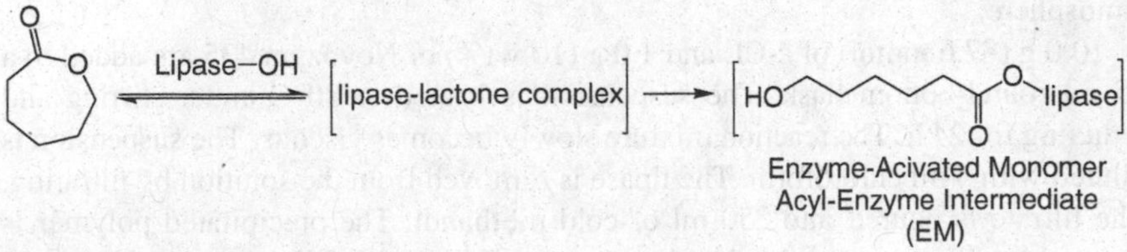

Enzyme-Acivated Monomer
Acyl-Enzyme Intermediate
(EM)

Initiation: nucleophilic attack of water or alcohol

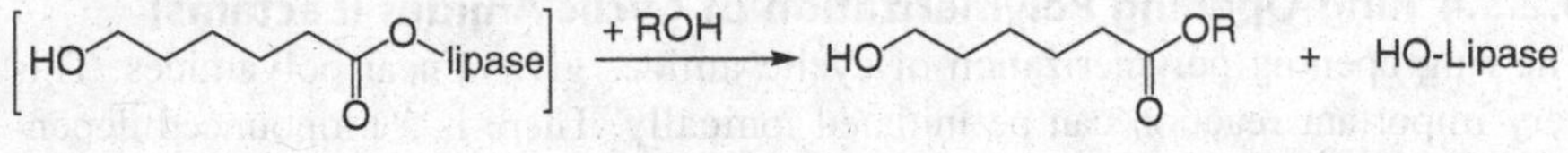

Propagation

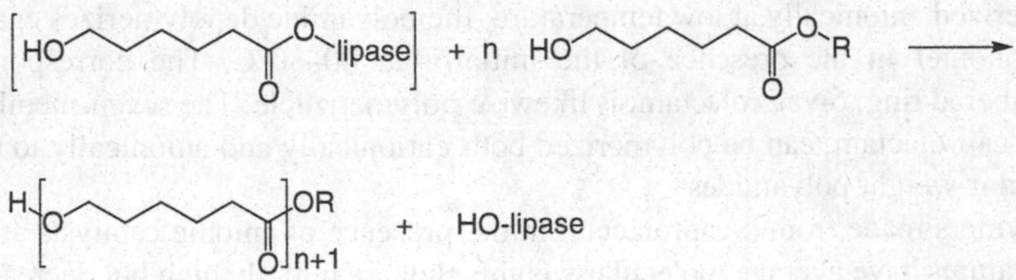

The reaction can be either performed in solvent or in bulk. By considering the principal reaction course involving an acyl-enzyme intermediate, the postulated mechanism for the lipase-catalyzed ROP of ε-CL is as follows. The polymerization via an "activated monomer mechanism". The reaction of lactone with lipase catalyst involving an enzyme-lipase complex and its ring-opening to give an acyl-enzyme intermediate (EM) is the key step of this reaction. In the initiation stage, is a nucleophilic attack of a nucleophile, such as water or an alkohol, which is present in the reaction mixture, on the acyl carbon of the intermediate to produce ω-hydroxycaroxylic acid (n = 1). It is followed by the propagation, in which EM is nucleophilically attacked by the terminal hydroxyl group of a propagation chain end to produce a one-monomer-unit elongated polymer chain.

Example 3.25b Novozym 435 Catalyzed Ring-Opening Polymerization of ε-Caprolactone in Bulk

Safety precautions: Before this experiment is carried out, Sect. 2.2.5 must be read as well as the material safety data sheets (MSDS) for all chemicals and products used.

ε-Caprolactone was purchased from Aldrich, dried over calcium hydride, distilled under reduced pressure, stored over 4 Å molecular sieves and under an argon atmosphere.

10.0 g (87.6 mmol) of ε-CL and 1.0 g (10 wt%) of Novozym 435 are added to a 25 ml round-bottom flask. The suspension is heated to 80°C under stirring and refluxing for 24 h. The reaction mixture slowly becomes viscous. The suspension is diluted with 5 ml chloroform. The lipase is removed from the solution by filtrating. The filtrate is poured into 250 ml of cold methanol. The precipitated polymer is isolated by filtration and dried under vacuum. (Yield: 90–95%).

3.2.3.4 Ring-Opening Polymerization of Cyclic Amides (Lactams)

The ring-opening polymerization of cyclic amides gives linear polyamides. This very important reaction can be initiated ionically. There is a pronounced dependence of the polymerizability of lactams on the ring size and on the number and position of the substituents. The five-membered lactam (γ-butyrolactam) can be polymerized anionically at low temperature; the polyamide depolymerizes again to the monomer in the presence of the initiator at 60–80°C. The corresponding 6-membered ring, δ-valerolactam, is likewise polymerizable. The seven-membered ring, ε-caprolactam, can be polymerized both cationically and anionically to high-molecular-weight polyamides.

Polymers made from ε-caprolactam in the presence of anionic catalysts at high temperatures have average molecular weights that are initially high but decrease on prolonged heating of the molten reaction mixture, finally attaining an equilibrium value. This change of molecular-weight distribution is caused by a transamidation reaction of the growing chains with the dead polyamide molecules. The polymerization of ε-caprolactam with anionic initiators can be considered as a true addition polymerization reaction:

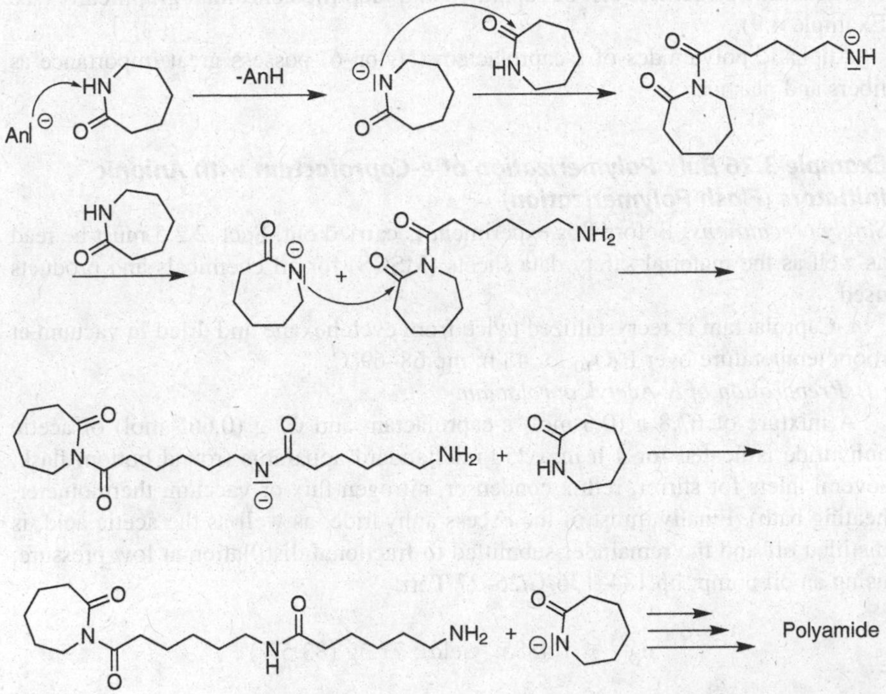

whereas the reaction initiated by catalytic amounts of water, e-aminocaproic acid, or benzoic acid as applied industrially, may be conceived as a stepwise addition polymerization involving migration of hydrogen atoms. The reaction starts with the addition of water or acid to ε-caprolactam; the resulting NH_2 or NH_3^+ end groups then add further lactam:

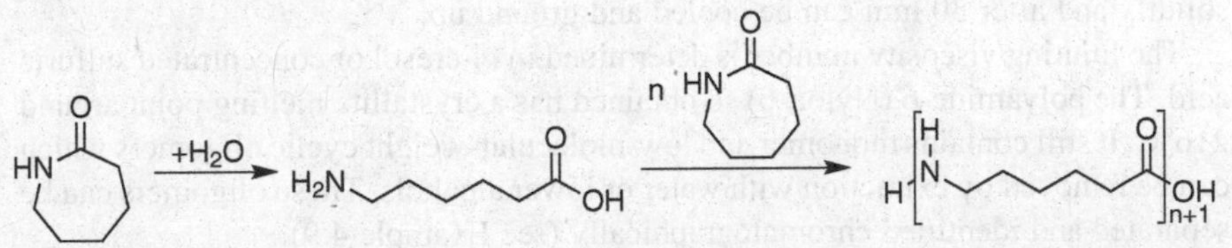

The molecular weight increases with increasing conversion. Regulation of the molecular weight can be achieved by adding small amounts of substances (e.g., benzoic acid) that can react with the polyamide chains by transamidation. Because of the transamidation reaction and hydrolysis of amide bonds, an equilibrium molecular-weight distribution is finally attained (see Sect. 4.1).

Polymers prepared at 250–270°C contain an equilibrium concentration of up to 10% of cyclic monomer and partly cyclic higher oligomers: after cooling, the monomer and oligomers can be recovered by extraction with water or lower alcohols. The oligomers can be separated and identified chromatographically (see Example 4.9).

Aliphatic polyamides of ε-caprolactam (Nylon-6) possess great importance as fibers and plastics.

Example 3.26 Bulk Polymerization of ε-Caprolactam with Anionic Initiators (Flash Polymerization)

Safety precautions: Before this experiment is carried out, Sect. 2.2.5 must be read as well as the material safety data sheets (MSDS) for all chemicals and products used.

ε-Caprolactam is recrystallized twice from cyclohexane and dried in vacuum at room temperature over P_4O_{10} for 48 h; mp 68–69°C.

(a) Preparation of N-Acetylcaprolactam

A mixture of 67.8 g (0.6 mol) ε-caprolactam and 67 g (0.665 mol) of acetic anhydride is heated for 4 h in a 250-ml standard apparatus (round bottom flask, several inlets for stirrer, reflux condenser, nitrogen flux or vaccum, thermometer, heating bath). Finally, most of the excess anhydride, as well as the acetic acid, is distilled off and the remainder submitted to fractional distillation at low pressure, using an oil pump; bp 134–136°C/26–27 Torr;

$$n_D^{25} = 1.4885; \text{ yield} : 77.5g \ (83.5\%)$$

(b) Polymerization Procedure

25 g (0.22 mol) of ε-caprolactam and 0.6 g (0.025 mol) of sodium hydride are placed in a flask, which is then evacuated and filled with nitrogen several times. The sample is melted to allow the sodium hydride to react. When the evolution of hydrogen has ceased, 0.33 g (0.002 mol) of *N*-acetylcaprolactam are added, the flask is well shaken, and then heated in an oil bath to 140°C. The contents rapidly solidify and after 30 min can be cooled and ground up.

The limiting viscosity number is determined in *m*-cresol or concentrated sulfuric acid. The polyamide-6 (Nylon-6) so obtained has a crystallite melting point around 216°C. It still contains monomer and low-molecular-weight cyclic oligomers which can be removed by extraction with water or lower alcohols. These oligomers can be separated and identified chromatographically (see Example 4.9).

3.2.3.5 Ring-Opening Polymerization of Oxazolines

The reaction of a nucleophilic monomer with an electrophilic monomer can, under suitable circumstances, lead to the formation of macromolecules that carry at least one charge at a chain end. Although the reaction has been known for a long time, it has gained importance for technical applications only recently. On the basis of the following scheme, the reaction mechanism can be explained by the homopolymerization of an oxazoline using methyl tosylate as an alkylating initiator.

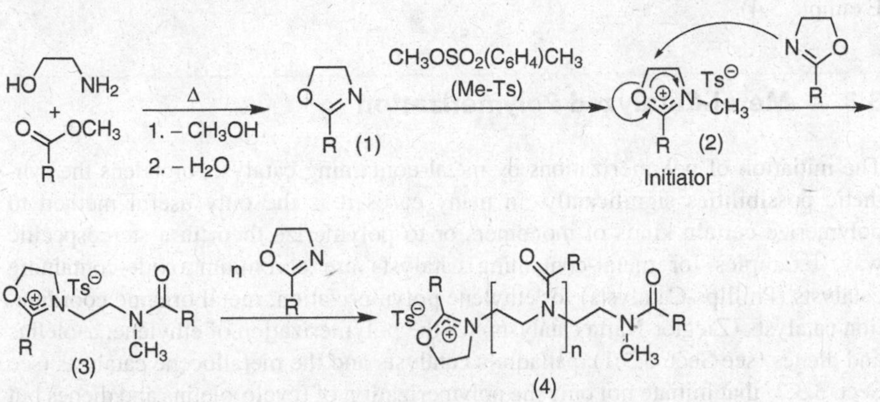

First of all a 2-substituted oxazoline (1) is formed by cyclocondensation of a carboxylic acid ester with 2-aminoethanol and a small amount of (1) is converted with an alkylating agent (e.g., methyl tosylate) to the activated, ionic form (2).

This *N*-alkylated heterocycle (2) acts as the actual initiator because it is attacked rapidly under ring-opening by an oxazoline molecule, present in excess. The newly formed dimer (3) contains an ionic ring function, which is subjected to the same attack as the initiator molecule. The molecular weight of the polymers is controlled by the amount of the alkylating agent. Other suitable initiators for the polymerization of oxazolines are Lewis acids, protic acids, and alkyl chloroformates.

In principle, one can obtain an unbranched polyethyleneimine by saponifying the amide groups that are located on the polymer (Example 5.4). It is important to note that the ring-opening polymerization of aziridine does not yield linear polyethyleneimines but rather highly branched polymer structures.

Example 3.27 Synthesis of a Linear, N-Acylated Polyethyleneimine Through Cationic Polymerization of 2-Methyl-2-Oxazoline in Bulk

Safety precautions: Before this experiment is carried out, Sect. 2.2.5 must be read as well as the material safety data sheets (MSDS) for all chemicals and products used.

In a 100 ml three-necked flask with stirrer and thermometer 1 mol% of *p*-toluenesulfonic acid methyl ester (methyl tosylate) are added to 3 g (0.03 mol) of anhydrous 2-methyl-2-oxazoline. The reaction mixture is stirred under nitrogen at 100–120°C. The bulk polymerization sets in immediately. After 30 min. the viscous polymer melt is poured in a dish where it solidifies within minutes. After cooling to room temperature about 2 g are dissolved in 20 ml ethanol and precipitated in 500 ml THF. The collected precipitate is dried under vacuum.

The polymer is fully amorphous with a glass transition temperature of 68°C. It is soluble in water, methanol, ethanol, propanol, $CHCl_3$, and DMF. Characterization can be carried out by IR-spectroscopy ($v_{C=O} = 1,633$ cm^{-1}) and by NMR spectroscopy (CDCl$_3$, $\delta = 3.47$ ppm, $-$N$-$CH$_2-$CH$_2-$, 2.13 ppm, CH$_3$).

The polymer can readily be hydrolized yielding a linear polyethyleneimine (see Example 5.4).

3.3   Metal-Catalyzed Polymerization

The initiation of polymerizations by metal-containing catalysts broadens the synthetic possibilities significantly. In many cases it is the only useful method to polymerize certain kinds of monomers or to polymerize them in a stereospecific way. Examples for metal-containing catalysts are chromium oxide-containing catalysts (Phillips-Catalysts) for ethylene polymerization, metal organic coordination catalysts (Ziegler-Natta catalysts) for the polymerization of ethylene, α-olefins and dienes (see Sect. 3.3.1), palladium catalysts and the metallocene catalysts (see Sect. 3.3.2) that initiate not only the polymerization of (cyclo)olefins and dienes but also of some polar monomers. More recently, progress in catalytic developments led to a number of new materials by Ring-Opening Metathesis Polymerization (ROMP) (see Sect. 3.3.3).

In all these cases and in contrast to starting a polymerization by initiators, there are no fragments of the starting molecule incorporated in the polymer chain. Consequently, the mechanisms are different to those of radical, anionic, or cationic polymerizations.

3.3.1   Polymerization with Ziegler-Natta-Catalysts

In 1953 K. Ziegler and coworkers discovered a class of heterogeneous catalysts that allowed ethylene to be polymerized at low pressures and low temperatures (low-pressure polyethylene = high-density polyethylene = PEHD).

These catalysts are made by mixing organometallic compounds of transition group elements IV–VIII with metal alkyls or metal hydrides of the groups I–III of the periodic system – for that reason they often are called metal organic mixed catalysts. Very efficient are combinations of the transition group compounds TiCl$_4$, TiCl$_3$, or VOCl$_3$ with alkylaluminum or alkylaluminum halide reagents, for example, Al(alkyl)$_3$, Al(alkyl)$_2$halogen, and Al$_2$(alkyl)$_3$halogen$_3$.

The transition group compound (catalyst) and the metal alkyl compound (activator) form an organometallic complex through alkylation of the transition metal by the activator which is the active center of polymerization (Cat). With these catalysts not only can ethylene be polymerized but also α-olefins (propylene, 1-butylene, styrene) and dienes. In these cases the polymerization can be regio- and stereoselective so that tactic polymers are obtained. The possibilities of combination between catalyst and activator are limited because the catalytic systems are specific to a certain substrate. This means that a given combination is mostly useful only for a certain monomer. Thus conjugated dienes can be polymerized by catalyst systems containing cobalt or nickel, whereas those systems do not work with ethylene and α-olefins. The latter can be polymerized with titanium-, vanadium-, or chromium-containing catalysts. A further restriction for the industrial process can be the activity of the catalyst and the stereoselectivity; in many cases both are not high enough. So-called carrier catalysts led to a considerable advance. They are obtained by precipitation of a transition metal compound onto an inert ($MgCl_2$) or reactive [$Mg(OH)_2$] solid. The polymerization activities achieved are very high (1,000 kg polyethylene per g titanium, corresponding to catalyst concentrations of a few ppm) so that the removal of the catalyst residue is not necessary, thus making the manufacturing process much easier (see Example 3.28). Further progress has been obtained by utilization of metallocenes (see Sect. 3.3.2).

By adding suitable electron donors (for example, ether or amines) the stereoselectivity of the propylene polymerization can be increased. Finally, a particular class of organometallic - catalysts should be mentioned thatcan bring about the ring-opening metathesis polymerization (ROMP) of cycloolefins to linear unsaturated polymers, e.g., of cyclopentene to poly(1-pentenylene) (see Example 3.33). Originally, tungsten compounds (e.g., $WCl_6/C_2H_5OH/C_2H_5AlCl_2$) have been applied for ROMP which are now largely substituted by the more effective Grubbs and Schrock catalysts.

The polymerization mechanism with Ziegler-Natta catalysts can be explained as follows:

The primary step consists in the formation of a π-complex between ethylene and the catalytic active center (Cat) that was already alkylated by the activator (R-M). This complexed monomer is then inserted into the metal-carbon bond with chain extension of two C-atoms. The propagation reaction consists of consecutive insertion steps. Termination occurs by β-elimination or by reaction with hydrogen. In both cases the active center Cat is maintained.

Formation of the catalyst (Cat):

Propagation:

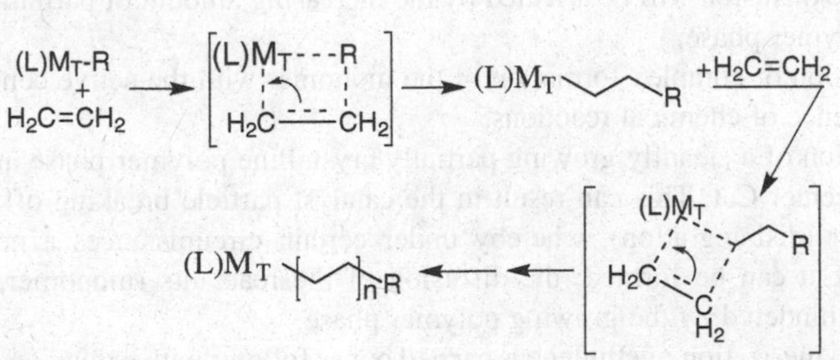

Termination:

(a) By β-elimination:

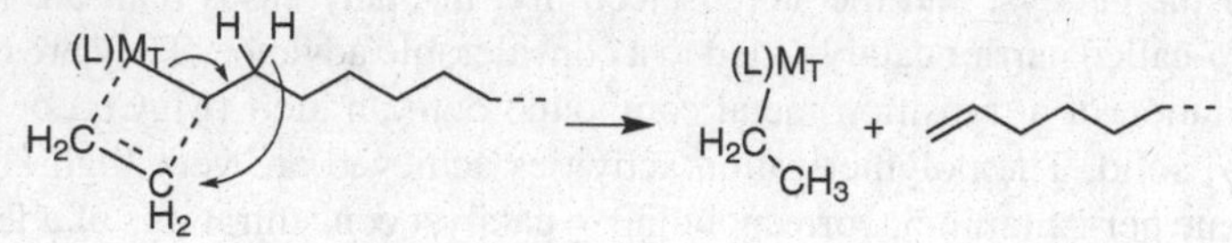

$$(L)M_T \quad + \quad$$

(b) By reaction with hydrogen:

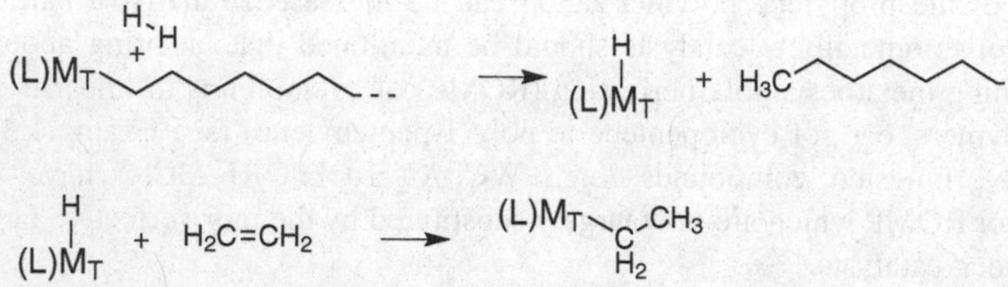

All factors that influence the stability of the transition metal-carbon bond (M_T-R) and/or the stability of the transition metal-ethylene bond (M_T-ethylene) are liable to affect the course of the reaction. Such factors are:
– Type and valence state of the transition metals M_T
– Type and number of ligands L at the transition metal M_T
– Type of the metal alkyl species (activator) R-M
– Catalyst morphology (crystallinity, porosity, external and internal surface).

The polymerization process consists of a series of consecutive and concurrent steps:
– Transfer of monomer and the molecular weight regulator, H_2, from the gas phase to the suspension medium,
– Diffusion of these components to the active center Cat; with increasing conversion, the diffusion will be affected by the increasing amount of partially crystalline polymer phase,
– Adsorption or complex formation of the monomer with the active center Cat,
– Occurrence of chemical reactions,
– Formation of a steadily growing partially crystalline polymer phase around the active center Cat. This can result in the catalyst particle breaking off from the polymer (disintegration), whereby under certain circumstances a new active center Cat can be formed; the diffusion of the reactants (monomer, H_2) can also be hindered by the growing polymer phase.

The polymerization of ethylene is carried out as follows with exclusion of air and water: the organometallic catalyst and the activator are suspended in an aliphatic hydrocarbon, the active centers Cat thereby being formed. Ethylene is passed in and the polymerization allowed to proceed at or slightly above atmospheric pressure, at a temperature below 100°C. The polyethylene precipitates as a swollen powder. As soon as the mixture has developed into a thick, dark-colored slurry, the reaction is terminated by destroying the catalyst (e.g., with butanol). The polymer is freed from most of the disperse medium by filtration, the remainder being removed by steam distillation. The damp polyethylene is finally dried.

Besides ethylene, higher olefins (propylene, 1-butylene), dienes, and a number of other monomers can be polymerized with organometallic mixed catalysts; the polymerization frequently proceeds stereospecifically, leading to tactic polymers (see Examples 3.29–3.31).

Example 3.28 Polymerization of Ethylene on a Supported Catalyst in Organic Suspension

Safety precautions: Before this experiment is carried out, Sect. 2.2.5 must be read as well as the material safety data sheets (MSDS) for all chemicals and products used.

Caution: The greatest possible care must be exercised when working with organoaluminum compounds since they ignite very easily on contact with air and water. All operations must, therefore, be carried out with the complete exclusion of air and moisture and pipettes must be flushed with nitrogen. Moreover, these substances cause wounds that are slow to heal, so that the wearing of safety goggles is mandatory and all contact with the skin must be avoided. The pipettes are cleaned as follows: After all the AlEt$_3$ has been run out, the pipette is filled with petroleum ether and allowed to drain again. It is then washed with acetone, dried and later cleaned with a solution of dichromate in concentrated sulfuric acid.

The polymerization apparatus (see Fig. 3.2) consists of a 1 l three-necked flask, fitted with stirrer, thermometer, gas inlet with tap, and gas outlet. On the inlet side the gas stream passes through three wash bottles: one as a safety bottle (A), one for the purification of ethylene, filled with 30 ml of petroleum ether (bp 100–140°C) and 5 ml of diethylaluminum chloride (B), and one filled with molecular sieves 5 Å (C). The last of these dries the ethylene further and also serves to trap aluminum hydroxide carried over from B. On the outlet side there are two wash bottles: the first is a safety bottle (D), and the second, (E), is filled with 50 ml of dry bis (2-hydroxyethyl) ether (diglycol), and isolates the apparatus from the external atmosphere.

(a) Preparation of the Supported Catalyst

1.1 g (10 mmol) of magnesium ethoxide (carrier) are suspended in 5 ml toluene. 20 ml of a 1 M TiCl$_4$ solution in toluene are added. The suspension is refluxed for 15 h. The precipitate is washed by decanting and stirring six times with 15 ml of toluene. The toluene above the solid should be free of titanium compounds. The volume of the suspension is filled up to 25 ml. The content of titanium of the suspension can be determined by colorimetric measurements with H$_2$O$_2$; 10 ml suspension contains approx. 2 mmol titanium compound.

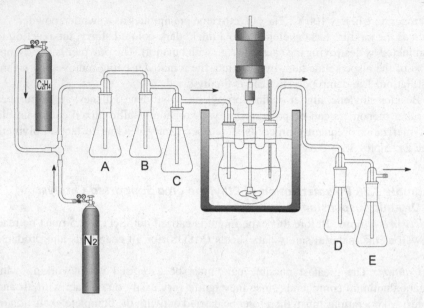

Fig. 3.2 Setup for the suspension polymerization of olefins in anhydrous medium

(b) Polymerization of Ethylene

Following the experimental arrangement in Fig. 3.2, the reaction vessel is a baked-out 500 ml four-necked flask with stirrer, gas inlet tube, and a condenser. A washing bottle with concentrated sulfuric acid, a washing bottle with KOH chips, and a washing bottle with toluene are attached to the gas inlet tube, so that the gas flow can pass through these bottles in the mentioned order. The condenser is attached to an empty washing bottle and to one filled with paraffin. Then 200 ml of absolutely dry toluene are placed into the flask under a nitrogen flow. The gas inlet tube has to reach into the liquid. Then, a solution of 0.5 g (3.1 mmol) triethylaluminum in 10 ml toluene and 2 ml of the supported catalyst suspension are added. The solution of the triethylaluminum is not allowed to come into contact with air and moisture. Now, the reaction flask is warmed up to 85°C and the nitrogen flow is stopped. Instead, ethylene from the pressure flask is passed into the solution. The gas flow is tuned in such a way that only a few or no gas bubbles can escape from the washing bottle. The start of the reaction can be observed by the fact that the initially finely distributed catalyst particles form a black coarse-grained precipitate and the toluene solution becomes colorless. In the course of the reaction the formed polyethylene grows around these grains as a dark-colored bulk. The stirrer has to work powerfully. After 4–6 h the polymerization is terminated by stopping the ethylene flow and by the addition of 100 ml of isopropanol. The resulting polyethylene is filtered off, repeatedly washed with methanol and dried in vacuum. Yield: 10–20 g polyethylene.

Polyethylene is, depending on the molecular weight, a waxy or solid, crystalline substance. Following the above-mentioned procedure, a high molecular crystalline product with a melting range around 130°C is obtained. At room temperature it is insoluble in all solvents. At higher temperatures (100–150°C) it can be dissolved in aliphatic and aromatic hydrocarbons. Viscosity measurements can be performed in xylene, tetralin or decalin at 135°C. To prevent the polymer from oxidative degradation, 0.2% of a commercially available stabilizer (for example, 2,6-di-*tert*-butyl-4-methylphenol or *N*-phenyl- b-naphthylamine) are added as antioxidant. Polyethylene can be formed to thin films by pressing between two metal plates (pressing time: 2 min at 180–190°C, see Sect. 2.5.2.1). After quenching with cold water, the films can be detached from the plates. They are convenient for recording infrared spectra, from which, for example, the crystallinity (see Sect. 2.3.5.5) or the degree of branching (see Sect. 2.3.2.2) can be determined.

Example 3.29 Stereospecific Polymerization of Propylene with Ziegler-Natta-Catalysts in Organic Suspension

Safety precautions: Before this experiment is carried out, Sect. 2.2.5 must be read as well as the material safety data sheets (MSDS) for all chemicals and products used.

Caution: Aluminum alkyls must be handled under total exclusion of oxygen and moisture (see Example 3.28).

(a) Preparation of Isotactic Polypropylene

The polymerization apparatus (Fig. 3.2) is flushed with nitrogen and the reaction flask filled with 500 ml of dry petroleum ether (distilled over sodium under nitrogen, bp 100–140°C). The flask is then heated to 60°C (±2°C) in a thermo-statted oil bath and the petroleum ether is saturated with propylene (about 15 min) while stirring (about 500 rpm). 1.26 ml (10 mmol) of diethylaluminum chloride and about 1 g (5 mmol) of $TiCl_3 \cdot 1/3\ AlCl_3$ (Stauffer AA, handled with exclusion of air and moisture) are added through the thermometer neck. The flow of propylene is adjusted such that very little escapes through the protective wash bottle E. The polymer separates out as a red-violet powder. The temperature is held steady by thermostatting at $60 \pm 2°C$. The stirring speed during polymerization is about 700 rpm.

After 2 h the polymerization is terminated by addition of 20 ml of 2-propanol, and stirring continued for another 30 min at 60°C; the reaction mixture decolorizes and becomes white. After filtration using a Büchner funnel and several washings with warm petroleum ether the polymer is dried to constant weight in vacuum at 70°C. Yield: 29.5 g.

The polypropylene so obtained has a high molecular weight and is crystalline. The proportion of isotactic polymer, determined by extracting with heptane for 10 h in a Soxhlet apparatus, is 98.5%. Isotactic polypropylene shows similar solubility behavior to polyethylene, but has a higher melting point (crystalline melting range 165–171°C).

(b) Effect of Heterogeneous Nucleation on the Crystallization of Isotactic Polypropylene

In the crystallization of isotactic polypropylene from the melt, the number and size of the spherulites (and hence the rate of crystallization) can be influenced by the addition of certain nucleating agents. The smaller the spherulites, the greater is the transparency of the polypropylene film. The mechanical properties can also be affected in some cases.

The effect of heterogeneous nucleation on the crystallization of isotactic polypropylene from the melt can be easily established as follows. A small amount of powdered polypropylene is well mixed with about 0.1 wt% of sodium benzoate in a mortar or by means of an analytical mill. Some of the mixture is transferred with a spatula to a microscope slide and melted at about 250°C on a hot block. A cover slip is pressed on to the melt with a cork to obtain as thin a film as possible. The sample is held at 200–250°C for some minutes and then allowed to crystallize at about 130°C on the hot stage of the microscope; an unadulterated polypropylene sample is crystallized in the same way. Both samples are observed under a polarizing microscope during crystallization, the difference in spherulite size between nucleated and untreated polypropylene can be seen very clearly. An ordinary microscope can also be used by placing polarizers on the condenser and eyepiece, and adjusting these to give maximum darkness.

Example 3.30 Stereospecific Polymerization of Styrene with Ziegler-Natta-Catalysts

Safety precautions: Before this experiment is carried out, Sect. 2.2.5 must be read as well as the material safety data sheets (MSDS) for all chemicals and products used.

Caution: Alkylaluminum reagents must be handled under total exclusion of atmospheric oxygen and moisture (see Example 3.28).

A dry 1 l three-necked flask, fitted with stirrer, thermometer, nitrogen inlet, and dropping funnel with pressure equalizer (according to Fig. 3.2), is evacuated and filled with nitrogen three times. 1.0 ml (9.1 mmol) of $TiCl_4$ is added by means of a syringe pipette under a brisk flow of nitrogen, and 27 ml (24 mmol) of triethylaluminum (0.9 M in hexane) (or triisobutylaluminum) are dropped in from the dropping funnel over a period of 10 min, with stirring. The reaction of the two catalyst components is initially strongly exothermic and the flask must therefore be cooled externally to about −20°C. As a precaution against possible breakage of the flask, the cooling bath must not contain water since it reacts extremely violently with triethylaluminium. A dry ice/1,2-dimethoxyethane bath can be used. When the additions are complete, the cooling bath is removed and stirring is continued for 5 min at room temperature, then 90 g of carefully dried styrene (seeExample 3.1) are added quickly from a second dropping funnel. The stirring rate is now increased and the oil bath heated to 50°C. After 1–2 h the contents of the flask become viscous and eventually gel-like after 2–3 h. The hot bath is removed and 30 ml of methanol are added over a period of 10 min through the dropping funnel, with vigorous stirring, in order to destroy the catalyst. The addition of methanol must be done very carefully; above all one must ensure immediate and thorough mixing. After the catalyst has been destroyed a further 350 ml of methanol are added, once again with vigorous stirring; the polystyrene then precipitates from the gel-like reaction mixture in the form of fine flakes. Stirring is continued for another 10 min, and the polymer is filtered off and washed with methanol. In order to remove catalyst residues the polymer is stirred for 1 h in a mixture of 500 ml of methanol and 5 ml of concentrated hydrochloric acid. It is then filtered off, washed with methanol and dried in vacuum at 60°C. Yield: 5–30%.

For the separation of the amorphous portion the dried polymer is stirred for 3 h in 500 ml of acetone, to which 2 ml of concentrated hydrochloric acid has been added. The insoluble portion is filtered off and dried in vacuum at 60°C. Yield: 85–95% of crystallizable polystyrene.

The crystallization of the acetone-insoluble polystyrene is completed by boiling for 2 h in freshly distilled butanone; it is then allowed to stand overnight at room temperature and finally filtered and dried in vacuum at 60°C. Yield of crystalline isotactic polystyrene: 95–100% of the acetone insoluble portion. The crystalline melting range and the density (see Sect. 2.3.5.1) are determined, as is also the limiting viscosity number in toluene at 20°C.

Crystallization and Characterization of Isotactic Polystyrene
The atactic polystyrene is precipitated by dropping the acetone/HCl solution into methanol and it is filtered through a sintered glass crucible; the atactic and crystalline portions are dried in vacuum at 50°C and finally weighed. The X-ray diffraction patterns of the two samples are compared with each other and with that of a polystyrene made by radical polymerization; likewise for the IR spectra (see Sects. 2.3.5.2 and 2.3.2.2).

The melting range of the isotactic, crystalline sample is determined with the aid of a hot-stage microscope; the following conditions of the sample can be distinguished:
1. Clearly defined particles
2. Blurred edges
3. Beginning of sintering
4. Beginning of melting
5. Melt runs together
6. Clear melt

The melting range is defined by the temperature interval between steps 4 and 5; for the crystalline polystyrene prepared above, this lies between 205°C and 215°C.

Example 3.31 Stereospecific Polymerization of Butadiene with Ziegler-Natta-Catalysts: Preparation of cis-1,4-Polybutadiene
Safety precautions: Before this experiment is carried out, Sect. 2.2.5 must be read as well as the material safety data sheets (MSDS) for all chemicals and products used.

Caution: Aluminum sesquichloride must be handled under total exclusion of atmospheric oxygen and moisture (see Example 3.28).

A trap, to be used for condensing the butadiene (see Fig. 3.3), is dried for 1 h at 120°C, evacuated, and filled with pure dry nitrogen. A butadiene cylinder is

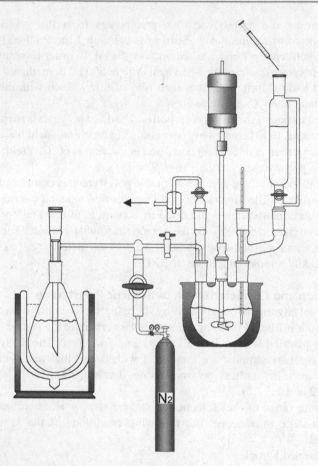

Fig. 3.3 Setup for the polymerization of butadiene

attached to the three-way tap via a P_2O_5 drying tube. The air in the tubing and drying tube between the three-way tap and cylinder is displaced by flushing with butadiene. The trap is then cooled in a dry ice/methanol bath and 20 g (33 ml) of butadiene condensed in.

Toluene is refluxed for a day over potassium. 200 ml are then distilled under nitrogen into a dry dropping funnel with pressure-equalizing tube. It is well stoppered for storage. 600 mg (1.68 mmol) of cobalt(III) acetylacetonate are weighed into a dry tube with attached stopcock. This is evacuated, filled with dry nitrogen, and 20 ml of dry toluene are then introduced through the bore of the tap with the aid of a hypodermic syringe. The closed tube is shaken until the cobalt compound has dissolved.

The polymerization is carried out in a 500 ml four-necked flask, fitted with stirrer, thermometer, adapter, and nitrogen inlet (see Fig. 3.3). The individual parts of the apparatus are previously dried for 1 h at 120°C, and while still hot are assembled as quickly as possible; the whole is then evacuated and filled with P_2O_5-dried nitrogen three times.

Finally, the dropping funnel containing the 200 ml of pure toluene is mounted and fitted with a self-sealing closure. Using a hypodermic syringe that has been dried at 120°C and flushed with dry nitrogen, 1.0 ml (4.6 mmol) of $Al_2Et_3Cl_3$ are injected through the self-sealing closure into the toluene in the dropping funnel; the syringe is washed out with the toluene by several strokes of the piston. The piston of the syringe should be smeared with a little paraffin in order to prevent its seizure in the cylinder. It should be washed with a mixture of 2-propanol and decalin (1:1 vol/vol) immediately after use. The toluene solution of the aluminum sesquichloride is now run into the flask, warmed on a water bath to 20–25°C, and 10 ml of the cobalt (III) salt solution added through the dropping funnel by means of a second hypodermic syringe, prepared in the same way as the first. The color now changes from green to gray-brown and the temperature rises to about 40°C. The two components of the catalyst are allowed to react with each other for 10 min. The cold bath round the butadiene is removed and the 20 g (0.37 mol) of butadiene are allowed to evaporate into the polymerization flask with stirring; this takes about 1 h. The reaction mixture is held at 20°C for another hour and a toluene solution of an antioxidant (e.g., 0.2% 2,6-di-*tert*-butyl-4-methylphenol) is then added to prevent crosslinking reactions during work-up. The polymer is precipitated by dropping the highly viscous solution into a fivefold amount of methanol. After settling, the supernatant liquid is decanted from the polymer which is then broken down into small pieces and stirred with fresh methanol for a few minutes; this purification process is repeated twice more, the polymer finally being filtered and dried in vacuum at 40°C. Yield: >90%. About 1 g of the polybutadiene is reprecipitated from toluene solution with methanol. The limiting viscosity number is determined in cyclohexane at 20°C and the molecular weight derived. The polymer may be further characterized as described in Example 3.21. Main applications of polybutadiene are in tires and in polystyrene/polybutadiene blends (high impact polystyrene, see Example 5.23).

3.3.2 Polymerization with Metallocene Catalysts

Metallocences are sandwich-like π-complexes derived, e.g., from dicyclopentadienylzirconium dichloride. At first, they were used only as soluble model systems (in combination with diethylaluminum chloride) in order to study Ziegler-Natta polymerizations. The discovery by Sinn and Kaminsky that, by using methylalumoxane (MAO), exceptionally high activities of polymerization were achieved and the results of Brintzinger that an additional bridging (linking) of the ligands in the metallocene results in outstandingly high regio- and stereoselectivities, make this class of catalysts very interesting, especially for industrial applications.

For example, it is possible to synthesize isotactic as well as syndiotactic polypropylene in high configurational purity and high yields. The same holds for syndiotactic polystyrene. Furthermore, metallocene catalysts open the possibility to absolutely new homopolymers and copolymers like, e.g., cycloolefin copolymers

(COC) and even (co)polymers of polar monomers. The simplest metallocene catalyst consists of two components. The first one is a π-complex (the actual metallocene) that can be bridged via a group X and therefore can become chiral:

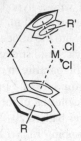

$$X = e.g., (CH_3)_2Si \text{ or } CH_2$$
$$M = e.g., Zr, Ti$$

The second component is a special alumina-organic compound, methylalumoxane (MAO), that is prepared by partial hydrolysis of trimethylaluminum and that contains linear as well as cyclic structures in the molecules.

$$n\,Al(CH_3)_3 \; + \; n\,H_2O \xrightarrow[-2n\,CH_4]{} \left[Al\underset{CH_3}{\overset{O}{|}}\right]_n$$

The mixture of metallocene and co-catalyst is soluble. Its active center, which is chiral, induces with a very low rate of defects only one type of monomer linkage ("single site catalysts"). That is why high activities (some 1,000 kg polymer/g zirconium), high molecular weights, narrow molecular weight distribution, and high steric homogeneity are achieved.

Example 3.32 Metallocene-Catalyzed Polymerization of Propylene to Highly Isotactic Polypropylene in Organic Suspension

Safety precautions: Before this experiment is carried out, Sect. 2.2.5 must be read as well as the material safety data sheets (MSDS) for all chemicals and products used.

(a) Solvent

Caution: Refluxing of toluene over sodium/potassium alloy must be done under permanent supervision.

Highly dried toluene is the most useful reaction medium. It is obtained according to the following procedure. Pre-dried toluene (over molecular sieve) is refluxed over a liquid sodium/potassium alloy (5–10 ml for 2 l of toluene) for 4–5 days. An alternative method is the addition of *n*-butyllithium in small portions (ca. 10 ml for 2 l toluene) which can be visualized with benzophenone as indicator. When the toluene is sufficiently dried (change of color) it is distilled off and stored under argon (The used syringes or pipettes have to be flushed prior to use with argon; after use they have to be cleaned immediately with toluene).

(b) Methylalumoxane (MAO)

Caution: Working with MAO has to be done under rigorous safety precautions! Wearing of safety goggles and protective gloves is a must. MAO is highly reactive! It reacts with moisture traces on the skin. If MAO is spilled, the contaminated area has to be covered with sand to prevent a fire. Then, MAO is hydrolyzed by the careful addition of 2-propanol or ethanol, followed by water (see also Example 3.28).

MAO is usually available as a 10% solution in toluene and is stable for approx. 2 months. If MAO should be stored over a longer period of time, the solvent has to be distilled off. Vacuum distillation at 40°C is therefore the method of choice (the cooling trap is cooled with liquid nitrogen!). The more and more viscous MAO solution should be stirred all the time. The solid MAO is then stored under argon.

(c) Polymerization of Propylene

A 250 ml flask with the equipment shown in Fig. 3.2 is flushed with argon. Then, 90 ml of the highly dried toluene are added with a syringe. Now are pipetted in under stirring: 5 ml of a 10% solution of MAO followed by 5 ml of a solution of rac-ethylene-bis(4,4,5,5′,6,6′,7,7-tetrahydro-1,1′-indenyl)zirconium dichloride (The used syringes or pipettes have to be flushed prior to use with argon; after use they have to be cleaned immediately with toluene).

Then the setup is evacuated until the vapor pressure of toluene is reached. Next, dry propylene is passed in until normal pressure is attained. The polymerization starts after approx. 15 min and the reaction mixture becomes turbid and might also warm up. A small flow of propylene into the reaction mixture is still maintained. After 1 h, the propylene flow is stopped and the polymerization is terminated by the addition of 10 ml ethanol. Then, the fluffy reaction mixture is added under stirring to 300 ml ethanol whereby the polymer precipitation is completed. MAO and metallocene residues are removed by the addition of 25 ml of a 10% HCl solution and stirring for 1 h. Finally, the solid polymer is filtered over a Büchner funnel, washed with ethanol and dried in vacuum at 40–60°C to constant weight (~12 h). Yield: 1–2 g. The obtained polypropylene is highly isotactic (determination by ^{13}C-NMR spectroscopy) and has a molecular weight of approx. 30,000.

3.3.3   Ring-Opening Metathesis Polymerization (ROMP)

Ring-opening metathesis polymerization (ROMP) is a transition metal alkylidene-triggered process in which cyclic olefins, whether mono-, bi- or multicyclic, undergo ring-opening and are concomitantly joined together to form a polymer chain. ROMP is thus a chain-growth polymerization and belongs, together with Ziegler-Natta-type polymerizations and group transfer polymerizations, to the family of polyinsertions. The mechanism is based on olefin metathesis. The ring-opening process occurs at the most stable site of the monomer, i.e. at the double bond.

Illustration of the ROMP process

It is important to note that, as all metathesis reactions, all steps are in principle reversible. Furthermore, the double bond of the monomer is formally preserved, resulting in one double bond per repeat unit. ROMP may therefore be regarded as an inversed ring-closing metathesis (RCM) reaction. ROMP is driven by the thermodynamics that are entailed with the reduction in ring-strain that occurs during incorporation of the monomer into the growing chain. In general, the ring-opening of 3-, 4-, 8- and larger-membered rings is energetically favored. Finally, since the polymer itself still contains double bonds, i.e., one per repeat unit, an intramolecular chain transfer reaction, i.e., a backbiting process may occur, leading to cyclic oligomers/polymers.

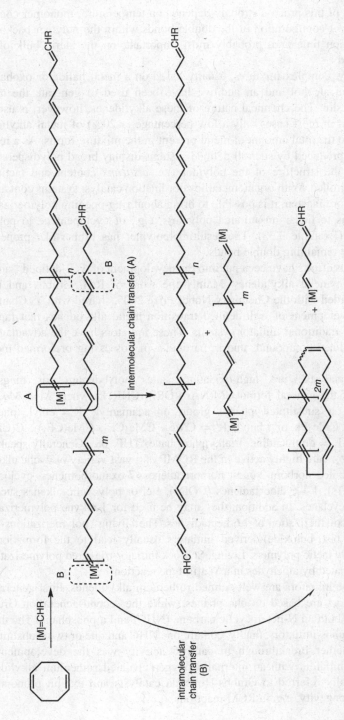

Inter- (A) and intramolecular (B) chain transfer reactions in ROMP.

The extent of this process strongly depends on temperature, monomer concentration, *cis/trans* configuration of the double bonds within the polymer backbone, solvent, reaction time and, probably most important, on the steric bulk of the monomer used.

Historically, complex mixtures, usually based on a metal halide or oxohalide, a tin alkyl, an alcohol and an additive have been used to generate the metal alkylidene in situ. The chemical nature of these alkylidenes, however, is usually ill defined and in most cases only a low percentage (<20%) of metal alkylidene with respect to the total amount of metal present in the mixture forms. As a result, the polymers produced by such ill-defined systems display broad polydispersities. Furthermore, the structure of the polymer (i.e. *cis/trans* content and tacticity) cannot be controlled. With organometallic coordination catalyst systems containing molybdenum or tungsten, it is possible to bring about ring-opening polymerization of cycloolefins to linear unsaturated polymers, e.g., of cyclopentene to poly(1-pentenylene) (Example 3.33). The resulting polymer has rubber-like properties because of the remaining double bonds.

Tremendous efforts have been put into the development of well-defined "single-site" transition metal alkylidenes. Mainly the work of R.H. Grubbs and R.R. Schrock (awarded with the Chemistry Nobel Prize 2005, shared with Y. Chauvin) led to the development of well-defined transition metal alkylidenes that rapidly outrivaled the traditional initiator systems. These initiators have the advantage of being well-defined compounds and in particular of possessing preformed metal-alkylidenes.

"Schrock-catalysts" are high-oxidation state molybdenum (or tungsten) alkylidenes of the general formula $M(NAr')(OR')_2(CHR)'L$, where $M = Mo, W$; $Ar' = $ phenyl, a substituted phenyl group or adamantyl; $R = $ ethyl, phenyl, trimethylsilyl, CMe_2Ph or t-butyl; $R' = CMe_3$, CMe_2CF_3, $CMe(CF_3)_2$, $C(CF_3)_2$, aryl, etc. and $L = $ quinuclidine, trialkylphosphane, THF, etc. Generally speaking, Schrock initiators are highly active in the ROMP of a vast variety of cyclic alkenes such as substituted norborn-2-enes, norbornadienes, 7-oxanorbornenes, cyclooctatetraenes (COTs), 1,4-cyclooctadienes (CODs), etc. or polycyclic alkenes such as certain quadricyclanes. In addition, they may be used for 1-alkyne polymerization and the cyclopolymerization of 1,6-heptadiynes. The "living" polymerizations triggered by Mo-bis(t-butoxide)-derived initiators usually lead to the formation of all-*trans*, highly tactic polymers. Living, Schrock initiator-triggered polymerizations are best terminated by aldehydes in a Wittig-type reaction.

Grubbs-type initiators are well-defined ruthenium alkylidenes. First-generation Grubbs initiators are based on phosphanes, while the second-generation Grubbs initiators bear both an N-heterocyclic carbene (NHC) and a phosphane. The third-generation Grubbs-initiators, finally, contain one NHC and one or two (substituted) pyridines. Another breakthrough in catalyst activity was the development of Grubbs-type initiators with an internally oxygen-chelated ruthenium alkylidene. They are usually referred to Grubbs-Hoveyda catalysts and exhibit pronounced stability and longevity, e.g. in RCM reactions.

1st-generation   2nd-generation   3rd-generation

Selection of Grubbs-type initiators

Compared to molybdenum- or tungsten-based Schrock catalysts, the reactivity of ruthenium-based Grubbs catalysts is somewhat different. Reactivity in $RuCl_2(PR_3)_2(CHPh)$ may efficiently be tuned rather via the use of different phosphanes than by the nature of the alkylidene moiety or by substitution of the chlorides by other, more electron-withdrawing groups. The stability as well as the reactivity order that can be deduced there from is $PPh_3 < PBz_3 < PCyPh_2 < PCy_2Ph < P\text{-}i\text{-}Bu_3 < P\text{-}i\text{-}Pr_3 < PCy_3$. In terms of polymer structure, ROMP of norborn-2-enes and norbornadienes using ruthenium-based systems generally results in the formation of polymers that, in most cases, predominantly contain *trans*-vinylene units. Polymerizations initiated by Grubbs-type initiators are best terminated by the use of ethyl vinyl ether, yielding methylidene-terminated polymers.

Example 3.33 Poly(1-Pentenylene) by Metathesis Polymerization of Cyclopentene with a Ziegler-Natta-Catalyst in Solution

Safety precautions: Before this experiment is carried out, Sect. 2.2.5 must be read as well as the material safety data sheets (MSDS) for all chemicals and products used.

Caution: Aluminum alkyls must be handled under total exclusion of oxygen and moisture (see Example 3.28).

(a) Preparation of $W(OCH_2CH_2Cl)_2Cl_4$

2.64 g (6.67 mmol) of WCl_6 and 50 ml of dry toluene are transferred under nitrogen to a 100 ml standard apparatus (round bottom flask, several inlets for stirrer, reflux condenser, nitrogen flux or vacuum, thermometer, heating bath) three-necked flask, equipped with thermometer, nitrogen inlet, dropping funnel with a pressure-equalizing tube, and magnetic stirrer. A solution of 1.07 g (0.9 ml, 13.33 mmol) of 2-chloroethanol in 15 ml of dry toluene is added dropwise at room temperature with stirring, over a period of about 30 min. The temperature should not be allowed to exceed 35°C. Stirring is continued for 1 h. The brown solution of the tungsten compound (0.1 M) is stored under nitrogen.

(b) Preparation of a 0.5 M Solution of (C$_2$H$_5$)$_2$AlCl in Toluene

30 ml of dry toluene and 2 ml (1.95 g, 16.13 mmol) of diethylaluminum chloride are placed in a 50 ml round-bottomed flask under nitrogen. The solution is well mixed and stored under nitrogen.

(c) Polymerization of Cyclopentene

450 ml of pre-dried toluene are placed in a 500 ml flask equipped with stirrer, thermometer, nitrogen inlet, and condenser for distillation. 125 ml of the toluene are distilled off under a gentle stream of nitrogen. The flask is cooled to −15°C under a slight excess pressure of nitrogen, 50 ml (38.6 g, 0.567 mol) of dry cyclopentene and, finally, 2 ml of the 0.1 M solution of W(OCH$_2$CH$_2$Cl)$_2$Cl$_4$ and 2 ml of the 0.5 M solution of (C$_2$H$_5$)$_2$AlCl are added. The polymerization commences immediately as can readily be seen from the marked increase in viscosity. The reaction temperature is kept at −10°C by external cooling. The polymerization is stopped after 5 h. To deactivate the catalyst a mixture of 0.5 g of 2,6 di-*tert*-butyl-4-methylphenol and 1 ml of ethanol dissolved in 15 ml of toluene are added with stirring; the solution is rapidly decolorized. The poly(1-pentenylene) is isolated by precipitation in about 1 l of propanol and drying to constant weight in vacuum at 50°C. Yield: about 70%. The solution viscosity is determined in toluene and the proportions of cis and trans double bonds estimated from the IR spectrum (see Table 3.9).

Example 3.34 ROMP of norborn-5-ene-2-methanol with a Grubbs-Type Initiator in Solution

Safety precautions: Before this experiment is carried out, Sect. 2.2.5 must be read as well as the material safety data sheets (MSDS) for all chemicals and products used.

All manipulations are carried out in a N$_2$-mediated dry box or under standard Schlenk conditions similar as described in example 3.19. Prior to use, CH$_2$Cl$_2$ is dried and deoxygenated by distillation from CaH$_2$ under Ar.

0.002 mmol of RuCl$_2$(PCy$_3$)$_2$(CHPh) (1st-generation Grubbs catalyst, Sigma-Aldrich or Materia Inc.) or RuCl$_2$(3-Br-pyridine)(IMesH$_2$)(CHPh) (3rd-generation Grubbs catalyst, IMesH$_2$ = 1,3-dimesitylimidazolin-2-ylidene, Sigma Aldrich or Materia Inc., or synthesized according to T. L. Choi, R. H. Grubbs, *Angew. Chem. Int. Ed.* **2003**, *42*, 1743–1746) are dissolved in 1 ml of CH$_2$Cl$_2$. Separately, 28 mg of norbornene-5-ene-2-ylmethanol (0.225 mmol, from Sigma-Aldrich) are dissolved in 2 ml of CH$_2$Cl$_2$ (monomer: initiator = 100:1). Under vigorous stirring, the initiator solution is quickly added to the monomer solution. Stirring is continued for another 20 min, then, ethyl vinyl ether is (0.5 ml) added. After another 20 min, the reaction mixture is poured onto 20 ml of pentane, the polymer is isolated by filtration and washed with diethyl ether. Isolated yield: >90%.

The *cis/trans* ratio of the double bonds can be estimated by IR-spectroscopy or determined quantitatively by ^{1}H-NMR spectroscopy. Here, the *cis:trans* ratio of the polymer can be estimated via integration of the olefinic protons. These can be found at $\delta = 5.6–5.3$ ppm and $\delta = 5.4–5.1$ ppm for the *trans-* and *cis*-double bonds, respectively. With the initiators used here, a *cis:trans* ratio of roughly 55:45 is obtained.

The solution viscosity can be determined in chloroform or THF. Additionally, the average degree of polymerization can be determined by ^{1}H-NMR-based end group analysis in THF-d$_8$ by determining the ratio of the phenyl protons over the protons of the double bond of the repeat unit.

3.4   Copolymerization

By copolymerization we understand the mutual polymerization of two or more monomers, with the resulting macromolecules containing repeating units of all the participating monomers. Depending on the distribution of the monomers in the macromolecules one differentiates four types of copolymers (nomenclature of copolymers see Sect. 1.2):
– Statistical copolymers
– Alternating copolymers
– Block copolymers
– Graft copolymers
Conventional polymerization methods yield macromolecules mostly with random (statistical) or nearly statistical and only very seldom with alternating distribution of the monomer units (see Sect. 3.4.1). Special methods are required in order to synthesize block and graft copolymers (Sect. 3.4.2).

3.4.1   Statistical and Alternating Copolymerization

When synthesizing random (statistical) and alternating copolymers two phenomena have to be kept in mind:
– Monomers that readily homopolymerize are not necessarily able to copolymerize well with another monomer. On the other hand, there are some monomers that are not able to homopolymerize but which can be induced to copolymerize.
– The tendency of monomers to copolymerize is strongly dependent on the nature of the growing chain end, e.g., on the type of the initiation. In nonradical polymerizations even the composition of the initiator or catalyst is important.
Copolymerization of two monomers has been very thoroughly investigated, but copolymerization of three or more compounds presents considerable difficulties on account of the multiplicity of variables. Nevertheless, terpolymers (from three monomers) are of technical importance and are produced on large scales. We limit ourselves here to the case of mutual polymerization of two monomers; there are then essentially four different possible propagation reactions: monomer M$_1$ can

react with a polymer chain whose growing chain end (radical or ionic) has been formed either from monomer M_1 or from monomer M_2; similarly for monomer M_2:

$$-M_1{}^{\bullet} + M_1 \xrightarrow{k_{11}} -M_1{}^{\bullet}$$

$$-M_1{}^{\bullet} + M_2 \xrightarrow{k_{12}} -M_2{}^{\bullet}$$

$$-M_2{}^{\bullet} + M_2 \xrightarrow{k_{22}} -M_2{}^{\bullet}$$

$$-M_2{}^{\bullet} + M_1 \xrightarrow{k_{21}} -M_1{}^{\bullet}$$

where k_{11}, k_{12}, k_{21}, k_{22} represent the rate constants of these propagation reactions, the first number in the subscript indicating the type of active center, and the second, the nature of the monomer that is adding to it. As long as the chain length is relatively large, the propagation reactions are rate-determining; as in homopolymerization, a quasistationary state is set up. It can then be shown that at low conversion ($<10\%$), the relative rates of consumption of the two monomers, and thus their relative amounts (in mol) in the copolymer, M_1/M_2, can be described by the following copolymerization equation:

$$\frac{m_1}{m_2} = \frac{[M_1]}{[M_2]} \cdot \left(\frac{r_1[M_1] + [M_2]}{r_2[M_2] + [M_1]}\right) \tag{3.18}$$

in which $r_1 = k_{11}/k_{12}$ and $r_2 = k_{22}/k_{21}$ are termed the reactivity ratios (copolymerization parameters). $[M_1]$ and $[M_2]$ represent the molar concentrations of monomer in the reaction mixture. The composition of a copolymer thus depends on the monomer feed ratio in the polymerization mixture. Since the parameters r_1 and r_2 are ratios of rate constants, they express the tendency of the growing chains to add either the same or the other monomer. If r is close to 1 it follows that a particular active chain end adds molecules of monomers M_1 and M_2 at random with approximately equal facility; $r > 1$ means that the addition of a monomer to a chain with the same end unit is strongly preferred. Reactivity ratios, being quotients of two rate constants, are not very temperature-dependent, but of course are strictly valid only for a particular polymerization temperature which must, therefore, always be indicated.

In some cases, the reactivity of the growing chain end depends on the nature of the last but one monomer unit. So, eight propagation constants have to be considered. This so-called penultimate effect can be the reason why the binary copolymerization cannot be described precisely enough by Eq. 3.18.

By rearranging Eq. 3.18 and inserting known values of the reactivity ratios r_1 and r_2, one can calculate the molar ratio of monomers that must be used in order to arrive at a copolymer with a chosen composition ($f = m_1/m_2$):

$$\frac{[M_1]}{[M_2]} = \frac{1}{2 \cdot r_1}\left\{(f-1) + \left[(f-1)^2 + 4 \cdot r_1 \cdot r_2 \cdot f\right]^{0.5}\right\} \tag{3.19}$$

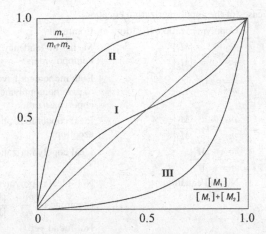

Fig. 3.4 Copolymerization diagram for the system styrene (M_1)/methyl methacrylate (M_2). I: radical copolymerization at 60°C; $r_1 = 0.52$, $r_2 = 0.46$; II: cationic copolymerization initiated by $SnBr_4$ at 25°C; $r_1 = 10.5$, $r_2 = 0.1$; III: anionic copolymerization initiated by Na in liquid NH_3 at −50°C; $r_1 = 0.12$, $r_2 = 6.4$

The dependence of the composition of the copolymer on the proportions of the monomers in the initial mixture can be portrayed graphically in a so-called copolymerization diagram (Fig. 3.4). The mole fraction of one of the two monomeric units in the resulting copolymer is plotted against the mole fraction of this monomer in the original reaction mixture; the curve can also be calculated from the reactivity ratios by means of Eq. 3.18.

As can be seen from Fig. 3.4, it is very rare for the polymer composition to correspond to that of the monomer mixture. For this reason the composition of the monomer mixture, and hence also that of the resulting polymer, generally changes as the copolymerization proceeds. Therefore, for the determination of the reactivity ratios one must work at the lowest possible conversion. In practical situations where, for various reasons, one is forced to polymerize to higher conversions, this leads to a chemical non-uniformity of the copolymers in addition to the usual non-uniformity of molecular weights.

If one wishes to attain high conversion at constant composition, the more reactive monomer must be added in a programmed manner. The procedure is as follows: from the copolymerization diagram (or from the reactivity ratios) one obtains the monomer composition that will lead, at low conversion, to the desired copolymer composition. A conversion/time curve is drawn up for this system and the composition of the copolymer determined from time to time. From this, one can find how much of the more reactive monomer is to be added at given times during the polymerization in order to maintain an approximately constant composition (see Example 3.39). Special computer software has already been developed for this.

From the copolymerization Eq. 3.18 and the copolymerization diagram (Fig. 3.4) some special cases can be derived (see Table 3.10). When the compositions of the

Table 3.10 Special cases of the copolymerization equation

r_1	r_2	$r_1 r_2$	Copolymerization equation	Comment
>0	~ 0	$\ll 1$	$\dfrac{m_1}{m_2} \approx r_1 \cdot \dfrac{[M_1]}{[M_2]} + 1$	M_2 has no tendency to give homopolymers
~ 0	~ 0	~ 0	$\dfrac{m_1}{m_2} \approx 1$	Both monomers have no tendency to give homopolymers→alternating copolymerization
$1/r_2$	$1/r_1$	~ 1	$\dfrac{m_1}{m_2} \approx r_1 \cdot \dfrac{[M_1]}{[M_2]}$	Ideal copolymerization without azeotrope
~ 1	~ 1	~ 1	$\dfrac{m_1}{m_2} \approx \dfrac{[M_1]}{[M_2]}$	Ideal copolymerization with azeotrope
<1	<1	<1	Normal equation	Non-ideal copolymerization, azeotrope at $\dfrac{m_1}{m_2} = \dfrac{[M_1]}{[M_2]} = \dfrac{r_2 - 1}{r_1 - 1}$
>1	>1	>1		Not found yet

monomer feed and the resulting copolymer are the same, one speaks of "azeotropic" copolymerization; Eq. 3.18 then reduces to:

$$\frac{m_1}{m_2} = \frac{[M_1]}{[M_2]} = \frac{r_2 - 1}{r_1 - 1} \tag{3.20}$$

In such cases the polymerization can be taken to relatively high conversion without change in composition of the copolymer formed (see Example 3.37). In the copolymerization diagram the azeotrope corresponds to the intersection point of the copolymerization curve with the diagonal. For example, from Fig. 3.4 it may be seen that in the radical copolymerization of styrene and methyl methacrylate the azeotropic composition corresponds to 53 mol% of styrene.

For the case where $r_2 = 0$, $k_{22} = 0$, M_2 does not add to growing chains having an M_2 end unit, so that homopolymerization of M_2 is also impossible. Eq. 3.18 then reduces to:

$$\frac{m_1}{m_2} = r_1 \cdot \frac{[M_1]}{[M_2]} + 1 \tag{3.21}$$

The resulting copolymer can contain at most 50 mol% M_2 units, even at high concentration of M_2 in the monomer mixture. This applies, for example, to maleic anhydride, and especially to such "monomers" as molecular oxygen or sulfur dioxide where, independent of the comonomer used, essentially alternating copolymers are obtained, with almost equal amounts of the two components.

Finally there are also cases in which two "monomers" copolymerize, both "monomers" being themselves incapable of homopolymerization ($k_{11} = k_{22} = 0$); this results in strict alternation of monomeric units of M_1 and M_2 in the copolymer (see Example 3.38).

If radical copolymerization is conducted in the presence of complexing agents it is sometimes possible to force monomers into forming alternating copolymers when they would otherwise give random copolymers. Lewis acids such as zinc chloride are especially suitable for this, also organoaluminum compounds. The reaction mechanism is not yet fully clarified; it is assumed that the additives form electron donor/acceptor complexes with the monomer or the active chain ends, leading to alternating insertion of the two monomers into the growing polymer chains.

It is important to note that the tendency of a monomer towards polymerization and therefore also towards copolymerization is strongly dependent on the nature of the growing chain end. In radical copolymerization the composition of the copolymer obtained from its given monomer feed is independent of the initiating system for a particular monomer pair, but for anionic or cationic initiation this is normally not the case. One sometimes observes quite different compositions of copolymer depending on the nature of the initiator and especially on the type of counterion. A dependence of the behavior of the copolymerization on the used catalyst is often observed with Ziegler-Natta or metallocene catalysts.

In the radical copolymerization of styrene and methyl methacrylate the reactivities are about the same, but for anionic initiation of an equimolar mixture of the two monomers the incorporation of methyl methacrylate is much preferred, while for cationic initiation of the same mixture the copolymer contains mostly styrene (see Fig. 3.4 and Example 3.35). Conversely, monomer pairs whose copolymerization behavior is well known can be used to test new initiator systems in order to draw conclusions about the polymerization mechanism.

As already indicated, values of reactivity ratios apply only to a given pair of monomers. There have been many attempts, especially for radical copolymerization, to derive parameters from the reactivity ratios, representing individual constants for each monomer which can be related to the structure of the monomers, and can be used to make predictions.

The basis of the scheme developed particularly by Alfrey and Price is the assumption that the activation energies of the propagation reactions, and hence the related rate constants and reactivity ratios, are governed primarily by resonance effects and by the interaction of the charges on the double bonds of the monomers with those in the active radicals. Accordingly, the rate constant of the reaction between a radical and a monomer is represented by:

$$k_{12} = P_1 \cdot Q_2 \cdot e^{-e_1 \cdot e_2} \tag{3.22}$$

where P_1 denotes the reactivity of a radical having an M_1 unit at the reactive end, Q_2 is the reactivity of the monomer M_2, and e_1 and e_2 are proportional to the charges on the corresponding species. It is assumed that the e value of the monomer is the same as that of the corresponding radical. Hence, it follows that:

$$r_1 = \frac{Q_1}{Q_2} \cdot e^{-e_1(e_1-e_2)} \qquad r_2 = \frac{Q_2}{Q_1} \cdot e^{-e_2(e_2-e_1)} \tag{3.23}$$

and hence

$$e_1 = e_2 \pm \sqrt{-ln(r_1 \cdot r_2)} \tag{3.24}$$

$$Q_1 = \frac{Q_2}{r_2} \cdot e^{\pm e_2 \sqrt{-ln(r_1 \cdot r_2)}} \tag{3.25}$$

On this basis, values of Q and e can be calculated for each monomer, so long as two arbitrary reference values are assumed. For this purpose Price took the values for styrene as $Q = 1.0$ and $e = -0.8$. Q and e values can then be obtained for all monomers that are copolymerizable with styrene. These monomers in their turn can serve as reference compounds for further determinations with other monomers that do not copolymerize with styrene. One of the main advantages of the so-called Q,e scheme is that the data can be presented in the form of a diagram instead of very complex tables of reactivity ratios.

Obviously the precision of this procedure is not very great, since the assumptions underlying the calculations of Q and e values can be regarded as best as semiquantitative. However, it has been shown that when the reactivity ratios are back-calculated from the Q,e values, quite good agreement is obtained with the experimental values, so that it is possible to make useful predictions of reactivity ratios for monomer pairs not previously investigated. On the other hand, it is questionable whether any theoretical conclusions should be drawn from the Q and e values concerning the behavior of different monomers. Thus, the e value of a monomer as a measure of the charge should be dependent on the nature of the solvent used for polymerization; but it has been shown, for example, that the copolymerization of styrene with methyl methacrylate in solvents of different dielectric constant (benzene 2.28, methanol 33.7, acetonitrile 38.8) give the same reactivity ratios, which should not really be the case if the foregoing assumptions are correct. A broadening of the basis of the Q,e scheme has therefore been suggested by various people. These further considerations permit somewhat deeper theoretical conclusions.

In most cases one of the two monomers is preferentially incorporated into the copolymer, so that the composition of the monomer mixture changes with increasing conversion, as also does the composition of the polymer chains. Therefore it is important, when determining reactivity ratios, to work at low conversions, so that at the end of the experiment the ratio of monomer concentrations is essentially the same as at the beginning. However, if high conversions are needed for preparative purposes, constant composition can be achieved by adding the more reactive monomer in a programmed manner (see Example 3.39).

The determination of the reactivity ratios requires a knowledge of the composition of the copolymers made from particular monomer mixtures; numerous analytical methods are available (see Sect. 2.3.2). In principle, it is possible to calculate r_1 and r_2, using Eq. 3.18, from the composition of only two copolymers that have been obtained from two different mixtures of the two monomers M_1 and M_2. However, it is more precise to determine the composition of the copolymers from several monomer mixtures and to calculate, for each individual experiment, values of r_2 that would correspond to arbitrarily chosen values of r_1 from the rearranged copolymerization equation:

$$r_2 = \frac{[M_1]}{[M_2]} \cdot \left\{ \frac{m_2}{m_1} \left(1 + \frac{[M_1]}{[M_2]} \cdot r_1 \right) - 1 \right\} \tag{3.26}$$

r_2 is then plotted against r_1, for each experiment to obtain a series of lines intersecting at the actual values of r_2 and r_1. In practice the lines do not intersect precisely at a point so that r_1 and r_2 are taken as the center of the smallest area that is cut or touched by all the lines. The size of this area allows an estimate of the limits of error.

A simpler method for determining the reactivity ratios is that of Fineman and Ross, in which the copolymerization Eq. 3.18 is rearranged to:

$$\frac{F}{f} \cdot (f - 1) = r_1 \cdot \frac{F^2}{f} - r_2 \tag{3.27}$$

or

$$\frac{f - 1}{F} = -r_2 \cdot \frac{f}{F^2} + r_1 \tag{3.28}$$

where $f = \frac{m_1}{m_2}$ and $F = \frac{[M_1]}{[M_2]}$

$(F/f)(f-1)$ is plotted against (F^2/f) to give a straight line (Eq. 3.27), of slope r_1 and intercept $-r_2$ on the ordinate. Alternatively, a plot of $(f-1)/F$ against f/F^2 according to Eq. 3.28 gives a line of slope $-r_2$ and intercept r_1 on the ordinate. The limits of error of r_1 and r_2 can be determined from the scatter of the points by the method of least squares. Besides these two classical methods, a number of further publications has recently appeared concerning the evaluation of copolymerization data. Among these, worthy of mention are not only the method of Kelen and Tüdös which is a graphical linear procedure, and perhaps the most ambitious but also the most exact method of Tidwell and Mortimer.

Under the condition that the reaction capability is only affected by the nature of the last monomer unit of the growing polymer chain end (terminal model, Bernoulli statistics), the copolymerization equation can be transformed according to Kelen and Tüdös:

$$\eta = r_1 \cdot \xi - \frac{r_2 \cdot (1 - \xi)}{\alpha} \tag{3.29}$$

where α is a arbitrarily chosen constant ($\alpha > 0$). The variable ξ can only adopt values between 0 and 1; η is plotted against ξ to a straight line with $(-r_2/\alpha)$ as the ordinate ($\xi = 0$) and the reactivity ratio r_1 where $\xi = 1$. Choosing the constant as the geometric middle of the smallest (F_{min}) and the largest (F_{max}) F value, a homogeneous distribution of the ξ-values of the interval (0.1) is achieved:

$$\alpha = \sqrt{F_{min} \cdot F_{max}} \qquad (3.30)$$

Because Eq. 3.26 is based on the differential form of the copolymerization equation, it is strongly valid only for infinitely low conversions, but this cannot be realized in real life. For higher conversions one has to start with an integrated form of the copolymerization equation. Fortunately, Kelen and Tüdös developed an elegant method of iteration. It allows the use of the earlier suggested method without the loss of graphical clearness.

In the literature one can find extensive compilations of reactivity ratios for numerous monomer pairs. For evaluation of the copolymerization experiments and for calculating the reactivity ratios, there is now extensive software available.

In copolymerizations of three monomers there are nine growing steps to be taken into account. From these, six reactivity ratios can be derived. They are difficult to obtain from terpolymerizations and are therefore taken from binary copolymerizations.

The statistical (random) copolymerization is an often used possibility in industry to change the properties of a polymeric compound. By incorporation of small amounts of a comonomer, the mechanical properties (hardness, stiffness) are almost maintained but other properties can intentionally be changed. Thus, by incorporation of small amounts of vinyl acetate into poly(vinyl chloride), the glass transition temperature is slightly lowered and an internal plasticization is achieved. The ability of synthetic fibers to take up dyes can be improved by incorporation of small amounts of comonomers with functional groups that confer a greater affinity towards organic dyestuffs. Polar groups also improve the printability of films and the stability of aqueous copolymer lattices. Significant differences in solubility between copolymers and the corresponding homopolymers are usually observed. The incorporation of larger amounts of comonomers leads then to larger changes in the mechanical properties. Thus, while polyethylene and isotactic polypropylene are crystalline polymers with the characteristic properties of a thermoplastic material, copolymers of ethylene and propylene are over a wide range of composition amorphous, rubber-elastic products.

Alternating copolymerization is a rare event. But when it occurs, the resulting copolymers exhibit unusual properties provided a 100% alternation was achieved (see Sect. 2.4)

Crosslinking copolymerization is also of great practical importance. Such copolymerization is achieved by taking compounds with two or more polymerizable double bonds and copolymerizing them with simple unsaturated monomers to form three-dimensional network materials. For example, crosslinked polystyrene is prepared by copolymerizing styrene with small amounts of divinylbenzene (Example 3.41). These crosslinked polymers are insoluble and nonfusible; however, depending on the degree of crosslinking, they swell to a limited extent in organic solvents and find application, for example, in the preparation of ion-exchange resins (see Sect. 5.2). The copolymerization of unsaturated polyesters (made from maleic or fumaric acid) with styrene is also a crosslinking copolymerization. Some non-conjugated dienes can polymerize radically via a cyclization propagation step to yield linear chains containing cyclic repeat units (so-called cyclopolymerization). Thus, in the polymerization of acrylic anhydride this tendency to ring formation is so great that the intramolecular reaction occurs almost to the exclusion of the intermolecular crosslinking reaction:

The rates and degrees of polymerizations in radical copolymerizations conform essentially to the same laws as for radical homopolymerization (see Sect. 3.1). Raising the initiator concentration causes an increase in the rate of polymerization and at the same time a decrease in the molecular weight; a temperature rise has the same effect. However, these assertions are valid only for a given monomer composition; in many cases the copolymerization rate depends very much on monomer composition and can pass through a pronounced minimum or maximum.

As already discussed for homopolymerization, radical copolymerizations can be carried out in bulk, in solution, and in dispersion. The composition of the copolymer obtained in suspension or emulsion may be different from that obtained by polymerization in bulk or solution if one of the monomers is more soluble in water than the other. In such a case the composition of the monomer mixture in the organic phase, or in the micelles where the copolymerization takes place, is not the same as the original composition.

There are no essential differences in experimental technique required for ionic copolymerizations, as compared with ionic homopolymerizations. However, the type of initiator and the solvent have a potential influence on the course of ionic copolymerizations as well as on the composition of the copolymers so that the optimum conditions for each monomer pair must be individually determined.

Finally, it should be mentioned that there exist two other routes for the synthesis of copolymers. First the partial chemical conversion of homopolymers (see Sect. 5.1), for example, the partial hydrolysis of poly(vinyl acetate). Secondly, by homopolymerization of correspondingly built monomers. An example for these macromolecular compounds, sometimes called pseudo-copolymers, is the alternating copolymer of formaldehyde and ethylene oxide synthesized by ring-opening polymerization of 1,3-dioxolane.

It should be pointed out that in both cases the copolymers cannot be obtained via "normal" copolymerization of the corresponding monomers.

Example 3.35 Copolymerization of Styrene with Methyl Methacrylate (Dependence on Type of Initiation)

Safety precautions: Before this experiment is carried out, Sect. 2.2.5 must be read as well as the material safety data sheets (MSDS) for all chemicals and products used.

Styrene and methyl methacrylate are destabilized and dried. A mixture of 8.00 g (76.8 mmol) of styrene and 7.68 g (76.8 mmol) of methyl methacrylate is then prepared in a dry receiver (see Sect. 2.2.5.3) under nitrogen and kept in a refrigerator until required.

(a) Radical Copolymerization

16 mg (0.1 mmol) of AIBN are weighed into a 50 ml flask and the air is displaced by evacuation and filling with nitrogen, using a suitable adapter (see Sect. 2.2.5.3). 5 ml of the prepared monomer mixture are then pipetted in, the flask removed under a slight positive pressure of nitrogen and immediately closed with a ground glass stopper secured with springs. The flask is now placed in a thermostat at 60°C, after 4 h it is cooled quickly in ice, and the reaction mixture is diluted with 25 ml of toluene. The solution is added dropwise to 200–250 ml of petroleum ether; the copolymer precipitates in the form of small flakes that tend to stick together somewhat (for this reason methanol is less suitable as precipitant). The copolymer is filtered off into a sintered glass crucible, washed with petroleum ether, and dried to constant weight in vacuum at 50°C. Yield: 10–20% with respect to the monomer mixture.

(b) Anionic Copolymerization

The initiator solution is prepared as follows. Phenylmagnesium bromide is prepared from 12 g (76.5 mmol) of bromobenzene and 1.86 g of magnesium in 30 ml of pure dry THF (see Example 3.19). The apparatus must be carefully dried (all openings being protected by drying tubes); exclusion of atmospheric oxygen is not absolutely necessary. The magnesium should be converted as completely as possible by gentle warming towards the end of the reaction.

The copolymerization is carried out as follows: 5 ml of the monomer mixture (preparation see above) are pipetted into a 50 ml round-bottomed flask that has previously been flamed under vacuum and filled with nitrogen using an adapter. The flask is closed with a self-sealing closure (see Sect. 2.2.5.3) and, after cooling to −50°C, 2 ml of freshly prepared phenylmagnesium bromide solution are injected by means of a hypodermic syringe. After 90 min at −50°C the mixture is diluted with 25 ml of toluene and the copolymer precipitated by dropping the solution into 200 ml of methanol containing about 10 ml of 2 M hydrochloric acid or sulfuric acid. Further treatment is as described under (a). Yield: 10–20% with respect to the monomer mixture.

The anionic polymerization can only be carried out at low temperature, since at higher temperature the Grignard compound can also react with the ester group of methyl methacrylate.

(c) Cationic Copolymerization

A 100 ml flask is fitted with an adapter (see Sect. 2.2.5.3), flamed under vacuum using an oil pump, and filled with nitrogen. 5 ml of the prepared monomer mixture (see above) are pipetted in, followed by 40 ml of an initiator solution prepared from 50 ml of pure dry nitrobenzene (see Example 3.40) and 300 mg (2.25 mmol) of anhydrous aluminum trichloride. The flask is now removed from the adapter under a slight positive pressure of nitrogen and immediately closed with a ground glass stopper. The flask is briefly shaken and allowed to stand at room temperature for 1 h. The solution is then dropped into methanol and the copolymer worked up as described above. Yield: 40–50% with respect to the monomer mixture.

Cationic copolymerization precedes under the above conditions at such a rate that even at much lower initiator concentration almost half the monomer mixture is polymerized in less than 1 h. The reaction comes to an end after 50% conversion since with cationic initiators, the copolymer consists almost entirely of styrene units.

(d) Characterization of the Copolymers

From the behavior on precipitation of the copolymers prepared by the three different methods, it may already be suspected that, in spite of the common starting mixture, one is dealing with different kinds of polymer. In order to prove this the following solubility tests are carried out.

50 mg of each of the three copolymers, as well as a mixture of equal parts polystyrene and poly(methyl methacrylate), are warmed with about 5 ml of the following solvents:
1. Mixture of acetone and methanol (volume ratio 2:1),
2. Acetonitrile,
3. Mixture of cyclohexane and toluene (volume ratio 4:1).

If any material remains undissolved, the supernatant liquid is decanted and dropped into methanol in order to precipitate the dissolved portion.

Solvents 1 and 2 are known to be good solvents for poly(methyl methacrylate); solvent 3 readily dissolves polystyrene. The solubility tests show that the radically polymerized sample is insoluble in all three solvents. The solubility is thus different from that of both poly(methyl methacrylate) and polystyrene. The anionically polymerized product dissolves on warming in the acetone/methanol mixture and also in acetonitrile; it is insoluble in cyclohexane/toluene. The solubility is thus similar to that of poly(methyl methacrylate). For the cationically initiated polymerization the product is only slightly soluble in acetone/methanol, insoluble in acetonitrile, but very readily soluble in cyclohexane/toluene. The solubility thus resembles that of polystyrene.

In addition to the solubility tests, the structures of the polymers can be determined easily from the IR and ^{1}H-NMR spectra.

Example 3.36 Radical Copolymerization of Styrene with 4-Chlorostyrene (Determination of the Reactivity Ratios)

Safety precautions: Before this experiment is carried out, Sect. 2.2.5 must be read as well as the material safety data sheets (MSDS) for all chemicals and products used.

20 mg (0.12 mmol) of AIBN are placed in each of nine test tubes fitted with joints, together with the approximate amounts of destabilized styrene and

Table 3.11 Test series for Example 3.36

Batch #	Styrene [g]	[mmol]	p-Chlorostyrene [g]	[mmol]
1	3.0	29	0	0
2	2.6	25	0.4	3
3	2.3	22	0.7	5
4	1.6	15	1.4	10
5	1.1	11	1.9	14
6	0.7	7	2.3	17
7	0.3	3	2,7	19
8	0.1	1	2.9	21
9	0	0	3.0	22

4-chlorostyrene indicated in Table 3.11, weighed accurately to three decimal places. If the samples cannot be polymerized on the same day they must be stored in a refrigerator.

The vessels are cooled in a methanol/dry ice bath, evacuated through an adapter (see Sect. 2.2.5.2) and, after thawing, filled with nitrogen; this procedure is repeated twice more. The tubes are now withdrawn from the adapter under a slight positive pressure of nitrogen, immediately closed with ground glass stoppers (secured with springs), and brought to 50°C in a thermostat. After 8 h the tubes are quickly cooled, the contents diluted with 5 ml of toluene and dropped into about 80 ml of stirred methanol. The copolymers are filtered off, reprecipitated twice and dried to constant weight in vacuum at 50°C. The yield under these conditions is about 300 mg (10%).

The chlorine content of the dried copolymers can be determined gravimetrically according to the method of Wurtzschmitt, and their composition derived. The copolymerization diagram is drawn and the reactivity ratios are calculated.

Example 3.37 Radical Copolymerization of Styrene with Acrylonitrile (Azeotropic Copolymerization)

Safety precautions: Before this experiment is carried out, Sect. 2.2.5 must be read as well as the material safety data sheets (MSDS) for all chemicals and products used.

The following experiment is designed to show the independence of the composition of a copolymer on the yield in the copolymerization of an "azeotropic" mixture. 3.23 g (31 mmol) of styrene, 1.01 g (19.0 mmol) of acrylonitrile, and 12.1 mg (0.05 mmol) of BPO are weighed into each of five tubes with joints, degassed, and polymerized under nitrogen at 60°C. The tubes are removed successively from the thermostat after 2, 4, 6, 8, and 10 h, and immediately quenched in a cold bath. The contents are dissolved in DMF. This can best be done as follows: The samples that are still fluid are washed out with 20 ml of DMF to which a little hydroquinone has been added; where the samples have solidified the tubes are broken open and the polymer is dissolved in 500 ml of DMF with vigorous stirring.

The copolymers are precipitated by dropping into 500 ml or 1 l of methanol, respectively, filtered and dried. The yield in wt% is plotted against polymerization time; the reasons for the deviation from linearity should be considered. The copolymers are reprecipitated once more from DMF into methanol, washed several times with methanol and dried. The compositions are determined by nitrogen analysis or spectroscopically.

Example 3.38 Radical Copolymerization of Styrene with Maleic Anhydride (Alternating Copolymerization)

300 ml of distilled toluene, 10.4 g (0.1 mol) of destabilized styrene (see Example 3.1), 9.8 g (0.1 mol) of pure maleic anhydride, and 0.1 g (0.4 mmol) of dibenzoyl peroxide are placed in a 500 ml standard apparatus (round bottom flask, several inlets for stirrer, reflux condenser, nitrogen flux or vaccum, thermometer, heating bath) and stirred at room temperature until a clear solution is obtained. The reaction mixture is continuously stirred and heated to boiling on a water bath; the copolymer gradually precipitates. After 1 h the mixture is cooled, the solid polymer filtered off, and dried to constant weight in vacuum at 60°C. Yield: 19–20 g.

The 1:1 copolymer so obtained has alternating monomeric units of styrene and maleic anhydride. This can be verified by NMR spectroscopy. It is insoluble in carbon tetrachloride, chloroform, toluene, and methanol, but soluble in THF, 1,4-dioxane, and DMF. It can be hydrolyzed to a polymeric acid (see Example 5.3)

Example 3.39 Radical Copolymerization of Methacrylic Acid with n-Butyl Acrylate in Emulsion (Continous Monomer Addition)

Safety precautions: Before this experiment is carried out, Sect. 2.2.5 must be read as well as the material safety data sheets (MSDS) for all chemicals and products used.

200 ml of distilled water are filled into a 500 ml standard apparatus (round bottom flask, several inlets for stirrer, reflux condenser, nitrogen flux or vaccum, thermometer, heating bath) three-necked flask equipped with dropping funnel, stirrer, thermometer, and nitrogen inlet tube. Then, 400 mg potassium peroxodisulfate (1 wt%) and 400 mg dodecyl sulfate are added. Oxygen is removed by passing nitrogen through the solution and heating to 80°C. Under a slight flow of nitrogen, 40 ml of a mixture made from 16 g methacrylic acid and 24 g n-butyl acrylate are added dropwise over a period of 1 h. A slight increase of temperature is observed. The contents of the flask appear milky turbid. Looking from the side, a pale blue color, induced by daylight scattering, is observed. On the other hand, a reddish color is observed by direct view from a high intensity, colorless background illumination. The copolymer can be isolated by adding dropwise the emulsion to sodium chloride solution or to dilute hydrochloric acid. Additionally, the emulsion should become almost clear with a strong increase of viscosity in alkaline medium. From this aqueous solution, the copolymer can be precipitated by the addition of Ca salts. Yield: quantitative.

Please note: due to differences in the copolymerization parameters of *n*-butyl acrylate and methacrylic acid a continuous addition of the monomer mixture is necessary in order to achieve a homogeneous composition of the copolymer product.

Copolymer emulsions with free carboxylic acid groups are useful, for example, for paper refinement as well as for coating of polar particles like pigments, fillers, or fibers. For this purpose, the particles to be coated are added under stirring to the weakly alkaline emulsion. Then, polyvalent salts or acids are added to separate out the copolymer dispersion which deposits almost quantitatively on the surface of the particles.

Example 3.40 Cationic Copolymerization of 1,3,5-Trioxane with 1,3-Dioxolane (Ring-Opening Copolymerization)

Safety precautions: Before this experiment is carried out, Sect. 2.2.5 must be read as well as the material safety data sheets (MSDS) for all chemicals and products used.

1,3,5-Trioxane and 1,3-dioxolane are purified as in Example 3.24 or by refluxing for a day over calcium hydride followed by fractional distillation. Nitrobenzene is refluxed over P_4O_{10} and distilled.

90 g (1 mol) of 1,3,5-trioxane and 9 g (0.12 mol) of 1,3-dioxolane are dissolved in 300 ml of nitrobenzene in a 500 ml flask that has been flamed under vacuum and filled with dry nitrogen or air. 0.18 ml (1.4 mmol) of boron trifluoride etherate (dissolved in 10 ml of nitrobenzene) are injected through a self-sealing closure (see Sect. 2.2.5.2). The mixture is now warmed to 45°C. The copolymer begins to precipitate within a few minutes. After 2 h the product is stirred with acetone, filtered off and washed well with acetone. In order to remove initiator and nitrobenzene residues, the polymer is boiled for 30 min with 1 l of ethanol, containing 1 wt% of tributylamine. It is then filtered off, washed with acetone, and sucked dry. About 90 g of a pure white copolymer are obtained containing about 30 $-O-CH_2-$ units for every $-O-CH_2-CH_2-$ unit. It melts at 156–159°C and has a molecular weight of about 30,000 (corresponding to an hsp/c value of about 0.04 l/g in DMF at 140°C). The thermal stability at 190°C (see Example 5.13) is compared with that of a homopolymer of 1,3,5-trioxane (see Example 3.24).

Example 3.41 Radical Copolymerization of Styrene with 1,4-Divinylbenzene in Aqueous Suspension (Crosslinking Copolymerization)

Safety precautions: Before this experiment is carried out, Sect. 2.2.5 must be read as well as the material safety data sheets (MSDS) for all chemicals and products used.

Styrene and 1,4-divinylbenzene (the latter as 50–60% solution in ethylbenzene) are destabilized and distilled as described in Example 3.1.

A standard apparatus (round bottom flask, several inlets for stirrer, reflux condenser, nitrogen flux or vaccum, thermometer, heating bath) fitted with a stirrer (preferably with revolution counter) is evacuated and filled with nitrogen three times. 250 mg of poly(vinyl alcohol) are placed in the flask (see Example 5.1) and dissolved in 150 ml of deaerated water at 50°C. A freshly prepared solution of 0.25 g (1.03 mmol) of dibenzoyl peroxide in 25 ml (0.22 mmol) of styrene and 2 ml (7 mmol) of 1,4-divinylbenzene is added with constant stirring in order to produce an emulsion of fine droplets of monomer in water. This mixture is heated to 90°C on a water bath while maintaining a constant rate of stirring and passing a gentle stream of nitrogen through the reaction vessel. After about 1 h (about 5% conversion) the crosslinking becomes noticeable (gelation). Stirring is continued for another 7 h at 90°C, the reaction mixture then being allowed to cool to room temperature with stirring. The supernatant liquid is decanted from the beads which are washed several times with methanol and finally stirred for another 2 h with 200 ml of methanol. The polymer is filtered off and dried overnight in vacuum at 50°C. Yield: practically quantitative. The crosslinked copolymer of styrene and 1,4-divinylbenzene so obtained can be used for the preparation of an ion-exchange resin (see Examples 5.9 and 5.10).

The swellability is determined by placing 1 g of the dried polymer in contact with toluene for 3 days in a closed 250 ml flask. The swollen beads are collected on a sintered glass filter (porosity G1), suction is applied for 5 min and the filter immediately weighed. The percentage increase in the weight of the beads can then be calculated. The size of the swollen and unswollen beads can be compared under the microscope.

Example 3.42 Copolymerization of Styrene with Methyl Acrylate (Internal Plasticization)

Safety precautions: Before this experiment is carried out, Sect. 2.2.5 must be read as well as the material safety data sheets (MSDS) for all chemicals and products used.

16.4 mg (0.1 mmol) AIBN are added to 4 ml of toluene in a thick-walled test tube with ground joint. Then, 0.01 mol of a monomer mixture consisting of styrene and methyl acrylate are added according to Table 3.12. To remove oxygen, the solution is bubbled with nitrogen via a Pasteur pipette for 30 s. The test tubes are sealed and maintained at 65°C for 3 h in a thermostatted water bath. The polymerization is terminated by cooling the test tubes in a cooling bath (ice/water). Then, 5 ml of toluene are added to the content of each test tube.

Samples 1 and 2 are precipitated by adding 50 ml of methanol containing 3 drops of concentrated hydrochloric acid. All other samples are precipitated by the addition of n-hexane. All of the precipitants should be cooled with an ice/water mixture. The yields have to be calculated.

To investigate the internal plasticization of polystyrene ($T_g = 105°C$) by insertion of methyl acrylate ($T_g = 4°C$), the samples are run on a DSC. Therefore, approximately 15 mg of each of the well dried polymers are weighed into small aluminum pans and measured by two heat-cool runs in a DSC apparatus. The glass transition is found as a characteristic jump in the heat capacity in the system. The glass transition temperature is evaluated after the second heating from the DSC plot.

Table 3.12 Test series for Example 3.42

Sample#	$\frac{[M_{MA}]}{[M_{MA}]+[M_S]}$	MA [g]	S [g]
1	0.00	0	5.208
2	0.10	0.430	4.687
3	0.25	1.076	3.906
4	0.35	1.507	3.385
5	0.45	1.937	2.864
6	0.55	2.367	2.343
7	0.65	2.798	1.823
8	0.75	3.228	1.302
9	0.85	3.659	0.781
10	0.90	3.874	0.521
11	0.95	4.089	0.260
12	1.00	4.305	0

Example 3.43 Three-Step Synthesis of Core/Double Shell Particles of Methyl Methacrylate/Butyl/Acrylate/Methyl Methacrylate

Safety precautions: Before this experiment is carried out, Sect. 2.2.5 must be read as well as the material safety data sheets (MSDS) for all chemicals and products used. This example describes the concept of core/shell impact modifiers for thermoplastic polymers (see Sect. 5.51).

(a) PMMA Core Synthesis by Crosslinking Copolymerization

In a reaction vessel equipped with mechanic propeller stirrer, argon inlet, and reflux condenser an emulsion is prepared of distilled water (300 ml), methyl methacrylate (MMA; 9.5 g), allyl methacrylate (ALMA; 0.5 g) and sodium dodecyl sulfate (SDS; 50 mg). The emulsion is heated (75°C), and ammonium peroxodisulfate (APS; 0.6 g) dissolved in water (5 ml) is added. After 50 min a mixture of MMA (28.5 g) and ALMA (1.5 g) is added dropwise (within 1 h). Subsequently, a "monomer emulsion" formed by water (250 ml), SDS (1.1 g), MMA (252 g), ALMA (8 g), and APS (0.5 g) is added continuously (within 225 min) to the reaction mixture. After complete addition, stirring is continued for further 1 h at 75°C and for further 2 h at 85°C. Finally, the emulsion is allowed to cool down to room temperature.

(b) Synthesis of the First (Elastomeric) Shell via Crosslinking Copolymerization

In a reaction vessel, the above PMMA latex (163 g) and water (175 g) are stirred and heated (75°C). A solution of APS (0.5 g) in water (5 ml) is added. Then, a mixture of butyl acrylate (BA) and ALMA (0.4 g) is added continuously (within 30 min). Subsequently, a "monomer emulsion" of water (90 ml), SDS (0.57 g), BA (120.5 g) ALMA (2.5 g), and APS (0.3 g) is added continuously, within 200 min. After complete addition, the polymerization is finished by stirring the mixture for further 1 h at 75°C and for further 2 h at 85°C. Finally, the mixture is allowed to cool down to room temperature.

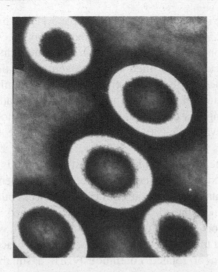

Fig. 3.5 TEM image of a cut through a SAN matrix with incorporated core/double shell particles. dark core: PMMA; bright area: PBA shell; diameter of the particles of the first shell: about 220 nm; the second PMMA shell is compatible with SAN matrix, therefore the contrast is very weak (Details see Bibliography)

(c) Synthesis of the Second (Thermoplastic) Shell by Homopolymerization of MMA

In a reaction vessel, the above PMMA-PBA latex (400 g) and water (75 ml) are stirred and heated to 75°C. A solution of APS (0.35 g) in water (5 ml) is added. Then, MMA (10 g) is added to the reaction mixture slowly, within 45 min. followed by a "monomer emulsion" of water (35 ml), SDS (0.25 g), MMA (50 g) and APS (0.15 g) which is added slowly and continuously as well (within 2 h). After complete addition, stirring is continued for a further 1 h at 75°C and for further 2 h at 85°C. Finally, the emulsion is allowed to cool down to room temperature.

Isolation of the polymer particals see Example 3.2 and 3.3.

Figure 3.5 shows a TEM picture of the core/double shell latex particles incorporated into an styrene/acrylonitrile (SAN) copolymer matrix (thin cut through the particle-filled matrix). The particles are very homogeneous in size and can also be used to prepare artifical opals.

Example 3.44 Radical Copolymerization of Butadien with Styrene in Emulsion

Safety precautions: Before this experiment is carried out, Sect. 2.2.5 must be read as well as the material safety data sheets (MSDS) for all chemicals and products used.

Caution: All work is to be undertaken behind a protective screen; hands should also be protected with safety gloves in order to avoid cuts should one of the flasks explode (internal pressure 3 bar).

Monomeric styrene is destabilized (see Example 3.1) and distilled into a suitable receiver (see Sect. 2.2.5.3) under nitrogen. Butadiene from a cylinder is condensed under nitrogen atmosphere into a trap cooled in a methanol/dry ice bath. A 500-ml pressure bottle is evacuated and filled with nitrogen. A solution of 5 g of sodium oleate (or sodium dodecyl sulfate) is then made up in 200 ml of deaerated water, and 0.25 g (0.93 mmol) of potassium peroxodisulfate is added; the mixture is shaken until everything has dissolved. The pH is adjusted to a value of 10–10.5 with dilute sodium hydroxide, 30 g (0.29 mol) of styrene containing 0.5 g of dodecylmercaptane as regulator, and 70 g (1.30 mol) of butadiene are added under nitrogen, and the bottle is sealed. The best procedure is to cool the pressure bottle in an ice/salt mixture, place it on a balance scale and then, under a hood, pour in a small excess of butadiene from the cold trap; the excess butadiene is allowed to evaporate on the balance until the correct weight is reached. The sealed bottle is allowed to warm to room temperature behind a safety screen, wrapped with a towel and vigorously shaken to emulsify the contents. It is then placed in a bath at 50°C and should be shaken or rotated continuously. (If a suitable apparatus is not available, the bottle can be shaken again vigorously after 1 h). The bottle is held at 50°C for 15 h behind a safety screen, then allowed to cool to room temperature and finally cooled in ice. It is weighed again to check whether any butadiene has escaped. The latex is then slowly poured into about 500 ml of stirred ethanol contained in a beaker under a hood; to the ethanol has previously been added 2 g N-phenyl-β-naphthylamine, this being the amount required to protect the copolymer against oxidation. Most of the unconverted butadiene evaporates and the copolymer is obtained as loosely coherent crumbs that are dried for 1–2 days in vacuum at 50–70°C. The composition of the copolymer can be calculated from the analytically determined double bond content or from the spectroscopically determined content of styrene units (see Sect. 2.3.2); the arrangement of the butadiene monomeric units can be found from the IR spectrum (see Example 3.21). By vulcanization (see Example 5.8) the copolymer can be converted into an insoluble, highly elastic product. It finds application in tires.

Example 3.45 Radical Copolymerization of Butadiene with Acrylonitrile in Emulsion

Safety precautions: Before this experiment is carried out, Sect. 2.2.5 must be read as well as the material safety data sheets (MSDS) for all chemicals and products used.

Monomeric acrylonitrile is distilled under nitrogen into a suitable receiver (see Sect. 2.2.5.3). Butadiene from a cylinder is condensed under nitrogen atmosphere into a trap cooled in a methanol/dry ice bath.

As described in Example 3.44, a mixture of 10 g (0.19 mol) of acrylonitrile containing 0.1 g dodecylmercaptane as regulator and 25 g (0.46 mol) of butadiene is polymerized in 50 ml of a 5% aqueous solution of sodium oleate (or sodium dodecyl sulfate) containing 0.25 g (0.93 mmol) of potassium peroxodisulfate as initiator. After 18 h the pressure bottle is allowed to cool to room temperature and is finally cooled in ice. It is now weighed again to confirm that no butadiene has been lost by leakage. The bottle is carefully opened, the latex poured into a beaker and

Table 3.13 Test series for Example 3.46

	Batch 1	Batch 2	Batch 3	Batch 4
Distilled water	23 ml	23 ml	23 ml	23 ml
Emulsifier[a]	684 mg	450 mg	684 mg	594 mg
Methacrylic acid	756 mg (8.8 mmol)	756 mg (8.8 mmol)	756 mg (8.8 mmol)	1.26 g (14.6 mmol)
Styrene	31.5 g (0.30 mol)	31.5 g (0.30 mol)	40.95 g (0.39 mol)	31.5 g (0.30 mol)
Butyl acrylate	31.5 g (0.27 mol)	31.5 g (0.27 mol)	22.05 g (0.17 mol)	31.5 g (0.27 mol)

[a]Commercial alkylbenzene sulfonic acid sodium salt (linear C10–C13 alkyl chains, 70% in water), (e.g., from Sasol)

0.5 g of *N*-phenyl- β-naphthylamine stirred in as antioxidant. A solution of 5 wt% of sodium chloride in 2% sulfuric acid is now added dropwise with stirring until the copolymer flocculates. It is filtered off, stirred vigorously with water several times in a beaker, filtered again and dried in vacuum at 65–70°C. The composition of the copolymer is determined from the nitrogen content determined according to the Kjeldahl method or from the double bond content. The arrangement of the butadiene monomeric units is found by IR spectroscopy (see Example 3.21).

Copolymers of butadiene and acrylonitrile are soluble in aromatic or chlorinated hydrocarbons but insoluble in aliphatic hydrocarbons. Therefore they find use in oil-resistant rubber articles (hoses, sealings etc.)

Example 3.46 Preparation of a Styrene/Butyl Acrylate/Methacrylic Acid Terpolymer Dispersion (Influence of Emulsifier)

Safety precautions: Before this experiment is carried out, Sect. 2.2.5 must be read as well as the material safety data sheets (MSDS) for all chemicals and products used.

The following experiment should demonstrate the influence of particle size and minimal film-formation temperature (which is connected with the glass transition temperature and therefore with the chemical structure of the polymers) on the properties of films, prepared from aqueous dispersions.

Four 250 ml standard apparatus (round bottom flask, several inlets for stirrer, reflux condenser, nitrogen flux or vaccum, thermometer, heating bath) are filled each with 38 ml of distilled water and 405 mg (1.8 mmol) ammonium peroxodisulfate. To the fourth flask, additionally 90 mg of an alkylbenzene sulfonic acid sodium salt emulsifier (linear C10–C13 alkyl chains, 70% in water) are added. After reaching an internal temperature of 80°C, 3 g of a monomer emulsion (prepared according to Table 3.13) are added to each of the four batches. Within an induction phase of 15 min the reactions start. This is visualized by the pale blue coloration of the content of the flasks. Then the remaining monomer emulsion is added over a period of 3 h. The internal temperature has to be maintained at 80°C during the whole reaction. After all monomer emulsion has been added, the batches are heated at 80°C for another 30 min. Then, the flasks are allowed to cool to room temperature. Now the reaction mixtures are adjusted to pH 8.5 by addition of 5% NaOH solution.

The contents of batch 2 and batch 4 are spread on a glass plate and dried at 35°C for 24 h. Then, the glass plates with the dried films are placed into water. Run 2 (particle size approx. 270 nm) becomes turbid very rapidly, whereas run 4 (particle size approx. 150 nm) stays clear.

Batch 1 and batch 3 are spread on a glass plate and dried at room temperature. Batch 1 gives a clear film without cracks (minimal film-forming temperature below room temperature) whereas batch 3 gives no coherent, crack-free film (minimal film-forming temperature above room temperature).

3.4.2 Block and Graft Copolymerization

Conventional copolymerizations yield macromolecules mostly with random distribution and only very seldom with alternating distribution of the monomer units. In order to synthesize block or graft polymers, special methods must be used, of which some are described in the following sections.

3.4.2.1 Block Copolymers

In the simplest case, block copolymers consist of successive series (blocks or segments) of A and B units. Depending on the number of linked blocks, one distinguishes diblock, triblock and multiblock copolymers:

diblock (A-A-A) – (B-B-B)

triblock (A-A-A) – (B-B-B) – (A-A-A)

multiblock (-A-A-A) – (B-B-B) – (A-A-A) – (B-B-B-)...

For the synthesis of block copolymers chain addition polymerization (ionic or radical) as well as condensation polymerization and stepwise addition polymerization can be used.

Well developed is the *anionic polymerization* for the preparation of olefin/diolefin – block copolymers using the techniques of "living polymerization" (see Sect. 3.2.1.2). One route makes use of the different reactivities of the two monomers in anionic polymerization with butyllithium as initiator. Thus, when butyllithium is added to a mixture of butadiene and styrene, the butadiene is first polymerized almost completely. After its consumption stryrene adds on to the living chain ends, which can be recognized by a color change from almost colorless to yellow to brown (depending on the initiator concentration). Thus, after the styrene has been used up and the chains are finally terminated, one obtains a two-block copolymer of butadiene and styrene:

Oxidative degradation by splitting off the double bonds in the butadiene blocks allows the styrene blocks to be isolated. The degradation of the chains can be followed by molecular weight determinations or viscosity measurements (see Examples 3.47, 5.12, 5.14, and 5.15).

A second route is termed sequential anionic polymerization. More recently, also controlled radical techniques can be applied successfully for the sequential preparation of block copolymers but still with a less narrow molar mass distribution of the segments and the final product. In both cases, one starts with the polymerization of monomer A. After it is finished, monomer B is added and after this monomer is polymerized completely again monomer A is fed into the reaction mixture. This procedure is applied for the production of styrene/butadiene/styrene and styrene/isoprene/styrene triblock copolymers on industrial scale. It can also be used for the preparation of multiblock copolymers.

A variation of the sequential anionic polymerization is the use of dianions as initiator, like sodium naphthalene. One starts with the polymerization of monomer A. Then monomer B is fed to the reaction mixture which adds immediately to the "living" anions at each end of block A and thus leads to a triblock copolymer with an A-middle block and two B-outer blocks. This triblock copolymer is still "alive" and repetition of the above procedure results in a multiblock copolymer (see Example 3.49).

Condensation polymerization and stepwise addition polymerization are, for example, applied for the preparation of block polyesters. The synthesis concepts are different from those of chain polymerization in that at least one "monomer" is an oligomer with one or two functional end groups, for example polytetrahydrofurane with a molecular weight of several hundred and OH–end groups (see Example 3.23). If this oligomer partially replaces butandiol in the condensation polymerization with terephthalic acid (compare Examples 4.1 and 4.2), a poly(ether ester) is obtained with hard polyester segments and soft polyether segments and with the properties of a thermoplastic elastomer.

Stepwise addition polymerization is used in the preparation of segmented polyurethanes (compare Sect. 4.2.1), e.g., poly(ester ether) urethanes which also find applications in thermoplastic elastomers. Here, both blocks are preformed separately and are linked together by reaction with isocyanates:

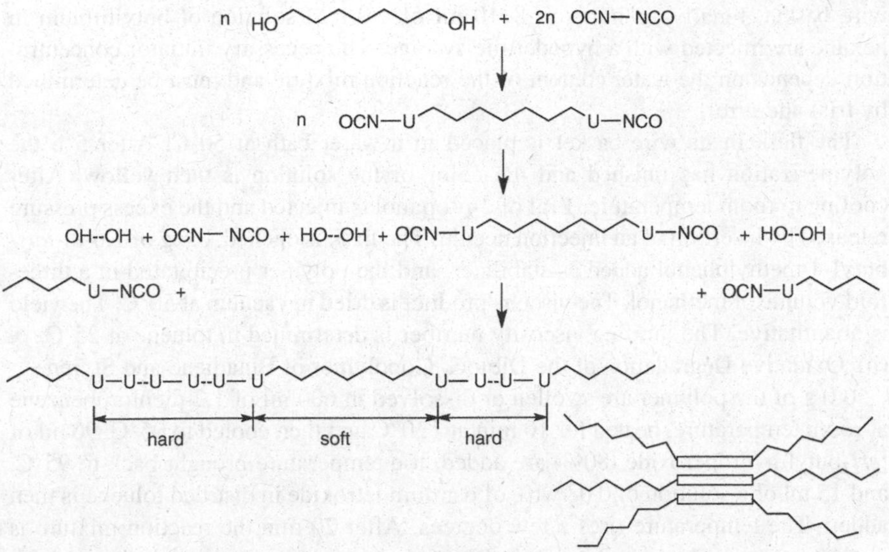

n HO~~~~~OH + 2n OCN—NCO

n OCN—U~~~~~U—NCO

OH--OH + OCN— NCO + HO--OH + OCN—U~~~~~U—NCO + HO--OH

~~~U—NCO +   + OCN—U~~~

~~~U—U--U— U--U— U~~~~~U— U--U— U~~~

|←—— hard ——|←—— soft ——|←—— hard ——|

In segmented polyurethanes as well as in many other block copolymers incompatible polymer segments are combined. This incompatibility of different CRUs means that block copolymers often form multiphase systems (see TEM-picture in Sect. 2.3.5.14) that are distinguished by special properties. They are already used in so-called thermoplastic elastomers (TPE). They consist of thermoplastic hard segments and of rubber-elastic soft segments, for example, styrene/butadiene/styrene block copolymers or segmented polyurethanes. They are not crosslinked via covalent bonds, but by formation of domains by the hard segments. On the other hand, they fulfill the requirements for rubber-elastic properties. This physical crosslinking is thermally reversible, so that these elastomers – in contrast covalently crosslinked elastomers – can be thermally processed. It must be stressed that in case of styrene/butadiene/styrene, the triblock composition is absolutely mandatory for obtaining TPE properties. Both styrene blocks must be terminal.

Example 3.47 Preparation of a Butadiene/Styrene Diblock Copolymer

Safety precautions: Before this experiment is carried out, Sect. 2.2.5 must be read as well as the material safety data sheets (MSDS) for all chemicals and products used.
(a) Preparation

A 500 ml two-necked flask, fitted with ground glass stopper and tap, is dried for 2 h at 110°C. Into the still warm flask are distilled 200 ml of toluene under nitrogen, a moderate stream of nitrogen (3 l/min) is then being passed through the solvent for 3 min, using a delivery tube with a sintered glass disk. The flask is immediately closed with a self-sealing rubber cap and the pressure reduced to 0.3–0.5 bar nitrogen. With the aid of a dry hypodermic syringe, 6 g (6.7 ml) of dry, stabilizer-free styrene are added, together with 14 g (23 ml) of dry, stabilizer-free butadiene from a vessel provided with an injection needle (the materials must be cooled to a suitable temperature). The flask is well shaken and placed in a suitable wire basket. Finally, as initiator, 8–10 ml of a 0.1 M solution of butyllithium in hexane are injected with a hypodermic syringe. The necessary initiator concentration depends on the water content of the reaction mixture and must be determined by trial and error.

The flask in its wire basket is placed in a water bath at 50°C. After 5 h the polymerization has finished and the color of the solution is then yellow. After cooling to room temperature, 1 ml of 2-propanol is injected and the excess pressure released by insertion of an injection needle. The flask is opened, 0.5 g of 2,6-di-*tert*-butyl-4-methylphenol added as stabilizer, and the polymer precipitated in a three-fold volume of methanol. The viscous product is dried in vacuum at 50°C. The yield is quantitative. The limiting viscosity number is determined in toluene at 25°C.

(b) Oxidative Degradation of the Diblock Copolymer of Butadiene and Styrene

6.0 g of the polymer are swollen or dissolved in 600 ml of 1,2-dichlorobenzene at room temperature, heated for 10 min at 120°C and then cooled to 95°C. 90 ml of *tert*-butyl hydroperoxide (80%) are added, the temperature brought back to 95°C, and 15 ml of a solution of 0.08 wt% of osmium tetroxide in distilled toluene is then added. The temperature rises a few degrees. After 20 min the reaction mixture is cooled to room temperature and shaken three times in a separating funnel with 600 ml of a mixture of 1,500 ml of methanol and 500 ml of water. The lower phase is the 1,2-dichlorobenzene solution. The 1,2-dichlorobenzene is distilled off under vacuum at 60°C (oil pump) and the residue weighed. About 2.4 g of a deep yellow, highly viscous mass are obtained. This can be decolorized by taking it up in toluene and warming with activated charcoal. The polymer then precipitates well in methanol. The reaction products of the degraded butadiene sequences, and also the styrene sequences having a molecular weight less than 500, are removed by this treatment. The limiting viscosity number of the residue is now determined in toluene at 25°C and the molecular weight of the styrene sequences determined. The IR spectrum of the residue from degradation corresponds very closely to that of polystyrene. The amount of styrene sequences is only slightly less than the amount of styrene in the block copolymer since only a small proportion of the styrene is present within short chain lengths containing both monomeric units, which form the junction points of the two-block sequences.

Example 3.48 Preparation of a t-Butyl Methacrylate/Styrene/t-Butyl Methacrylate (→ Acrylic Acid/Styrene/Acrylic Acid) Triblock Copolymer

Safety precautions: Before this experiment is carried out, Sect. 2.2.5 must be read and the material safety data sheets (MSDS) for all chemicals and products examined.

General: The triblock copolymer is prepared by anionic polymerization. As in Example 3.19, the greatest care must be taken to exclude air and moisture. Also all transfers have to be carried out under rigorous exclusion of air, either in a glove box or using Schlenck techniques. Tetrahydrofurane is dried as described in Example 3.19.

The monomers are dried for 24 h over calcium hydride and then distilled under vacuum at room temperature. 1,1-Diphenyl ethylene (DPE) is dried by dropwise addition of *n*-butyl lithium until a stable dark-red color appears. The DPE is then distilled in vacuo at 120°C. Naphthalene (best purity) can be used as obtained.

Lithium naphthalene is prepared by adding, under argon, lithium metal to a solution of naphthalene (3 wt%) in THF (1 equiv. Li with respect to naphthalene). The mixture is stirred for 7 h at room temperature until all lithium has disappeared and an intensively green solution is obtained. The true concentration of the thus obtained solution of lithium naphthalene can be determined by carrying out a model styrene polymerization and determining the resulting degree of polymerization. When carefully sealed and put under an atmosphere of argon, this solution is stable for several months at −30°C.

Lithium naphthalene (0.5 mmol) in THF is added quickly to a cooled (−78°C) solution of styrene (5 g) in THF (50 ml). The mixture immediately turns orange. Stirring at −78°C is continued for 30 min. A small sample may be taken from the mixture with a syringe and quenched in degassed methanol. This sample can be used to measure the molar mass of the styrene block (by GPC). The mixture is allowed to warm to −18°C, then a solution of 1,1-diphenylethylene (108 mg, 0.6 mmol) in THF (1 ml) is added whereupon the color immediately changes to an intensive red. Now *tert*-butylmeth-acrylate (*t*-BMA; 5 g) is added quickly. The color disappears, and stirring is continued at −18°C for further 60 min. Degassed methanol (1 ml) is added to terminate the reaction. After allowing the mixture to warm to room temperature, the obtained polymer is precipitated by pouring the whole mixture into methanol (500 ml). The precipitate is filtered off, washed with methanol, and dried in vacuo to constant weight. The molar mass of this triblock copolymer may again be measured by GPC.

Hydrolysis of the poly(*t*-BMA) blocks is achieved by dissolving the triblock copolymer (1 g) in dioxane (100 ml) at room temperature. Conc. hydrochloric acid (1.5 ml) is added, and the resulting mixture is heated (100°C) and stirred for 18 h. After cooling to room temperature the solution is concentrated down to approx. 10 ml which are poured into cold (0°C) diethylether (100 ml). The precipitate is filtered off, washed with methanol and dried in vacuo to constant weight. The result is an acrylic acid-*b*-styrene-*b*-acrylic acid triblock copolymer.

The parent triblock copolymer t-BMA-*b*-S-*b*-t-BMA and the hydrolyzed triblock copolymer acrylic acid–*b*-styrene-*b*-acrylic acid should be analyzed by IR spectroscopy. The ester and acid carbonyl peaks indicate the saponification (see Sect. 2.3.2.2)

The two products differ in polarity and thus in hydrophilicity. The acrylic acid-*b*-styrene-*b*-acrylic acid triblock copolymer exhibits amphiphilic properties and forms an emulsion in dilute sodium hydroxide solution where the acid groups are neutralized.

Example 3.49 Preparation of a Multiblock Copolymer of 4-Vinylpyridine and Styrene by Anionic Polymerization

Safety precautions: Before this experiment is carried out, Sect. 2.2.5 must be read as well as the material safety data sheets (MSDS) for all chemicals and products used.

The block copolymer is prepared by anionic polymerization. As in Example 3.19, the greatest care must be taken to exclude air and moisture.

Monomeric styrene is destabilized as in Example 3.1 and pre-dried with calcium chloride. The monomer is now allowed to stand over calcium hydride for 24 h and then distilled under reduced pressure of nitrogen into a previously flamed out

Schlenk tube. Pure 4-vinylpyridine is distilled twice over KOH pellets in vacuum. It is then vacuum distilled under nitrogen through a column packed with Raschig rings into a previously flamed out Schlenk tube (bp 62°C/12 Torr). The closed Schlenk tubes containing the monomers are stored in a refrigerator until required. The preparation of the initiator solution (sodium naphthalene) is described in Example 3.19.

The polymerization is carried out as follows: a Schlenk tube that has been flamed out and filled with nitrogen, is charged with 50 ml of pure THF and 1 ml of sodium naphthalene (see Example 3.19). Using a nitrogen-filled syringe, 4.6 ml (40 mmol) of styrene are added to this solution at room temperature with vigorous agitation, while nitrogen is passed through the tube. The closed tube containing the red polymerizing mixture is allowed to stand for 15 min at room temperature.

To 5.3 g of 4-vinylpyridine is added to THF up to a volume of 50 ml; 5 ml of this solution (containing 5 mmol 4-vinyl pyridine) are added in the same way to the above solution containing the "living" polystyrene, with vigorous agitation. After 15 min another 40 mmol of styrene are added, followed 15 min later by another 5 mmol of 4-vinylpyridine; this operation is repeated once more. 15 min after the last addition of monomer the block copolymer is precipitated by dropping the solution into a mixture of 300 ml of diethyl ether and 300 ml of petroleum ether. The polymer is filtered, washed with ether, filtered again, and dried in vacuum at room temperature.

The blocks of 4-vinylpyridine must not be too long, otherwise the polymer is no longer completely soluble in THF. This can easily be observed when 20 ml lots of 4-vinylpyridine are used instead of 5 ml; the solution then becomes cloudy, but after fresh addition of styrene it turns clear again.

For comparison, polystyrene and poly(4-vinylpyridine) are prepared by anionic polymerization with sodium naphthalene as initiator. Poly(4-vinylpyridine) precipitates from THF; the mixture is poured into 200 ml of diethyl ether and the polymer filtered off. The polymer is then reprecipitated from pyridine solution into a tenfold amount of diethyl ether and dried in vacuum.

The IR spectra of all three polymers are recorded and compared with one another. The incorporation of monomeric units of 4-vinylpyridine can also be demonstrated by nitrogen analysis of the block copolymer. The solubility behavior is also determined. Poly(4-vinylpyridine) is soluble in pyridine, methanol, and chloroform, but insoluble in toluene and diethyl ether; it swells considerably in water. On the other hand, the block copolymer, like polystyrene, is soluble in pyridine, chloroform, and toluene; but unlike polystyrene, it swells significantly in methanol.

3.4.2.2 Graft Copolymers

Block copolymers are linear, but *graft* copolymers are branched, with the main chain generally consisting of a homopolymer or a random copolymer, while the grafted side chains are composed of either the same or another monomer or several monomers:

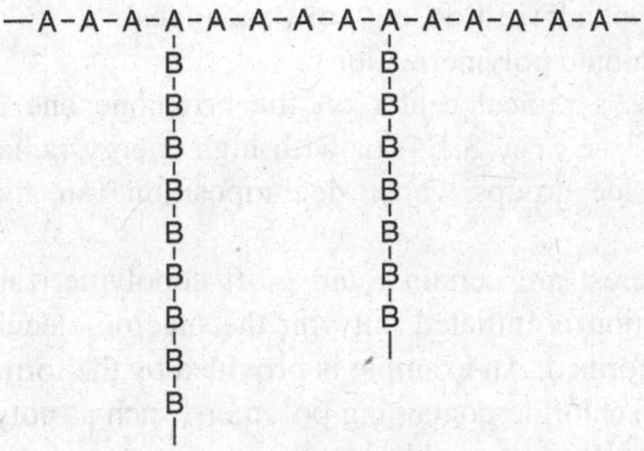

$$—A–A–A–A–A–A–A–A–A–A–A–A–A–A–A—$$

The numerous ways for the synthesis of *graft* copolymers can be divided into three categories.

To the first category belong the homo- and copolymerization of macromonomers. For this purpose, macromolecules with only one polymerizable end group are needed. Such macromonomers are made, for example, by anionic polymerization where the reactive chain end is modified with a reactive vinyl monomer. Also methacrylic acid esters of long-chain aliphatic alcohols or monofunctional polyethylene oxides or polytetrahydrofurane belong to the class of macromonomers.

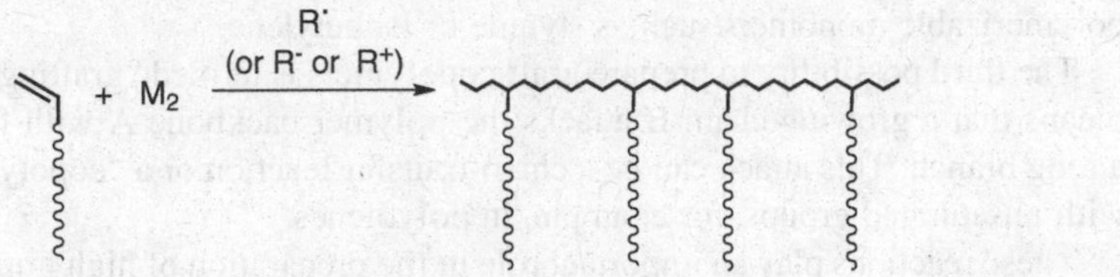

An example for polystyrene macromonomer synthesis:

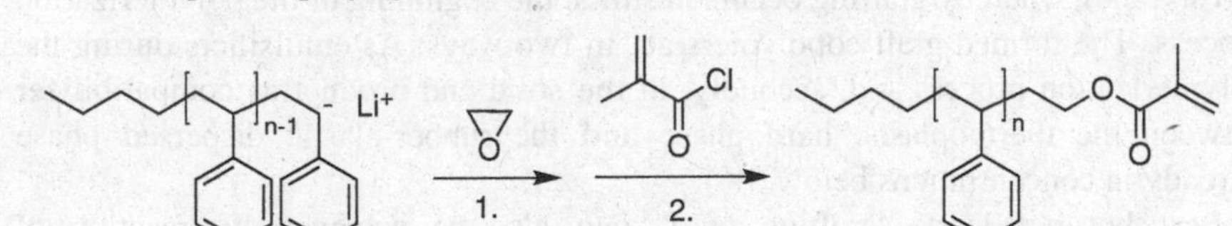

The second possible route is called "grafting from". This means that active sites are generated at the polymer backbone A which initiate the polymerization of monomer B, thus leading to long-chain branches:

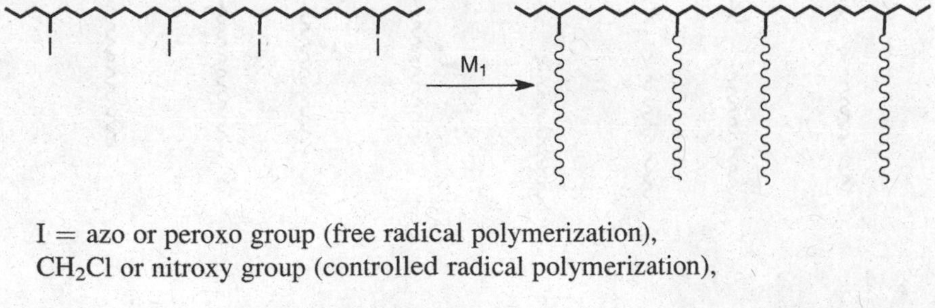

I = azo or peroxo group (free radical polymerization),
CH$_2$Cl or nitroxy group (controlled radical polymerization),

CH_2Cl (cationic polymerization, e.g., of oxazolines),

halogens + Li (anionic polymerization).

In order to create a radical center on the backbone one may irradiate with ultraviolet radiation (see Sect. 3.1.4) or with high energy radiation. Autoxidation leads to hydroperoxide groups whose decomposition can also lead to suitable radical centers.

Of particular interest are certain ionic graft copolymerizations in which the polymerization reaction is initiated only on the macromolecular framework and no homopolymer is formed. An example is provided by the formation of polymeric carbonium ions from chloride-containing polymers, such as poly(vinylchloride), in the presence of diethylaluminum chloride:

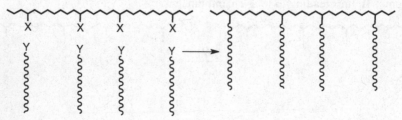

The polymer cations can then initiate the polymerization of cationically polymerizable monomers such as styrene or isobutylene.

The third possibility to prepare graft copolymers is termed "grafting onto". This means that a growing chain B attacks the polymer backbone A with formation of a long branch. This attack can be a chain-transfer reaction or a "copolymerization" with unsaturated groups, for example, in polydienes.

These reactions play an important role in the preparation of high impact polystyrene (see Example 5.23) and of ABS-polymers (made from acrylonitrile, butadiene, and styrene), whereby grafting occurs in situ at the beginning of the polymerization process. The formed graft copolymers act in two ways: As emulsifiers during the polymerization process and, secondly, in the solid end product as compatibilizer between the thermoplastic hard phase and the rubber-elastic dispersed phase (already in concentrations below 3%).

Last but not least, "grafting onto" can also be achieved by reaction of monofunctional oligomers with the reactive side groups of a polymer backbone. The esterification of (co)polymers of methacrylic acid chloride with polyethylene oxides or polytetrahydrofuranes bearing only one OH-group belongs to this category.

It has to be kept in mind that in most of the recipes for the preparation of graft copolymers, mentioned above, larger amounts of homopolymers B are formed.

In general, graft copolymers play often a role as compatibilizer in polymer blends (see Sect. 5.5).

The characterization of block or graft copolymers is generally much more difficult than that of random copolymers (see Sect. 2.3.2.7). Especially, DSC measurements are useful for the characterization of the different segments (determination of T_g). Also dynamic-mechanical measurements are used to distinguish statistical copolymers from those with block or graft structure.

The properties of these types of copolymer depend markedly on the number and length of the blocks and side chains, as well as on the structure of the monomeric units and their molar ratio. In general, graft and block copolymers combine additively the properties of the corresponding homopolymers, while random copolymers normally exhibit the average of the properties of the two homopolymers. This can be recognized very well from the glass transition temperatures: Statistical copolymers show only one T_g which is often found in the region between the T_g values of the two corresponding homopolymers. In contrast to this, block and graft copolymers show, because of the microphase-separation, two T_g values that correspond to the individual homopolymers.

There is, therefore, the possibility of preparing copolymers with desired combinations of properties. This can generally not be achieved by blending the corresponding homopolymers, since chemically different polymers are rarely compatible with one another.

Example 3.50 Graft Copolymerization of Styrene on Polyethylene

Safety precautions: Before this experiment is carried out, Sect. 2.2.5 must be read as well as the material safety data sheets (MSDS) for all chemicals and products used.

A dry, weighed polyethylene film (length 50 mm, breadth 25 mm, thickness 0.1–0.2 mm) is placed in a tube (about 70 ml capacity) fitted with a ground glass joint and stopcock. After addition of some crystals of benzophenone (as sensitizer) the sample is heated for 1 h on a water bath at 60°C. The outside of the tube is dried, the stopcock closed, and the tube exposed to 15 min irradiation from a mercury lamp (maximum of emission in the region of $\lambda = 253$–254 nm).

While the sample is being irradiated a second tube is filled with 50 ml of destabilized styrene (see Example 3.1) and 0.1 ml of tetraethylenepentamine, the tube then being evacuated and filled with nitrogen. With the aid of a large syringe 40 ml of this solution are injected through the bore of the stopcock into the tube containing the film. The tube is now evacuated, filled with nitrogen, and the closed vessel maintained at 60°C in a thermostat for 2 h. The grafted polyethylene foil is then extracted in a Soxhlet apparatus for 1 h with ethyl acetate, dried to constant weight in vacuum at 50°C and weighed. The extract is added dropwise to about 400 ml of methanol; the precipitated homopolymer of styrene is filtered off through a sintered glass filter, dried in vacuum at 50°C, and weighed.

The amount of grafted styrene is given by the increase in weight of the film. It may also be calculated quite well from the densities of polyethylene (d_1), polystyrene (d_2), and the graft copolymer (d_3) according to the following Eq. 3.31:

$$\frac{100 - x}{d_1} + \frac{x}{d_2} = \frac{100}{d_3} \tag{3.31}$$

where x is the weight percentage of grafted polystyrene, relative to the weight of the grafted film. The density of polystyrene (d_2) may be taken as 1.05 g/cm^3; the densities d_1 and d_3 can be determined on small pieces (0.2–0.5 cm^2) of the polyethylene and grafted films by the flotation method using ethanol/water mixtures (see Sect. 2.3.5.1). In addition, the ratio of grafted polystyrene and homopolymer of styrene is determined.

Example 3.51 Radical Graft Copolymerization of Vinylpyrrolidone onto Poly(vinylalcohol)

Safety precautions: Before this experiment is carried out, Sect. 2.2.5 must be read as well as the material safety data sheets (MSDS) for all chemicals and products used.

An initiator solution is prepared by dissolving 0.625 g of 2,2-azobis (2-aminopropene) dihydrochloride in 6 ml water. The solution is transferred under argon into a syringe.

A standard apparatus (250 ml, three-neck round-bottom flask, thermometer, heating bath) is equipped with magnetic stirrer, reflux condenser, dropping funnel with pressure release, and one neck is sealed with a septum which can be used for thermometer inlet but also addition of initiator solution using a syringe; the apparatus is flushed with argon. Under argon atmosphere one adds 8.3 g of a 30-wt% polyvinylalcohol aqueous solution (2.5 g of PVA, molar mass about 30,000 g/mol), 1/5 of the vinylpyrrolidone amount (4.5 g), and about 3 drops of the initiator.

Under slow argon stream (inlet via dropping funnel, outlet via condensor) the mixture in the flask is heated to 80°C. Within 3 h the remaining portion of vinylpyrrolidone (18.5 g) is added through the dropping funnel, and within 4 h, half of the remaining initiator solution (about 3 g water solution) is added through the septum using the syringe. The mixture is stirred for 1 h at 80°C and then the remaining initiator solution is added within 45 min. The mixture is polymerized to full conversion by keeping it for additional 3.5 h at 80°C. After cooling to room temperature one obtains a light yellow, viscous polymer solution in water (about 30 wt%).

Characterization of the Graft Copolymer

Two solutions have to be prepared:
- Polymer solution A: 3 g of the obtained graft polymer solution + 6 ml water (=10 wt% solution)

- Polymer solution B: 9 g 10 wt% aqueous polyvinylpyrolidone (M_n about 50,000 g/mol, e.g., Luviskol K30 from BASF AG) solution + 1 g 10 wt% aqueous polyvinylalcohol solution (M_n about 30,000 g/mol).

About 10 drops of solution A and B, respectively, are added dropwise to 30 ml of an acetone/ethanol mixture (2:1). While the ungrafted mixture causes a visible clouding, the graft product is fully soluble.

Bibliography

Allen G, Bevington J (eds) (1989) Comprehensive polymer science, vols 3 and 4. Pergamon, Oxford

Amajjahe S, Choi S, Munteanu M, Ritter H (2008) Angew Chem Int Edit 41:716

Benaglia M, Rizzardo E, Alberti A, Guerra M (2005) Macromolecules 38:3129–3140 (RAFT, see Example 3.15b)

Brunelle DJ (ed) (1996) Ring opening polymerization: mechanisms, catalysis, structure, utility. Hanser, Munich

Buchmeiser MR (2000) Chem Rev 100:1565 (ROMP)

Buchmeiser MR (2005) Adv Polym Sci 176:89 (ROMP)

Cherdron H, Brekner M-J, Osan F (1994) Cycloolefin copolymers. Angew Makromol Chem 223:121

Chiefari J, Chong YK, Ercole F, Krstina J, Jeffery J, Le TPT, Mayadunne RTA, Meijs GF, Moad CL, Moad G, Rizzardo E, Thang SH (1998) Macromolecules 31:5559

Corpart P, Charmont D, Biadatti T, Zard S, Michelet D (1998) WO 0.858.974 to Rhodia Chimie, Invs (MADIX)

Cowie JMG (ed) (1985) Alternating copolymers. Plenum, New York/London

Fink G, Mülhaupt R, Brintzinger HH (1995) Ziegler catalysts. Springer, Berlin/Heidelberg/New York

Fouassier JP (1995) Photoinitiation and photocuring – fundamentals and applications. Hanser, Munich

Grubbs RH (2003) In: Grubbs RH (ed) Handbook of metathesis, vol 1–3, 1st edn. Wiley VCH, Weinheim

Hawker CJ, Bosman AW, Harth E (2001) New polymer synthesis by nitroxide mediated living radical polymerization. Chem Rev 101:3661–3688

Houben-Weyl (1987) Methoden der organischen Chemie. Makromolekulare Stoffe, vol E20. Thieme, Stuttgart/New York

Ivin KJ, Saegusa T (eds) (1984) Ring opening polymerization, vol 3. Elsevier, London/New York

Jeromin J, Noll O, Ritter H (1998) Macromol Chem Phys 199:2641 (Cyclodextrin in Polymerization)

Kaminsky W, Arndt M (1997) Metallocenes for polymer catalysis. Adv Polym Sci 127:143

Komber H, Erber M, Däbritz F, Ritte H, Stadermann J, Voit B (2011) Macromolecules 44(9):3250

Kronganz VV, Trifunac AD (eds) (1995) Processes in photoreactive polymers. Chapman and Hall, New York

Matsumura S, Kobayashi S, Ritter H, Kaplan D (2006) Adv Polym Sci 194:95; Kobayashi S (2009), Macrol Rapid Commun 30:237

Matyjaszewski K, Xia J (2001) Atom transfer radical polymerization. Chem Rev 101:2921–2990

Moad G, Solomon DH (1995) The chemistry of free radical polymerization. Pergamon-Elsevier, Oxford

Quinn JF, Barner L, Barner-Kowollik C, Rizzardo E, Davis TP (2002) Macromolecules 35:7620 (RAFT)

Ritter H, Tabatabai M (2002) Advanced macromolecular and supramolecular materials and processes. Kluwer Academic/Plenum publishers (Cyclodextrin in polymerization), New York, pp 41–53

Ruhl T, Spahn P, Hellmann GP (2003) Polymer 44:7625 (Artifical opals prepared by melt compression; see Example 3.43)

Scheirs J, Kaminsky W (1999) Metallocene-based polyolefins, vol 2. Wiley-VCH, Weinheim

Schrock RR (2009) Chem Rev 109:3211–3226 (ROMP)

Swarc M (1996) Ionic polymerization fundamentals. Hanser, Munich

Szejtli J (1998) Cyclodextrin technology. Kluwer, Dordrecht

Togni A, Haltermann RL (1998) Metallocenes, vol 2. Wiley-VCH, Weinheim

Trnka TM, Grubbs RH (2001) Acc Chem Res 34:18 (ROMP)

Wenz G (1994) Angew Chem 106:851 (Cyclodextrin in polymerization)

Synthesis of Macromolecules by Step Growth Polymerization

4

4.1 Condensation Polymerization (Polycondensation)

Condensation polymerizations (polycondensations) are stepwise reactions between bifunctional or polyfunctional components, with elimination of small molecules such as water, alcohol, or hydrogen and the formation of macromolecular substances. For the preparation of linear condensation polymers from bifunctional compounds (the same considerations apply to polyfunctional compounds which then lead to branched, hyperbranched, or crosslinked condensation polymers) there are basically two possibilities. One either starts from a monomer which has two unlike groups suitable for polycondensation (AB type), or one starts from two different monomers, each possessing a pair of identical reactive groups that can react with each other (AABB type). An example of the AB type is the polycondensation of hydroxycarboxylic acids:

$$n \; HO-(CH_2)_x-COOH \xrightarrow{-(n-1) H_2O} H[O-(CH_2)_x-CO]_nOH$$

An example of the AABB type is the polycondensation of diols with dicarboxylic acids:

$$n \; HO-(CH_2)_x-OH + n \; HO-CO-(CH_2)_y-COOH \xrightarrow{-(2n-1) H_2O} H[O-(CH_2)_x-O-CO-(CH_2)_y-CO]_nOH$$

The formation of a condensation polymer is a stepwise process. Thus, the first step in the polycondensation of a hydroxycarboxylic acid (AB type) is the formation of a dimer that possesses the same end groups as the initial monomer:

$$2 \; HO-(CH_2)_x-COOH \xrightarrow{-H_2O} HO-(CH_2)_x-CO-O-(CH_2)_x-COOH$$

D. Braun et al., *Polymer Synthesis: Theory and Practice*,
DOI 10.1007/978-3-642-28980-4_4, © Springer-Verlag Berlin Heidelberg 2013

The end groups of this dimer can react in the next step either with the monomeric compound or with another dimer molecule, and so on. The molecular weight of the resulting macromolecules increases continuously with reaction time, unlike many addition polymerizations, e.g., radical polymerizations. The intermediates that are formed in independent, individual reactions, are oligomeric and polymeric molecules with the same functional end groups as the monomeric starting compound. In principle, these intermediates can be isolated without losing their capability for further growth. Thus, all reactions

$$M_i + M_j \rightarrow M_{j+i}$$

can occur, where M_i, M_j denote oligomeric or polymeric species containing i, j monomeric units, respectively. The formation of a reaction scheme that takes into account the multitude of possible reactions between i-mers and j-mers, with different rate constants k_{ij}, and with the prevailing concentrations $[M_i]$ and $[M_j]$, would be extraordinarily complicated. If, however, one assumes that in each single step a reaction between an end group of the molecule M_i with an end group of the molecule M_j takes place in such a way that it is independent of i or j, the kinetic treatment of polycondensation is considerably simplified.

Thus, it is assumed that the reactivity is independent of molecular weight. This "principle of equal reactivity" holds true for both condensation and stepwise addition polymerizations. It means that there is no difference in the reactivity of the end groups of monomer, dimer, etc. and, therefore, that the rate constant is independent of the degree of polymerization over the total duration of reaction.

On this basis the kinetics of polycondensation were worked out a long time ago. The following points are of particular interest:
- Dependence of the average molecular weight on conversion,
- Dependence of the average molecular weight on the molar ratio of reactive groups,
- Dependence of the conversion and average molecular weight on the condensation equilibrium, and
- Exchange reactions such as transesterification or transamidation.

The progress of a polycondensation can be followed in a simple manner by analysis of the unreacted functional groups. If the reactive groups are present in equimolar amounts, which is generally desired (see below), it is sufficient to analyze for one of the two groups, for example, the carboxyl groups in polyester formation. If the number of such functional groups initially present is N_0, and the number at time t is N, the extent p of condensation is defined as the fraction of functional groups that have already reacted at that time:

$$p = \frac{N_0 - N}{N_0} \tag{4.1}$$

Multiplying p by 100 yields the conversion in %.

Table 4.1 Degree of polymerization P_n and number-average molecular weight M_n (assuming a structural element of molecular weight 100) as a function of conversion in condensation and stepwise addition polymerizations

Conversion (%)	P_n	M_n
50	2	200
75	4	400
90	10	1,000
95	20	2,000
99	100	10,000
99.5	200	20,000
99.95	2,000	200,000

The number-average degree of polymerization P_n is defined as the number of structural units per polymer chain (N_0 = number of monomer molecules originally present, N = total number of molecules at the appropriate stage of reaction, including the as yet unconverted monomer molecules). Hence from Eq. 4.1 one obtains:

$$P_n = \frac{N_0}{N} = \frac{1}{1-p} \tag{4.2}$$

P_n is thus dependent on the conversion ($p \cdot 100$). This is illustrated by some numerical examples in Table 4.1.

From Table 4.1 it is clear that to attain the commercially interesting molecular weights of 20,000–30,000, the reaction must be driven to conversions of more than 99%.

The degree of polymerization P_n is affected not only by the yield, but also by the molar ratio of the reacting functional groups A and B. If N_A and N_B are the numbers of such groups originally present, and r is their ratio ($r = N_A/N_B$; $r \leq 1$), then the average degree of polymerization P_n can be expressed by modifying Eq. 4.2 to give Eq. 4.3:

$$P_n = \frac{1+r}{2 \cdot r \cdot (1-p) + 1 - r} \tag{4.3}$$

On equivalence of both functional groups A and B, which means $r = 1$, Eq. 4.3 reduces to Eq. 4.2.

If all the A groups have reacted (i.e., $p = 1$), Eq. 4.3 simplifies to:

$$P_n = \frac{1+r}{1-r} \tag{4.4}$$

In general, the effect of an excess of one component is greater, the higher the conversion. At 99% conversion the molecular weight is reduced to half its value by a 2% excess of one component; at 99.5% conversion the same effect is produced by

Table 4.2 Degree of polymerization P_n as a function of the ratio of the functional groups A and B in condensation and stepwise addition polymerizations (conversion 100% with respect to A, i.e., component A has completely reacted)

Excess of component B (mol%)	$r = N_A/N_B$	P_n
10	0.09	19
1	0.99	199
0.1	0.999	1,999
0.01	0.9999	19,999

a 1% excess. Some numerical examples are given in Table 4.2 for the effect of non-equivalence of components on the degree of polymerization (at 100% conversion of the lesser component).

Table 4.2 shows the importance of the exact equivalence of functional groups in polycondensation reactions, since even a 1 mol% excess of one of the two groups limits the maximum attainable degree of polymerization P_n to less than 200. For polycondensations of the AB type, e.g., hydroxycarboxylic acids or amino acids, this equivalence is automatic since the monomer contains both groups. On the other hand, for polycondensations of the AABB type, e.g., between diols and dicarboxylic acids, a small excess of one component causes the reaction to come to a halt when only the end groups of the component present in excess are left because these are unable to react with each other.

Even if both functional groups are present in equivalent amounts at the beginning of reaction, this equivalence can be disturbed during the course of reaction by evaporation, sublimation, or side reactions of one of the reaction partners. A monofunctional compound that can react with one of the bifunctional reaction partners acts in the same way as an excess of a bifunctional component. Therefore, high purity of the monomers is absolutely essential in the preparation of high-molecular-weight polymers by condensation and stepwise addition polymerization. On the other hand, this also means that it is possible to regulate precisely the average molecular weight of the polymer formed either by using a controlled excess of one component or by addition of a monofunctional compound.

The effects described by Eqs. 4.1, 4.2, 4.3, and 4.4 apply to both stepwise addition and condensation polymerizations. In polycondensation, two further factors must be considered: The condensation equilibrium and the exchange reactions. The *condensation equilibrium* limits the conversion and hence the average molecular weight. As in the case of esterification of monofunctional compounds, the corresponding polycondensations are to be treated as equilibrium reactions, governed by the law of mass action. For example, in the case of polyesterification, if 1 mol of hydroxy groups (1/2 mol of diol) reacts with 1 mol of carboxylic acid groups (1/2 mol of dicarboxylic acid), this takes the form (according to Flory and Schulz):

$$K = \frac{p \cdot n_w}{(1 - p)^2} \tag{4.5}$$

As before, p is the fraction of functional groups that have already reacted (see Eq. 4.1), i.e., the mols of ester groups formed (N_0 and N are the numbers of functional groups present at the start and a given time of the reaction, respectively), $(1-p)$ the molar quantity of unreacted hydroxy and carboxy groups, and n_w is the mole fraction of water present in the reaction mixture. Solving Eq. 4.5 for p we obtain the upper limit of conversion as a function of the ratio $\beta = K/n_w$:

$$p = \frac{1}{2\beta}\left(1 + 2\beta - \sqrt{1 + 4\beta}\right) \tag{4.6}$$

Since the number-average degree of polymerization P_n is reciprocally proportional to the free functional groups $1-p$ still present (see Eq. 4.2), the upper limit of the degree of polymerization, when governed by the condensation equilibrium, is given by Eq. 4.7:

$$P_n = \frac{2\beta}{\sqrt{1 + 4\beta} - 1} \tag{4.7}$$

Since, in practically all cases, one aims at as high a conversion as possible, i.e., a value of p tending towards unity, Eqs. 4.6 and 4.7 may be simplified to obtain:

$$p \approx 1 - \sqrt{\frac{1}{\beta}} \tag{4.8}$$

and

$$P_n \approx \sqrt{\beta} = \sqrt{\frac{K}{n_w}} \tag{4.9}$$

If the equilibrium constant K has a value between 1 and 10, less than a thousandth of the total amount of water formed in the reaction mixture is sufficient to prevent the formation of really high-molecular-weight condensation polymers. Hence it follows that it is extremely important to remove as completely as possible the low-molecular-weight reaction products, for example, water, eliminated during a polycondensation. In principle, these equilibriums are also known in stepwise addition polymerizations (polyaddition) like the back-reactions of urethane groups. Since they mostly occur at higher temperatures only, they can be neglected.

The second factor that additionally effects polycondensations are *exchange reactions* which can occur between free end groups and junction points in the chain, for example, between OH end groups and ester groups of a polyester (transesterification):

Neither the number of free functional groups, nor the number of molecules, nor the number-average degree of polymerization, is thereby altered. Thus, two equal-sized macromolecules could react with one another to give one very large and one very small macromolecule; conversely a very large and a very small macromolecule can be converted into two macromolecules of similar size. Independent of the initial distribution these exchange reactions will in each case lead to a state of dynamic equilibrium in which the rates of formation and consumption of molecules of a given degree of polymerization are equal. This results in an equilibrium distribution of molecular weights which is formally the same as that obtained by purely random polycondensation. Therefore, in normal polycondensations the exchange reactions will not affect the molecular-weight distribution. On the other hand by mixing a low molecular weight and a high-molecular-weight polyester in the melt, especially if a catalyst is still present, the molecular-weight distribution very soon adjusts to an equilibrium distribution with a single maximum, instead of the two in the original mixture. Such exchange reactions between the end groups and the junction positions are known for polyesters, polyamides, polysiloxanes, and polyanhydrides.

Finally, it has to be considered that also the formation of macrocyclic oligomers and polymers cannot be fully neglected. For example, by use of mass spectrometry, the existence of ring-shaped polycondensates is definitely elucidated in all types of polycondensates.

Although the procedures for the preparation of condensation polymers are analogous in many respects to those for the condensation reactions of monofunctional compounds, some additional factors must be taken into account if one wishes to attain high molecular weights in polycondensations. These are essentially consequences of Eqs. 4.1, 4.2, 4.3, 4.4, 4.5, 4.8, and 4.9. Firstly, the condensation reaction must be specific and proceed with the highest possible yield, otherwise only mixtures of oligomers will be obtained. Furthermore, for polycondensation reactions in which two or more components may participate, care must be taken to ensure strict equivalence in the proportions of the reacting groups through-out the reaction. The equilibrium position of the reaction must be displaced as far as possible towards condensation. By analogy with the condensation of monofunctional compounds this may be achieved by removing the low-molecular-weight reaction product, e.g., water, as completely as possible from the reaction mixture. This can be done by distillation under high vacuum or by azeotropic distillation. It is advantageous to pass very dry inert gas through the well-stirred reaction mixture in order to facilitate the diffusion of the eliminated component from the viscous solution formed during polycondensation. High demands are placed on the purity of the starting materials. They must be especially free from monofunctional compounds since these block the end groups of the resulting macromolecules and so prevent further condensation. Only bifunctional compounds can be used for the preparation of linear condensation polymers, since polyfunctional compounds give rise to branching, hyperbranching, and crosslinking. Finally, polycondensation reactions should be carried out under exclusion of oxygen, since oxidative decomposition can easily occur at the high reaction temperatures that are frequently needed.

4.1.1 Polyesters

Polyesters are macromolecules whose monomeric units are linked with an ester group:

$$H \left[O-(CH_2)_x-O-\overset{\displaystyle O}{\overset{\displaystyle \|}{C}}-(CH_2)_y-\overset{\displaystyle O}{\overset{\displaystyle \|}{C}} \right]_n OH$$

Their properties depend markedly on their chemical composition. Pure aliphatic polyesters are generally crystalline and have melting points below 100°C (with the melting point increasing with the number of methylene groups between the ester groups). At room temperature they are soluble in formic acid, methylene chloride, and dichlorobenzene, and they are easily hydrolyzed. In contrast, polyesters made from aromatic or cycloaliphatic dicarboxylic acids and diols, like polyesters made from terephthalic acid and ethylene glycol, or terephthalic acid and 1,4-cyclohexylenedimethanol, have different properties. They also crystallize but have significantly higher melting points. At room temperature, they are insoluble and hard to hydrolyze. Additionally, they have good mechanical properties, which make them suitable for the preparation of fibers, films, and moldable plastics.

Unusual properties of fully aromatic polyesters are observed if they have at least partially a rigid planar chain structure. In particular, they can form thermotropic liquid crystalline states (see Example 4.5). As already discussed in Sect. 1.2.4 an important structural prerequisite for LCPs of Type A in order to attain the liquid crystalline state of aromatic polyesters (and aromatic polyamides, see Example 4.14), is a rigid main chain according to the following construction principle:

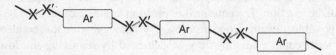

The flat-structured aromatic units (Ar) are responsible for the formation of the liquid crystalline domains. The two-atom linking group –X–X′– should be stabilized by additional π-bonds that can contribute to the rigidity of the chain. Suitable are azo, azomethine, ethenyl, ester, and amide groups. For polyaramides, processed from solution, a strong *para*-linkage is favorable (see Example 4.14). In contrast to this, an all *para*-linkage in thermotropic polyesters leads to high melting points and high melt viscosities, so that a thermoplastic processing is mostly impossible. Depression of the melting point by maintaining the possibility to form liquid crystalline states is mainly achieved by the following two concepts:
– Insertion of appropriate side groups,
– Insertion of discontinuity in the main chain.

The effect of the incorporation of side groups into aromatic polyester is demonstrated in Example 4.5.

Discontinuities in the main chain can be obtained by different means. Example 4.6 describes the incorporation of flexible aliphatic segments. Of special interest is the overall orientation of incorporated mesogens by mechanically or electrically applied external forces.

From a practical point of view, the deviation from main-chain continuity is the most interesting one. Especially the concept of the so-called parallel offset of the main chain is favored. It is realized in a commercial product made from 4-hydroxybenzoic acid and 2,6-hydroxynaphthalenecarboxylic acid. The incorporation of small amounts of 2,6-hydroxynaphthalenecarboxylic acid causes a discontinuity in the main chain but only in form of a parallel offset of some chain segments. Thus, the ability to form liquid crystalline states is largely maintained as well as the anisotropic properties of the molten and solid polyester. Instead, the melting point is significantly reduced, as desired.

The preparation of fully aromatic polyesters is mostly performed by melt condensation. Complex phase equilibriums and states of aggregation are formed in this manner. This is in most cases not easy to control under laboratory conditions. Example 4.5 describes a polyester that is easy to prepare under laboratory conditions and allows the observation of typical properties of some thermotropic liquid crystalline polymers.

From an industrial point of view, not only the high-molecular-weight linear polyesters are of interest. Also, a series of low-molecular-weight linear or branched polyesters (Example 4.1) find application in surface coating systems (alkyd resins), as coreactants in unsaturated polyester resins (Example 4.8), or in polyurethane foams (Examples 5.26 and 5.27).

Polyesters are prepared by the following methods:
- Polycondensation of hydroxycarboxylic acids,
- Polycondensation of diols with dicarboxylic acids,
- Polycondensation of diols with derivatives of dicarboxylic acids (e.g., cyclic dicarboxylic acid anhydrides, dicarboxylic acid chlorides, or dicarboxylic acid esters),
- Polycondensation of dicarboxylic acids with derivatives of diols (e.g., acetates of bisphenols), or
- Ring-opening polymerization of lactones (see Sect. 3.2.3.3)

The establishment of the equilibrium is often accelerated by acidic or basic catalysts, for example, by strong acids (p-toluenesulfonic acid), metal oxides (antimony trioxide), Lewis acids (titanium tetrabutoxide, tin acetates or tin octoates), weak acid salts of alkali metals or alkaline earth metals (acetates, benzoates), or by alcoholates.

4.1.1.1 Polyesters from Hydroxycarboxylic Acids

The synthesis of polyesters from hydroxycarboxylic acids belongs to the polycondensation of the AB type. Both reacting groups are part of the same monomer. Polycondensation of aliphatic hydroxycarboxylic acids, such as glycolic acid, lactic acid, or 12-hydroxy-9-octadecenoic acid (ricinoleic acid), are mostly carried out in the melt and yield high-molecular-weight polymers. However, these reactions often result in the formation of cyclic esters (lactones). The tendency towards ring formation depends on the number of methylene groups between the hydroxy and carboxylic acid groups; it is greatest for γ-hydroxybutyric acid, δ-hydroxyvaleric acid, and ε-hydroxycaproic acid. The polycondensation of aromatic hydroxycarboxylic acids, e.g., 4-hydroxybenzoic acid and 2,6-hydroxynaphthalenecarboxylic acid, which are very useful components in the synthesis of thermotropic liquid-crystalline polyesters, has gained importance in recent years.

4.1.1.2 Polyesters from Diols and Dicarboxylic Acids

Polycondensation of diols with dicarboxylic acids is often performed in the melt. However, it does not always lead to high-molecular-weight polyesters. Sometimes, the starting materials or the resulting polyester are thermally unstable at the high condensation temperatures. If the reactants and the polyester are well soluble, one can carry out the polycondensation in solution (see Example 4.2). The elimination of water from diols and dicarboxylic acids frequently occurs rather slowly. In such cases suitable functional derivatives of the diols and dicarboxylic acids (esters or anhydrides) can be used instead of the direct condensation, as described in Sect. 4.1.1.3.

Example 4.1 Preparation of a Low-Molecular-Weight Branched Polyester from a Diol, a Triol and a Dicarboxylic Acid by Melt Condensation

Safety precautions: Before this experiment is carried out, Sect. 2.2.5 must be read as well as the material safety data sheets (MSDS) for all chemicals and products used.

(a) *Preparation of a Slightly Branched Polyester*

A 250 ml three-necked flask is fitted with a nitrogen inlet, a stirrer, a thermometer that reaches deep into the flask, a fractionating column of about 20-cm length packed with Raschig rings, at the upper end the another thermometer, and a condenser for distillation, with vacuum adapter and graduated receiver. 73 g (0.5 mol) of adipic acid, 55.3 g (0.52 mol) of anhydrous bis(2-hydroxyethyl) ether, (diglycol), and 4.5 g (0.0335 mol) of anhydrous 1,1,1-tris-(hydroxymethyl) propane are weighed in, some antibump granules added, and the air is displaced by evacuation and filling with nitrogen. The flask is then heated slowly under a gentle stream of nitrogen; when the contents become fluid between 80°C and 110°C, the stirrer is switched on. Polyesterification sets in at 130–140°C as evidenced by the formation of water. The internal temperature is now raised to 200°C at such a rate that the temperature at the head of the column does not exceed 100°C. During this period (normal pressure phase) most of the water (18 g) is eliminated; it is important not to interrupt the stirring because of the danger of bumping.

As soon as the temperature at the head of the column drops much below 100°C, after the internal temperature has reached 200°C, the vacuum pump is attached and the pressure reduced to 12–14 Torr. At this stage of the experiment (vacuum phase), distillation of the diglycol must be completely avoided; if necessary one must evacuate more slowly. The reaction is now allowed to continue at 12–14 Torr and 200°C, samples being taken at intervals of 5 h in order to determine the acid number (see below). In order to take the samples, the apparatus is momentarily filled with nitrogen and then evacuated again. When the acid number is smaller than 2 (total time about 35–40 h) the experiment is terminated. Addition of a catalyst, e.g., 5 ppm $SnCl_2$ before the vacuum phase reduces the reaction time about 25–30%. The slightly branched polyester remains, after cooling under nitrogen, as a viscous, pale yellow mass. It has an OH number of about 60 and can be directly processed and converted to an elastic polyurethane foam (see Example 5.26).

(b) *Preparation of a Highly Branched Polyester*

8.45 g (0.06 mol) of adipic acid, 21.3 g (0.075 mol) of oleic acid, 22 g (0.15 mol) of phthalic anhydride, and 52.5 g (0.39 mol) of anhydrous 1,1,1-tris-(hydroxymethyl)propane are placed into the apparatus as described under (a) and the air is displaced by evacuation and filling with nitrogen. The mixture is slowly heated under a stream of nitrogen; at 120°C the contents of the flask have melted and the stirrer can be started. The internal temperature is raised to 190°C over a period of 2 h; as soon as the temperature at the top of the column drops below 70°C, the pump is attached and the apparatus slowly evacuated to 40 Torr over the course of 2 h. At this pressure and an internal temperature of 190°C, the mixture is stirred for a further 8 h. The pump is then switched off and the product allowed cooling under nitrogen. The highly branched polyester, which remains as a viscous liquid, has an acid number of 2 and an OH number of 350; it can be used directly for the preparation of a rigid polyurethane foam (see Example 5.27). Addition of 5 ppm $SnCl_2$ as catalyst reduces the reaction time by 10%.

(c) *Determination of the Acid Number*

1–2 g of the polyester are dissolved by warming with 50 ml of acetone and, after cooling, are titrated as quickly as possible with 0.1 M alcoholic potassium hydroxide using phenolphthalein as indicator, until the red color remains for a second. The alkali requirement of the solvent is determined in a blank experiment. The acid number is given by the weight of KOH in mg, required to neutralize 1 g of substance:

$$\text{acid number} = 5.61 \cdot f \cdot (A - B)/E$$

where
A = titer for the sample in ml
B = titer for the blank in ml
E = weight of sample in g
f = concentration of the alcoholic KOH in mol/l

(d) Determination of the Hydroxy Number

To acetylate the free hydroxy groups of a polyester, a solution of 15 g of freshly distilled acetic anhydride (bp 140°C) in 35 g of freshly distilled dry pyridine is prepared. The solution, which turns slightly yellow with time, is stored in a dark bottle.

To determine the hydroxy number, about 2.5 g of polyester are weighed to the nearest 10 mg into each of three conical flasks with ground joints. The samples are each dissolved with gentle warming in 10 ml of pyridine. After cooling, 10 ml of the acetylating solution are added with a pipette to each of the flasks. 10 ml of pyridine and 10 ml of acetylating solution are placed in each of two conical flasks with ground joints, for the determination of the blank value. All five flasks are closed with stoppers fastened with adhesive tape. They are then placed in an oven at 110°C for 70 min. (Caution when removing: Use safety goggles and safety gloves!). After cooling, a few drops of 0.1% aqueous Nile blue chloride solution are added to each of the samples which are then titrated with 1 M NaOH solution until the color changes suddenly to violetred. The hydroxy number indicates how many mg KOH are equivalent to the free hydroxy groups present in 1 g of substance.

$$\text{OH number} = 56.1 \cdot f \cdot (B - A)/E$$

where
A = titer for the sample in ml
B = titer for the blank in ml
E = weight of sample in g
f = concentration of the NaOH in mol/l.

The average value is taken of those values which do not differ by more than ±2 from each other.

Example 4.2 Preparation of a High-Molecular-Weight Linear Polyester from a Diol and a Dicarboxylic Acid by Condensation in Solution

Safety precautions: Before this experiment is carried out, Sect. 2.2.5 must be read as well as the material safety data sheets (MSDS) for all chemicals and products used.

The recycling apparatus shown in Fig. 2.4 is charged with 29.55 g (0.25 mol) of 1,6-hexanediol (purified by vacuum distillation), 29.55 g (0.25 mol) of recrystallized succinic acid, 100 ml of dry toluene, and 0.75 g of pure p-toluenesulfonic acid; at the same time, the siphon is also filled with toluene so that the circulation of the solvent can begin immediately. After adding some antibump granules the solution is heated on an oil or air bath to boiling; the toluene should flow quickly through the drying tube filled with soda-lime back into the flask. After some hours, when about three-quarters of the theoretical amount of water has collected in the separator, the soda-lime is renewed for the first time; it is renewed again after another 10 h.

The viscosity of the solution gradually increases and so does the temperature in the flask. In order to maintain a quick rate of distillation (and therefore polycondensation), each time the internal temperature reaches 130°C about 25 ml of pure toluene are added. After about 25 h the flask is cooled to room temperature and the solution is added drop-wise to a tenfold amount of methanol; the polymer is filtered off and dried to constant weight in vacuum at 40°C. Yield: about 90%.

The linear polyester so obtained is of higher molecular weight than that produced by melt polycondensation. It is soluble in toluene, chloroform, and formic acid, and readily hydrolyzable (see Example 5.1); the melting point is about 100°C. The increase of molecular weight during polycondensation can be followed by taking samples from time to time (initially at short intervals) and determining the limiting viscosity numbers of the precipitated and dried samples; the carboxyl and hydroxy end groups can also be quantitatively determined (see Example 4.1).

Hyperbranched Polymers

Hyperbranched polymers are a relatively new type of highly branched materials, which in contrast with dendrimers, often can be prepared in a one-step synthesis of AB_x (x ≥ 2) monomers. The growth proceeds in an uncontrolled fashion and the development of the structure is statistic, as for linear polymerizations. The resulting macromolecules, as dendrimers, have a high density of functional groups. The number of these groups can be directly connected with the degree of polymerization (DP): if the reaction is performed with an AB_2 monomer, each molecule will have exactly one functional group more than the number of repetition units (number of B groups: DP + 1). Dendritic, linear, and terminal units can be identified, depending on the number of reacted functional groups: in a dendritic unit both B groups have been involved in the reaction; while in linear and terminal units, only one, and none of the B groups respectively, have been employed, and can therefore still be used for successive modifications. The impossibility of forming "entanglements" and the highly compact structure are most certainly responsible for their lower solution viscosities and the better solubilities, compared to the linear analogues. Furthermore, most hyperbranched polymers are amorphous and, therefore, no crystallite melting point T_m can be detected.

Schematic representation of hyperbranched polymer (A = focal unit)

By modifying the functional groups they can be used, for example, as crosslinkers in high solid or powder coatings and in thermosets. Because of their good miscibility and low melt viscosity, they find applications as melt modifiers and as blend components. Modified hyperbranched polymers, like alkyl chain substituted poly(ether)s and poly(ester)s sometimes exhibit amphiphilic behavior. They can, therefore, be used as carriers for smaller molecules, for example, dyestuffs into polypropylene.

Example 4.3 Preparation of a Hyperbranched Polyester by Polycondensation of 4,4-bis(4'-hydroxyphenyl)Valeric Acid

Safety precautions: Before the experiment is carried out, Sect. 2.2.5 must be read as well as the material safety sheets (MSDS) for all chemicals and products used. Due to the high temperature of the oil bath needed in this experiment, special care has to be taken and one has to work in a closed hood. In addition, a safety shield is required when working under high vacuum.

4,4-Bis(4'-hydroxyphenyl)valeric acid (15.1 g, 52.7 mmol) is a commercial monomer but has to be used in high purity (not colored). If the purity of the chemical as received is not sufficient it has to be purified by column chromatography.

4,4-Bis(4'-hydroxyphenyl)valeric acid and 3 drops of dibutyl tin diacetate are placed in a dry standard apparatus with mechanical stirrer connected to a vacuum line. The polycondensation is carried out in two steps in order to obtain high conversion and high molar mass: First, the monomer is stirred in a nitrogen stream for 70 min at 200°C; during this time period an increase in viscosity of the melt should be observed. After that, vacuum is applied at 225°C for 3 h with stirring and then for 1 h without stirring until a final vacuum of 0.010 mbar is reached. The crude product is dissolved twice in approximately 100–150 ml THF and precipitated into 500 ml water. After filtration the product is dried over phosphorus pentaoxide for several days.

The molar mass of the products is very dependent on the final conversion achieved and thus, on the final quality of the vacuum. Under the described conditions number average molar masses as determined by GPC in DMAc/LiCl (3 g/l)/water (2 vol%) will reach values between 2,000–5,000 g/mol.

The polymers are easily soluble in polar solvents like THF, DMAc, and DMSO, indicating that the solubility is enhanced by the large number of polar end groups and the branched structure. It is fully amorphous with a glass transition temperature between 90°C and 110°C, depending on molar mass.

The degree of branching is determined by NMR analysis of the product in DMSO-d6. The different structural isomers (linear l, dendritic d, terminal t units) can be identified best by analyzing the quarternary carbon found at about 148 ppm (up to 6 signals for this carbon for a sample containing oligomers). However, quantitative analysis of the ^{13}C NMR requires special conditions. Similarly, the proton NMR gives three signals for the methyl group corresponding to the different structural units at about 1.5 ppm which can be quantified (d = 1.63 ppm, l = 1.58 ppm, t = 1.53 ppm). The degree of branching is calculated by $(I_t + I_d)/(I_t + I_d + I_l)$ and should be close to 0.5 (I = intensity of the proton NMR signal).

4.1.1.3 Polyesters from Diols and Dicarboxylic Acid Derivatives

The polycondensation of a diol and the diester of a dicarboxylic acid (e.g., the dimethyl ester) can be carried out in the melt at a considerably lower temperature than for the corresponding reaction of the free acid. Under the influence of acidic or basic catalysts a transesterification occurs with the elimination of the readily volatile alcohol (see Example 4.3). Instead of diesters of carboxylic acids one can also use their dicarboxylic acid chlorides, for example, in the synthesis of high-melting aromatic polyesters from terephthaloyl dichloride and bisphenols. The commercially very important polycarbonates are obtained from bisphenols and phosgene, although the use of diphenyl carbonate as an alternative component is of increasing interest (see Example 4.4). Instead of free acids, cyclic carboxylic acid anhydrides (e.g., maleic anhydride or phthalic acid anhydride) are suitable for the synthesis of polyesters. Unsaturated polyester resins, which are synthesized from diols and maleic anhydride (together with other dicarboxylic acid anhydrides) possess great importance. The maleic ester units, which are incorporated into the polyester chain (or fumaric ester units which are formed by thermal rearrangement) can be crosslinked with styrene by radical copolymerization (see Example 4.8).

Example 4.4 Preparation of Polyester from Ethylene Glycol and Dimethyl Terephthalate by Melt Condensation

Safety precautions: Before this experiment is carried out, Sect. 2.2.5 must be read as well as the material safety data sheets (MSDS) for all chemicals and products used.

Pure ethylene glycol is dried by refluxing for 1 h in the presence of 2 wt% metallic sodium and is then distilled. Dimethyl terephthalate is recrystallized from methanol and carefully dried in vacuum (mp 141–142°C). 9.7 g (0.05 mol) of dimethyl terephthalate, 7.1 g (0.115 mol) of ethylene glycol, 0.015 g of pure anhydrous calcium acetate, and 0.04 g of pure antimony trioxide are weighed into a 50 ml round-bottomed flask. The flask is then fitted with a Claisen stillhead, an air-cooled condenser, a vacuum adapter, and a graduated receiver (or a measuring cylinder). The air is removed by evacuating and filling with nitrogen, and the components melted on an oil or metal bath at 170°C. The Claisen head is now immediately fitted with a capillary tube (attached to a ground joint) that reaches to the bottom of the flask and a slow stream of nitrogen is passed through. The transesterification sets in almost at once; the progress of reaction is followed by the amount of methanol that has distilled over into the graduated receiver. When the methanol production decreases (after about 1 h) the temperature is raised to 200°C for 2 h, whereby the remainder of the methanol distills over. Finally, the excess ethylene glycol is distilled over during 15 min at 220°C, the temperature then being raised to 280°C. After 15 min the graduated receiver is replaced by a simple round-bottomed flask and the apparatus gradually evacuated while keeping the temperature steady at <0.5 Torr. After another 3 h the polycondensation is complete. The flask is allowed to cool under nitrogen and then broken carefully with a hammer to remove the poly(ethylene terephthalate). The polyester is soluble in *m*-cresol and can be reprecipitated in ether or methanol. The limiting viscosity number is determined in *m*-cresol or in a mixture of equal parts of phenol and tetrachloroethane (see Sect. 2.3.3.3); the softening range is also determined. Fibers can be spun from the melt or pulled out from the melt by means of a glass rod.

Example 4.5 Preparation of a Polycarbonate from 4,4-Isopropylidenediphenol (Bisphenol A) and Diphenyl Carbonate by Transesterification in the Melt

Safety precautions: Before this experiment is carried out, Sect. 2.2.5 must be read as well as the material safety data sheets (MSDS) for all chemicals and products used.

The reaction between diols and phosgene is especially useful for the preparation of polycarbonates. The commercially very important polycarbonates are obtained from aromatic dihydroxy compounds (bisphenols). In contrast to the products of the reaction of aliphatic diols and phosgene they show considerable resistance to water but they are degraded by strong alkali, ammonia, and amines. Due to their angled chain structure with relatively low mobility, the aromatic colorless polycarbonates exhibit high glass transition temperatures, and they are often amorphous and highly transparent. Furthermore, they have good mechanical and optical properties.

The reaction between a dihydroxy compound (bisphenol) and phosgene, which is performed on an industrial scale, proceeds even at room temperature. The reaction is generally carried out in a biphasic medium consisting of methylene chloride (with dissolved phosgene) and aqueous sodium hydroxide (with dissolved bisphenol sodium salt) and a phase transfer catalyst (e.g. triethylamine). The procedure is termed interfacial polycondensation (see Sect. 4.1.2.3 and Examples 4.5, 4.12, and 4.13).

Another useful method is the transesterification of diphenyl carbonate with bisphenols under elimination of phenol. However, the major disadvantages are that the process requires high temperature and vacuum. The polycondensation reaction is carried out stepwise, i.e., the temperature and vacuum are raised slowly, in order to avoid diphenyl carbonate loss by sublimation at the beginning of the reaction. However, an excess of diphenyl carbonate is necessary in order to obtain high molecular weight.

This method of transesterification is of high technical interest. Particularly the reaction of bisphenol A with diphenyl carbonate is a preferred phosgene-free process because biphenyl carbonate can be obtained directly from phenol and dimethyl carbonate. The latter is an industrial product made from CO and methanol.

The following example illustrates this synthetic method.

Experimental Procedure
This reaction should be carried out in a hood behind a protective shield. Bisphenol A powder is particularly irritating to the eyes. All contact with this substance is, therefore, to be avoided.

A 250-ml three-necked flask is fitted with metal stirrer, nitrogen inlet, Vigreux column (30 cm), and condenser for distillation with a vacuum adapter and round flask as receiver. The high-vacuum pump must be connected over a cold trap with the vacuum adapter.

9.12 g (0.04 mol) bisphenol A (4,4′-isopropylidenediphenol), 9.42 g (0.044 mol) diphenyl carbonate, and 1.6 mg sodium methoxide are placed in the reaction flask under a stream of nitrogen. The flask is then evacuated to about 30 Torr and placed in a preheated (150°C) metal bath (Woods metal), whereby the components melt. The mixture is now vigorously stirred and gradually (about 1°C/min) heated to 220°C under reduced pressure. After the larger part of segregated phenol is distilled off, the pressure is reduced to 7 Torr, and the mixture heated to 235°C over a period of 1 h. Subsequently, the temperature is increased to about 280–300°C under higher vacuum (<0.1 Torr) during a period of 2 h. It is held at this temperature until the reaction mixture becomes very viscous. The vacuum is then released and the molten mass poured into a porcelain dish.

The solidified polycarbonate is a colorless, glassy material, which is soluble in methylene chloride, chloroform, pyridine, 1,4-dioxane, and tetrahydrofuran.

For characterization, the polymer is dissolved in methylene chloride, precipitated again in methanol and dried at 50°C in vacuum. The limiting viscosity number is determined in tetrahydrofuran at 20°C and the molecular weight calculated (see Sect. 2.3.3.3). Additional information can be obtained by infrared spectroscopy. There should be no absorption of the OH-group at ~3,400 cm^{-1}.

Polycarbonates find use as amorphous thermoplasts in nonbreakable windshields, housings for electric and electronic equipment, and as substrate in optical storage devices (CD, DVD).

Example 4.6 Preparation of a Liquid Crystalline (LC), Aromatic Main-Chain Polyester by Polycondensation in the Melt

Safety precautions: Before this experiment is carried out, Sect. 2.2.5 must be read as well as the material safety data sheets (MSDS) for all chemicals and products used.

This example describes the lowering of the melting point by introduction of side groups.

The highest purity of monomers and exact control of the reaction conditions are necessary for the synthesis of aromatic polyesters.

Purification of the Monomers
Phenylhydroquinone: 20 g of phenylhydroquinone are purified by fractional vacuum distillation (p < 0.01 Torr) and stored under dry nitrogen.

Bromoterephthaloyl chloride: A mixture of 10 g bromoterephthalic acid, 20 g freshly distilled thionyl chloride, and 1 ml of DMF is refluxed under a stream of nitrogen for 3 h. Next, the excess thionyl chloride is removed by vacuum distillation (about 10 Torr). The remaining bromoterephthalic dichloride is converted to bromoterephthalic acid dimethyl ester by adding the fourfold amount of anhydrous methanol. The excess of methanol is taken off on a rotary evaporator or by distillation and the pure bromoterephthalic acid dimethyl ester can be isolated by fractional distillation (p < 0.01 Torr) over a short column. The purity of bromoterephthalic acid dimethyl ester can be checked by gas chromatography. The pure bromoterephthalic acid dimethyl ester is converted to the acid by saponification with the fourfold amount of sodium hydroxide in boiling methanol. The bromoterephthalic acid is precipitated by addition of half-concentrated hydrochloric acid. It is filtered after cooling and washed with demineralized water and dried. Pure bromoterephthalic acid is placed in an exactly weighed reaction flask (the polycondensation reaction is carried out in the same flask), then freshly distilled thionyl chloride and 1 ml of DMF are added and treated as described above. After the reaction is complete, the excess of thionyl chloride is distilled off by vacuum distillation. Now the air is removed from the apparatus by repeated evacuation and filling with nitrogen.

Preparation of Polyester by Melt Condensation
The exact amount of bromoterephthaloyl chloride is calculated by subtracting the weight of the empty flask from that of the reaction flask. Under a stream of nitrogen an equimolar amount of phenylhydroquinone is added. The further conversion has to be carried out according to a precise temperature program. The mixture is heated under a slow stream of nitrogen (important for the removal of hydrochloric acid; the pH is checked using moist pH paper) with stirring (with a sealed stirrer) for 2 h at 120°C, then 2 h at 180°C, 1 h at 200°C, 1 h at 220°C, and finally 2 h at 240°C. After cooling, the solid residue is ground, dissolved in a mixture of 1,1,2,2-tetrachloroethane (the dissolution takes place more rapidly with slight warming), and precipitated in methanol.

Characterization of the polyester can be done by infrared spectroscopy and inherent viscosity measurement ($c = 5$ g/l at 25°C) in a mixture of 1,1,2,2-tetrachloroethane/phenol (1:1). The inherent viscosity should be at least 0.1 l/g.

Example 4.7 Preparation of a Thermotropic, Main-Chain Liquid Crystalline (LC) Polyester by Interfacial Polycondensation

Safety precautions: Before this experiment is carried out, Sect. 2.2.5 must be read as well as the material safety data sheets (MSDS) for all chemicals and products used.

$$n \; HO-\!\!\bigcirc\!\!-\!\!\bigcirc\!\!-OH \; + \; n \; \underset{O}{Cl}\!-\!\!\!\overset{O}{\underset{}{C}}\!\!\!\cdots\cdots\!\!\!\overset{O}{\underset{}{C}}\!-Cl$$

$$\downarrow \; -2n \; HCl$$

$$\left[O-\!\!\bigcirc\!\!-\!\!\bigcirc\!\!-O-\!\!\overset{O}{\underset{O}{C}}\!\!\cdots\cdots\!\!\overset{O}{\underset{}{C}} \right]_n$$

In a 500 ml flat-bottomed three-necked flask, equipped with a powerful stirrer, dropping funnel, nitrogen inlet, and bubble counter are placed successively 0.93 g (5 mmol) of bisphenol A, 1.0 g of tetrabutylammonium bromide (phase transfer catalyst), and 2.5 g of KOH, suspended in 50 ml of water. A stream of nitrogen is now passed into the carefully stirred mixture at such a rate that the bubbles can be individually distinguished. The mixture is stirred until the initially cloudy reaction mixture becomes clear. A solution of 1.34 g (5 mmol) of dodecanedioic acid chloride in 50 ml of dry dichloromethane is introduced quickly through the dropping funnel. After the addition is complete, stirring is continued for a further 10 min and during this time the stirring rate should be adjusted to give good mixing. The polymer is precipitated from the white emulsion by pouring it dropwise into 400 ml of methanol. The polymer is separated by centrifugation, the upper organic layer decanted off, the polymer washed by slurrying several times with methanol, then with acetone, and dried under reduced pressure (water pump).

The liquid crystalline phases in polymeric materials can often be identified by polarized light microscopy and DSC measurements. The DSC diagram shows endothermic signals during heating because the molecular order of the polymer changes. The order of the above LC polymer is lost in two steps, therefore two peaks appear in the DSC curve. At the melting point (about 200°C) the transition of the crystalline, highly ordered structure to the smectic LC phase takes place, in which the rigid parts of molecules are arranged in layers, parallel to each other. The centers of gravity of the molecules possess a direction-dependent mobility. At the clearance point (about 260°C) this order is lost and the polymer chains are transformed into the fully irregular isotropic phase.

With polarizing microscopy the birefringence of a thin polymer film (between two glass plates) can be observed between two crossed polarizers provided that the polymer shows ordered regions in the same order of dimension as the wavelength of the light. A further characteristic of main-chain LC polymers is their low melt viscosity. Therefore, the glass plates can easily be moved against each other above the melting point.

Another class of crosslinkable polyesters is the so-called alkyd resins. Alkyd resins are defined as branched or crosslinked polyesters obtained, for example, by polycondensation of a dicarboxylic acid with a polyfunctional alcohol. Branching and crosslinking occur consecutively in a controllable manner. Thus, in the polycondensation of glycerol with phthalic acid anhydride, there is first formed a branched polyester that remains soluble and fusible as long as the polycondensation is interrupted before more than about 75 mol% of the hydroxy or carboxyl groups have reacted. If this degree of condensation is exceeded the branched polyester transforms by further polycondensation (self-crosslinking) into completely insoluble products. Since one is dealing with crosslinking by polycondensation here, a temperature of about 200°C is generally required (baking varnishes). Of course, if carboxylic acids containing double bonds are incorporated into the alkyd resins they can be crosslinked by the action of atmospheric oxygen at lower temperatures. However, the crosslinking then proceeds by a different mechanism.

The dicarboxylic acids normally employed are phthalic acid or its anhydride, or mixtures of them with, for example, adipic acid or unsaturated acids. The polyhydroxy compounds generally used are glycerol, trimethylolpropane, or pentaerythritol. In the preparation and crosslinking of alkyd resins (with the exception of some unsaturated alkyd resins) the reaction is practically always carried out in the melt, taking the precautions discussed in Sect. 4.1.

Example 4.8 Preparation of Unsaturated Polyesters

Safety precautions: Before this experiment is carried out, Sect. 2.2.5 must be read as well as the material safety data sheets (MSDS) for all chemicals and products used.

Polycondensation of a diol with a dicarboxylic acid, either of which may contain a double bond, results in an unsaturated polyester. For this purpose suitable starting compounds are maleic anhydride and 2-butylene-1,4-diol. These can also be used mixed with saturated dicarboxylic acids or diols (copolycondensation) in order to vary the number of double bonds per macromolecule and thereby the properties of the polyester. Unsaturated polyesters are generally prepared by melt condensation. The resulting products are often viscous or waxy substances of relatively low molecular weight.

The incorporation of double bonds into polyesters provides the possibility for subsequent reactions. Of particular interest in this context are reactions that lead to crosslinking.

One of the best examples of ionic crosslinking is the addition of polyhydroxy compounds to double bonds, which sometimes takes place during the actual preparation of the unsaturated polyester. However, radically induced crosslinking reactions are much more important. They may be brought about by the action of oxygen and light, especially in the presence of suitable catalysts such as cobalt(II) compounds ("air drying"). The possibilities are considerably widened if one carries out a crosslinking copolymerization: the unsaturated polyester is dissolved in a radically polymerizable monomer, e.g., styrene and the polymerization then initiated by addition of a radical-forming system. Their structure can be represented schematically as follows:

The temperature required depends on the initiator used. With peroxides such as dibenzoyl peroxide, cyclohexanone peroxide, or cumyl hydroperoxide the reaction is carried out at 70–100°C ("hot curing"); with redox systems it can be done at room temperature ("cold curing"). Suitable redox systems consist of a combination of peroxides with reducing agents that are soluble in organic media, for example, metal salts (cobalt- or copper-naphthenoates or -octanoates) and tertiary amines (*N,N*-dimethylaniline). The most commonly used monomer for this crosslinking graft copolymerization is styrene; allyl compounds such as diallyl phthalate can also be employed. In all cases transparent, insoluble, three-dimensionally crosslinked products are formed that exhibit good heat stability and, especially when mixed with glass fibers, good mechanical properties.

The crosslinking graft copolymerization (curing) is generally carried out by first dissolving the unsaturated polyester (about 70 parts) by stirring in the monomer to be grafted (about 30 parts). High temperatures (up to 130°C) must often be used for this purpose. In order to prevent premature polymerization it is best to add some inhibitor (0.1–0.5 wt% hydroquinone or 4-*tert*-butylpyrocatechol). As soon as a homogeneous solution is obtained, it is cooled to the desired polymerization temperature and the initiator (about 1–3 wt% with respect to the total solution) is stirred in. The onset of polymerization is evidenced by gelation of the reaction mixture; the initially added inhibitor does not upset the catalyzed polymerization. The fully polymerized mass (allow a reaction time of up to 24 h according to the reaction conditions) often contains a low proportion of soluble material (homopolymer of the vinyl compound) that can be determined quantitatively by extraction of a well-ground sample using a suitable solvent (e.g., toluene when the monomer is styrene) for a period of 30 min. In this way one obtains an indication of the relative proportions of grafting and crosslinking.

The crosslinking copolymerization of unsaturated polyesters with styrene is utilized industrially for the production of surface coatings. More extensive are applications in fiber-reinforced plastics where the unsaturated polyesters are used as a matrix for inorganic and organic fibers (see Example 5.20). Areas of application are (large) parts in boats, vehicles, and sport equipment construction.

(a) *Preparation of the Unsaturated Polyester*

In a 500 ml three-necked flask, fitted with stirrer, nitrogen inlet, thermometer, and condenser for distillation with vacuum adapter, are placed 40 g (0.5 mol + excess of 5 mol%) of propane-1,2-diol, 24.5 g (0.25 mol) of maleic anhydride, 37 g (0.25 mol) of phthalic acid anhydride, and 20 mg (0.01%) of hydroquinone (as polymerization inhibitor). The air is removed from the apparatus by evacuating and filling with nitrogen. The mixture is then heated under a slow stream of nitrogen; melting occurs at 80–90°C so that the stirrer can be set in motion. Esterification begins at 180–190°C as indicated by distillation of water. This temperature is maintained until the acid number of the polyester has fallen to 50; this takes about 5–6 h. The melt is now cooled to 140°C and 100 g of monomeric styrene (to which 0.02% hydroquinone has been added after distillation) are introduced with stirring over a period of 1 min. The mixture is cooled immediately to room temperature with a water bath in order to prevent premature polymerization. One thus obtains a viscous, colorless to pale-yellow solution of the unsaturated polyester in styrene.

Determination of the Acid Number

1–2 g of the polyester (without styrene) are dissolved by warming with 50 ml acetone and, after cooling, are titrated as quickly as possible with 0.1-N alcoholic potassium hydroxide using phenolphthalein as indicator, until the red color remains for a few seconds.

The alkali consumption of the solvent is determined in a blank experiment. The acid number is given by the weight of KOH in mg, required to neutralize 1 g of substance.

(b) *Crosslinking (Curing) of the Unsaturated Polyester with Styrene*

Cold curing: To crosslink the polyester at room temperature (cold curing), 10 g of the polyester solution prepared in (a) are placed in a small beaker or flat tin can and the components of a suitable redox system stirred in one after the other (on no account at the same time!), for example:

1. 0.06 ml of a 10% solution of cobalt(II) naphthenoate (or octanoate) in styrene and
2. 0.2 ml of cyclohexanone peroxide or 2-butanone peroxide (e.g., as 50% solution in dibutyl phthalate);

or

1. 200 mg of dibenzoyl peroxide and
2. 0.05 ml of pure dimethylaniline.

After 30–40 min the mixture warms up and gels, indicating that the crosslinking graft copolymerization has begun. After 1 h, the mixture has become almost solid, most of the styrene having polymerized.

Hot curing: To carry out the crosslinking at higher temperatures (hot curing), 0.1 g of dibenzoyl peroxide is dissolved in 10 g of the polyester solution prepared in (a) and heated to 80°C. The polymerization sets in after a few minutes (gelation) and is essentially complete after 15 min.

After curing: The sample obtained by cold or hot curing is not yet fully polymerized and to obtain optimum rigidity, an after-curing treatment is necessary. If it is performed at room temperature it takes 2–3 weeks, at 30–40°C a few days are sufficient and at 70–80°C some hours are adequate.

1 g each of the well-ground samples (see Sect. 2.5.1) obtained by cold, hot, and after curing is boiled with 10 ml of toluene for 30 min, filtered after cooling, and washed with toluene. The samples are thoroughly dried in vacuum at 60°C and the weight loss is determined. The toluene solutions are dropped into methanol and any extracted polystyrene is precipitated. The swellability of all samples is determined in an organic solvent.

Glass-fiber-reinforced plates can also be made in the above manner. For this purpose a mat of glass fibers is impregnated with the polyester solution already containing the radical initiator and cured at the appropriate temperature (see Example 5.20).

4.1.2 Polyamides

Polyamides are macromolecules whose constitutional repeating units are joined by amide groups, for example:

$$\left[\underset{H}{N}\text{-}(CH_2)_x\text{-}\underset{H}{N}\text{-}\overset{O}{\underset{\parallel}{C}}\text{-}(CH_2)_y\text{-}\overset{O}{\underset{\parallel}{C}} \right]_n$$

It has become the custom to name linear aliphatic polyamides according to the number of carbon atoms of the diamine component (first named) and of the dicarboxylic acid. Thus, the condensation polymer from hexamethylenediamine and adipic acid is called polyamide-6,6 (or Nylon-6,6), while the corresponding polymer from hexamethylenediamine and sebacoic acid is called polyamide-6,10 (Nylon-6,10). Polyamides resulting from the polycondensation of an aminocarboxylic acid or from ring-opening polymerization of lactams are indicated by a single number; thus polyamide-6 (Nylon-6) is the polymer from e-aminocaproic acid or from ε-caprolactam.

Many properties of polyamides are attributable to the formation of hydrogen bonds between the NH and CO groups of neighboring macromolecules. This is evidenced by their solubility in special solvents (sulfuric acid, formic acid, *m*-cresol), their high melting points (even when made from aliphatic components), and their resistance to hydrolysis. In addition, polyamides with a regular chain structure crystallize very readily.

The melting points of *aliphatic polyamides* depend on the chain length of the starting materials: the melting point falls with increasing distance between the amide groups in the macromolecule, but polyamides made from components with an even number of carbon atoms melt at a higher temperature than the neighboring odd-numbered members of the series. Alkyl side chains (on carbon or nitrogen) lower the melting point, but at the same time improve the solubility.

Aliphatic polyamides are used as fibers and as engineering thermoplasts in automotive, electrical, and consumer product applications.

Aromatic polyamides (polyaramides) melt at considerably higher temperatures, often above their decomposition temperature so that they are mostly not thermoplastically processable. But they can be spun from solution to fibers of high technical value. Polyaramide fibers from *m*-phenylene diamine and isophthaloyl dichloride are nonflammable and find use in fire-protecting clothing and upholstery. The para-linked polyaramides, due to their linear and stiff main chain are able to form liquid crystalline phases in solution. Fibers spun from these solutions exhibit extraordinary high values of Young's modulus and strength.

The special properties and differences between all-*meta* linked and all-*para* linked polyaramides have three reasons:
- Microlinearity of the chain as a result of all-*para* linkage,
- Stiffness and planarity of the chain because of the partial double bond characteristics of the amide groups, and
- Stability of the conformation in the solid state by intermolecular hydrogen bonds.

As a consequence of this almost perfect alignment of molecule structures, such polyamides are able to orientate in solution and to form liquid crystalline phases (see Sect. 1.3.4). Out of these solutions one obtains fibers of poly(*p*-phenylene terephthalamide) (PPTA) having 5–10-fold higher values for stiffness and strength as the all-*meta* linked polymers. In addition, PPTA crystallizes, whereupon the fibers achieve an extraordinary temperature resistance: in a nitrogen atmosphere they decompose at temperatures above 550°C without melting.

These facts can be explained by a simple comparison of the stretched conformations of poly(*p*-phenylene terephthalamide) and poly(*m*-phenylene *iso*-phthalamide). Whereas the all-*para* linked polyamide can build up many hydrogen bonds to the neighboring molecules, this is less possible for the all-*meta* linked polymer:

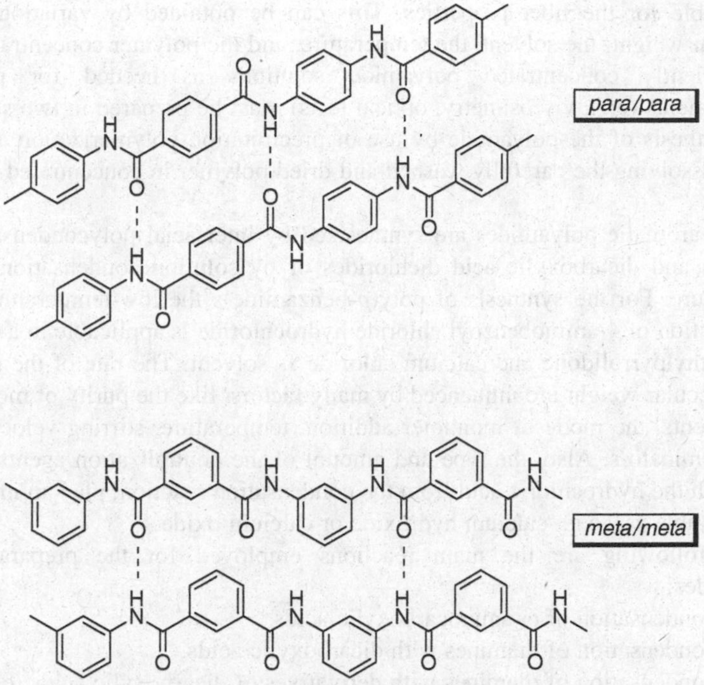

para/para

meta/meta

The extreme insolubility of PPTA in common organic solvents is a serious disadvantage for the production and processing of this type of polymer. The solubility in strong polar solvents, e.g., N-methylpyrrolidone is improved by complexation of the polar amide groups with alkali halides or alkaline earth halides. The solubility still amounts to merely 3% and is, therefore, not sufficient for a spinning process. However, concentrated polyamide solutions can be obtained in concentrated sulfuric acid without additional complexing agent. In both cases the intermolecular polymer–polymer hydrogen bonds are broken. Furthermore, an electrostatic repulsion of the complexed (i.e., charged) amide groups within the polymer chain causes an additional stretching and stiffening of the macromolecules. This is evidenced by an α-value larger than 1 in the Mark-Houwink equation.

If one follows the solution viscosity in concentrated sulfuric acid with increasing polymer concentration, then one observes first a rise, afterwards, however, an abrupt decrease (about 5–15%, depending on the type of polymers and the experimental conditions). This transition is identical with the transformation of an optical isotropic to an optical anisotropic liquid crystalline solution with nematic behavior. Such solutions in the state of rest are weakly clouded and become opalescent when they are stirred; they show birefringence, i.e., they depolarize linear polarized light. The two phases, formed at the critical concentration, can be separated by centrifugation to an isotropic and an anisotropic phase. A high amount of anisotropic phase is desirable for the fiber properties. This can be obtained by variation of the molecular weight, the solvent, the temperature, and the polymer concentration.

Sufficiently concentrated polyamide solutions as needed for physical measurements (e.g., viscosimetry, optical tests) must be prepared in two steps: At first, synthesis of the polyamide by use of precipitation polymerization and secondly, dissolving the carefully washed and dried polymer in concentrated sulfuric acid.

Fully aromatic polyamides are synthesized by interfacial polycondensation of diamines and dicarboxylic acid dichlorides or by solution condensation at low temperature. For the synthesis of poly(p-benzamide)s the low-temperature polycondensation of 4-aminobenzoyl chloride hydrochloride is applicable in a mixture of N-methylpyrrolidone and calcium chloride as solvent. The rate of the reaction and molecular weight are influenced by many factors, like the purity of monomers and solvents, the mode of monomer addition, temperature, stirring velocity, and chain terminators. Also, the type and amount of the neutralization agents, which react with the hydrochloric acid from the condensation reaction, play an important role. Suitable are, e.g., calcium hydroxide or calcium oxide.

The following are the main reactions employed for the preparation of polyamides:
- Polycondensation of ω-aminocarboxylic acids,
- Polycondensation of diamines with dicarboxylic acids,
- Polycondensation of diamines with derivatives of dicarboxylic acids (e.g., acid chlorides), and
- Ring-opening polymerization of lactams (see Sect. 3.2.3.4).

In principle, the attainment of chemical equilibrium can be accelerated by catalysts; however, in contrast to polyester formation, catalysts are not absolutely essential in the above-mentioned polycondensations. The first two types of reactions are generally carried out in the melt; solution polycondensations at higher temperature, e.g., in xylenol or 4-$tert$-butylphenol are of significance only in a few cases on account of the poor solubility of polyamides. On the other hand, polycondensation of diamines with dicarboxylic acid chlorides can be carried out either in solution at low temperature or as interfacial condensation (see Sect. 4.1.2.3).

4.1.2.1 Polyamides from ω-Aminocarboxylic Acids

The formation of polyamides by elimination of water from aminocarboxylic acids at high temperature (see Example 4.10) is generally only possible with acids having more than four methylene groups between the amino and carboxyl groups; under these conditions α- or β-aminocarboxylic acids undergo preferential ring closure (e.g., glycine/2,5-dioxopiperazine; γ-aminobutyric acid/γ-butyrolactam).

Example 4.9 Preparation of an Aliphatic Polyamide by Polycondensation of ε-Aminocaproic Acid in the Melt

Safety precautions: Before this experiment is carried out, Sect. 2.2.5 must be read as well as the material safety data sheets (MSDS) for all chemicals and products used.

The polycondensation can be performed in the apparatus described in Example 4.3. For the preparation of small amounts of Nylon-6 the following procedure is especially suitable, and at the same time permits a quantitative determination of the water eliminated.

5 g of ε-aminocaproic acid are weighed into a thick-walled tube of about 20 ml capacity, carrying a ground joint. The tube is fitted with as small a distillation adapter as possible and connected via a short tube to a weighed U-tube filled with anhydrous granular calcium chloride. The adapter and attached tube are wrapped with aluminum foil. A slow stream of nitrogen is passed through the distillation head which serves both to carry away the eliminated water and to provide a buffer against atmospheric oxygen, the presence of which leads to strongly colored products.

The tube is now immersed nearly up to its neck in an oil or metal bath and the ε-amino-caproic acid melted at a bath temperature of 220°C. Then it is heated quickly to 260°C and held at this temperature under a continuous stream of nitrogen for 15 min. Should any water condense on the wall of the connecting tube leading to the calcium chloride tube, it can be removed by warming with a hot-air blower before the heating bath is taken away and the melt allowed to cool under nitrogen. Finally, the tube is broken and the solid polyamide removed. The calcium chloride tube is weighed again to determine the amount of water eliminated. The experiment is repeated twice more, extending the reaction times to 30 and 60 min, respectively. The limiting viscosity numbers of the three samples are determined in concentrated sulfuric acid at 30°C ($c = 10$ g/l) using an Ostwald viscometer (capillary diameter 0.6 mm). The increase of the η_{sp}/c values with reaction time is a measure of the progress of the polycondensation.

The resulting Nylon-6 has a melting point of 215°C; fibers can be drawn from the melt. The product still contains small amounts of cyclic and linear polyamides that can be extracted from the finely ground product with methanol in a Soxhlet apparatus (12 h). The extract contains ε-caprolactam as well as cyclic and linear oligomers up to the pentamer; these can be quantitatively determined by evaporating of the methanol in vacuum. The ε-caprolactam is then dissolved out by digestion with anhydrous ether. The residue is taken up in methanol (1% solution) and passed through a column filled with a strongly acidic cation exchange resin (elute with the same volume of methanol). The linear oligomers are retained on the column, while the cyclic oligomers are quantitatively determined by evaporating the eluate to dryness. They can be separated by chromatography (e.g., HPLC).

4.1.2.2 Polyamides from Diamines and Dicarboxylic Acids

The polycondensation of diamines with dicarboxylic acids can be carried out simply by melting together the highly purified components under nitrogen at 180–300°C. However, considerable amounts of diamine can be carried over with the water that distills off, especially towards the end of the reaction when vacuum is applied. The equivalence of the reaction partners, which is a prerequisite for the attainment of a high molecular weight, is thereby disturbed. Therefore, it is advantageous to start with an excess of diamine; the optimum excess must be determined for each individual case. Accordingly, the technical synthesis of polyamides from diamines and dicarboxylic acids is carried out in autoclaves.

An elegant variation of this procedure is to carry out the polycondensation with the salt of a diamine and a dicarboxylic acid (see Example 4.10). In this case, the formation of salt (which is also the first step in the direct polycondensation of a diamine and a dicarboxylic acid) and the polycondensation are carried out as two separate steps. The salts can be obtained in good crystalline form most simply by mixing equimolar amounts of diamine and dicarboxylic acid in a solvent in which the salt formed is insoluble (e.g., ethanol). In order to attain high molecular weights by such polycondensations the salts should be as neutral as possible (exactly equivalent amounts of diamine and dicarboxylic acid) and very pure (recrystallize, for example, from mixtures of ethanol/water).

Example 4.10 Preparation of Polyamide-6,6 from Hexamethylenediammonium Adipate (AH Salt) by Condensation in the Melt

Safety precautions: Before this experiment is carried out, Sect. 2.2.5 must be read as well as the material safety data sheets (MSDS) for all chemicals and products used.

Preparation of the AH Salt

12 g (0.082 mol) of pure adipic acid (mp 152°C) are dissolved in 100 ml of 95% ethanol and 9.7 g (0.082 mol) of pure hexamethylenediamine are dissolved in a mixture of 27 ml of ethanol and 10 ml of water. Both solutions are filtered if they are not perfectly clear.

The adipic acid solution is now placed in a 250 ml beaker and the diamine solution added dropwise with stirring over a period of 8 min, during which the solution warms up to 40–45°C. After stirring for a further 30 min and allowing to cool, the crystallized AH salt is filtered off with suction, washed twice with 95% ethanol and dried in vacuum. Yield: 90–95%; mp 183°C (with loss of water); pH value of 9.5% aqueous solution: 7.62. Impure AH salt can be recrystallized from ethanol/water mixtures (volume ratio 3:1).

Polycondensation of the AH Salt

A 50 ml pear-shaped flask, fitted with distillation head, air condenser, vacuum adapter, and receiver is three-quarters filled with AH salt and the air removed by evacuation and filling with nitrogen. It is then heated under nitrogen for 1 h on a silicone oil bath at 220°C, and for further 3 h at 260–270°C. After cooling, the flask is broken carefully with a hammer. The polyamide from adipic acid and hexamethylenediamine melts at 265°C. It can be spun from the melt into threads which can be cold drawn. The viscosity number is determined in concentrated sulfuric acid or in 2 M KCl in 90% formic acid (see Sect. 2.3.3.3).

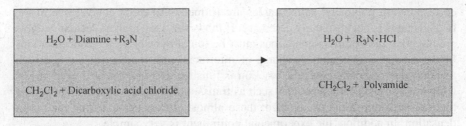

Fig. 4.1 Interfacial polycondensation of dicarboxylic acid and diamine

4.1.2.3 Polyamides from Diamines and Dicarboxylic Acid Derivatives

As in the preparation of polyesters, also in the preparation of polyamides, the reaction temperature can be considerably reduced by using derivatives of dicarboxylic acids instead of the free acids. Especially advantageous in this connection are the dicarboxylic acid chlorides which react with diamines at room temperature by the Schotten-Baumann reaction; this polycondensation can be carried out in solution as well as by a special procedure known as interfacial polycondensation (see Examples 4.11 and 4.12).

The solution polycondensation usually works at room temperature in approximately 10% solution in toluene, methylene chloride, N-methylpyrrolidone (NMP), or dry tetrahydrofuran (THF). Tertiary amines (triethylamine) or dispersed calcium hydroxide are added as acid acceptor. This procedure has the following advantages: The polycondensation is carried out at low temperature (0–40°C); it is nevertheless very fast, the reaction usually being over in a few minutes. At low temperatures practically no side reactions occur. Disadvantages are the following: relatively large amounts of solvent must be purified and handled; and large amounts of salts are formed as by-products. Condensation in solution at low temperature is, therefore, above all a laboratory method, in which these disadvantages are not so important. It is to be particularly recommended in that it yields a high-molecular-weight condensation polymer in a short time with simple equipment.

In interfacial polycondensation, the two components are separately dissolved in two immiscible solvents. The polycondensation can now take place only at the interface of the two liquids, whereby the practically instantaneously formed thin polyamide film prevents further diffusion of the two reactants. The polycondensation can only continue when this film is pulled carefully away from the interface; the process can thus be run continuously in a simple way (Fig. 4.1).

Interfacial polycondensation can be also performed in dispersion (Example 4.13): For this purpose the solution of acid dichloride is dispersed in the aqueous solution of diamine by vigorous stirring (if necessary in the presence of a water-soluble dispersion stabilizer). The polycondensation then takes place at the surface of the droplets. Water is especially suitable as solvent for the diamine component, while aliphatic chlorinated hydrocarbons are best for the dicarboxylic acid dichlorides.

Interfacial polycondensation can be carried out not only with aliphatic but also with aromatic dicarboxylic acid dichlorides (with disulfonic acid dichlorides the corresponding polymeric sulfonamides are formed). An exact equivalence of the two reactants is not absolutely necessary. If, however, very high molecular weights are the goal, the optimum conditions must be found by varying the concentrations of the reactants. A major advantage of interfacial polycondensation is the high reaction rate; in many cases the reaction is finished in a few minutes even at low temperature. Thus, side reactions, such as transamidation, as well as oxidative and thermal decomposition are avoided; these almost always occur during melt condensation. In addition, the experimental equipment is very simple.

In this way one can prepare successfully even high-molecular-weight, very high-melting polyamides that are obtainable by the usual methods only in low molecular weights, if at all. Furthermore, reactants can be used that still carry reactive groups (for example, hydroxy groups, C/C double or triple bonds). An advantage over solution condensation at low temperature is that the eliminated hydrogen chloride does not precipitate as a salt that needs to be separated later, but remains in solution as amine hydrochloride. The molecular weights attainable by interfacial polycondensation are at least as high as those obtained by melt condensation (10,000–30,000), and are often much higher.

Example 4.11 Preparation of Polyamide-6,10 from Hexamethylenediamine and Sebacoyl Dichloride in Solution and by Interfacial Polycondensation

Safety precautions: Before this experiment is carried out, Sect. 2.2.5 must be read as well as the material safety data sheets (MSDS) for all chemicals and products used.

(a) *By Polycondensation in Solution at Low Temperature (Precipitation Polycondensation)*

The alcohol-free chloroform required for this experiment is first prepared by running 70 ml of chloroform through a column filled with basic aluminum oxide.

Sebacoyl dichloride is obtained as follows: 20 g of sebacoic acid and 50 g of thionyl chloride are refluxed on a water bath for 2 h. The excess thionyl chloride is then distilled off and the sebacoyl dichloride fractionated in vacuum (bp 142°C/ 2 Torr).

1.77 g (16.7 mmol) of hexamethylenediamine and 5.1 ml of triethylamine are dissolved in 33 ml of alcohol-free chloroform in a 250 ml three-necked flask fitted with stirrer and reflux condenser. With vigorous stirring, 3.57 ml (16.7 mmol) of sebacoyl dichloride in 13.5 ml of alcohol-free chloroform are then added as quickly as possible (in about 10 s) at room temperature, stirring being continued for another 5 min. The polycondensation occurs instantaneously with considerable heat evolution and precipitation of the condensation polymer. The reaction mixture is cooled and filtered, the product being washed successively with chloroform, petroleum ether, 1 N hydrochloric acid, water, and 50% acetone and is finally dried in vacuum at 50°C. Yield: about 70%; a further low-molecular-weight fraction can be isolated from the filtrate by shaking with petroleum ether.

(b) *By Interfacial Polycondensation*

A solution of 3 ml (14 mmol) of freshly distilled sebacoyl dichloride (for preparation see above) in 100 ml of carbon tetrachloride is placed in a 250 ml beaker. A solution of 4.4 g (38 mmol) of hexamethylenediamine in 50 ml of water is carefully run on to the top of this solution, using a pipette. (The aqueous solution can be made more readily visible by coloring it with a few drops of phenolphthalein solution.) A polyamide film is immediately formed at the interface and can be pulled out from the center with tweezers or clamps and laid over some glass rods; it can now be pulled out continuously in the form of a hollow thread and wound up on to a spool driven by a slow-running motor. The polycondensation comes rapidly to a standstill if the motor is stopped, but immediately recommences, even after some hours, when the motor is restarted.

The drawn-off polyamide thread is thoroughly washed in 50% ethanol or in acetone, then with water and dried in vacuum at 30°C. The Nylon-6,10 so obtained has the same physical properties as that obtained by condensation in solution (see under above) or by melt condensation. It melts at 228°C and can be spun from the melt by pulling with a glass rod; the threads can be cold drawn.

Example 4.12 Synthesis of a Lyotropic Liquid Crystalline Aromatic Polyamide from Terephthalic Acid Dichloride and Silylated 2-chloro-1,4-Phenylenediamine by Polycondensation in Solution

Safety precautions: Before this experiment is carried out, Sect. 2.2.5 must be read as well as the material safety data sheets (MSDS) for all chemicals and products used.

In order to activate the aromatic amino groups, the synthesis is carried out with trimethylsilyl-substituted diamines:

$$+ 2n \ ClSi(CH_3)_3$$

Purification of the Monomers

During the synthesis of aromatic polyamides special attention must be paid to the purity of the monomers and an exact control of the reaction conditions.

2-Chloro-1,4-phenylenediamine sulfate is dissolved in water and the pH value adjusted to 10 by addition of a 50% potassium hydroxide solution. After evaporation at reduced pressure, the diamine is recovered by repeated extraction with ether. The ether layer is dried with Na_2SO_4, the ether is removed and the diamine sublimed in vacuum at 80°C, whereby a colorless product results which decomposes above 135°C. Melting point: 64–65°C.

Terephthaloyl dichloride is distilled in vacuum, Melting point: 82–83°C.
LiCl is dried in vacuum at 150°C for 24 h.

Polycondensation

In a 250 ml three-necked flask fitted with stirrer and internal thermometer 11.04 g
(0.078 mol) of 2-chloro-1,4-phenylenediamine are dissolved in 150 ml dry N,N-
dimethylacetamide (containing 2 wt% LiCl). 29.5 ml (0.233 mol) of highly pure
trimethylchlorosilane (>99%) are dropped into the solution under stirring at 20°C.
Then 15.71 g (0.078 mol) of terephthaloyl dichloride are added, whereupon the
temperature and the solution viscosity increase immediately. After 2 h opaque,
lytropic liquid crystalline solution is obtained. This solution is poured into a beaker
and water is slowly added to the solution, whereupon the polyamide precipitates. It
is washed with water to remove the salt-containing solvent. Finally, the product is
purified by extraction with propane-2-ol. The polymer is dried in a vacuum oven at
100°C. The polyamide is characterized by determination of the solution viscosity at
20°C (1.25 g of polymer in 50 ml N-methylpyrrolidone with 2 wt% of LiCl).

Example 4.13 Microencapsulation of a Dyestuff by Interfacial Polycondensation

Safety precautions: Before this experiment is carried out, Sect. 2.2.5 must be read
as well as the material safety data sheets (MSDS) for all chemicals and products
used.

Microencapsulation means the envelopment of liquid droplets or solid particles
with natural or synthetic polymers. The encapsulation of a substance with a
polymer membrane is undertaken for various reasons, for example, as protection
against moisture, or to obtain delayed dissolution of fertilizers, herbicides, or drugs
by microencapsulation with semipermeable membranes.

Various techniques have been developed for the preparation of microcapsules
with diameters of 1–5,000 μm; one of these involves the method of interfacial
polycondensation. The following example describes the microencapsulation of
a dyestuff, which has practical application in the manufacture of carbon-free copy
paper.

(a) *Preparation of Dye-Containing Microcapsules*

The following ingredients are prepared in parallel:

1. *Dispersion system*: 1 g of low-molecular-weight poly(vinyl alcohol) which
   serves both as dispersing agent and protective colloid, is dissolved in about
   150 ml of distilled water in a narrow 600 ml beaker. It is stirred with a paddle
   stirrer at room temperature for about 2 h until dissolved. The polymer used
   should be a partially (88%) hydrolyzed poly(vinyl acetate), a 4% aqueous
   solution of which has a viscosity of 4 cP.
2. *Acid chloride solution*: 7.7 g (0.04 mol) of terephthaloyl dichloride are finely
   powdered in a mortar and dissolved in 40 g of dibutyl phthalate by stirring at
   70°C, taking care to exclude moisture. Dissolution takes about 30 min; any
   insoluble material is filtered off.

3. *Dyestuff solution*: 1 g of crystal violet lactone [3,3-bis(*p*-dimethylaminophenyl)-6-dimethylaminophthalide] is dissolved in 10 g of dibutyl phthalate by stirring at 90°C for 15 min. The solution is brown/red.

4. *Amine solution*: 3 g (0.03 mol) of diethylenetriamine and 3.2 g of NaOH pellets are dissolved in 20 ml of distilled water with cooling; time required: 15 min. (*Caution*: use hood and avoid all contact of the amine with the skin or clothes!)

The paddle stirrer in the dispersion system [vessel (1)] is now replaced by a high-speed disperser with a strong shearing action. To minimize the formation of foam the contents are stirred at about 2,000 rpm, solutions (2) and (3) are then mixed together with gentle stirring (magnetic stirrer). This water-insoluble mixture is run dropwise into the dispersion vessel over a period of 30 s, the stirrer speed having been raised to about 7,000 rpm as the first drops enter. After a further 30 s the emulsion drop size has adjusted to about 6–10 μm. The particle size can be checked by examination of a sample under a transmission microscope (magnification about 500) with a built-in measuring scale.

The amine solution (4) is now immediately run in dropwise over a period of 30 s and the stirrer speed reduced to 2,000 rpm After 2–3 min the high-speed stirrer is replaced by a normal paddle stirrer and the dispersion stirred (500 rpm) for a further 30 min at room temperature to complete the interfacial polycondensation between terephthaloyl dichloride and diethylenetriamine.

(b) *Testing the Microcapsules*

The microcapsules can be seen under the microscope as individual spherical particles or as small agglomerates.

The use of the microcapsules to make carbon-free copy paper can be demonstrated in principle. The dispersion is painted on to a sheet of typing paper with a fine paint brush. To prevent the paper from curling up it is fixed to a glass plate with adhesive tape. After drying the paper with a hot air drier, the coated side is laid on a silica gel plate. The other side is then written on with gentle pressure. This breaks down the microcapsules, causing the colorless color-forming solution to flow out onto the underlying silica gel plate, developing a blue-colored imprint; thus a blue copy of the writing appears on the silica gel plate, e.g., a plate for thin layer chromatography. One must take care that the painted paper is completely dry, otherwise no copy appears. The resulting blue coloration on the silica gel plate can be removed reversibly by moistening.

4.1.3 Phenol-Formaldehyde Resins

The high-molecular-weight products formed by the condensation of phenols with carbonyl compounds (especially with formaldehyde) are known as phenolic resins. They are mixtures of structurally nonuniform compounds that are initially soluble and fusible but which can become crosslinked (cured) by subsequent reactions. One distinguishes between acid- and base-catalyzed condensations, since they lead to different end products; the properties of the condensation polymer are also affected by the mole ratio of phenol to formaldehyde.

Condensation in acid medium gives soluble, fusible phenolic resins, with an average molecular weight between 600 and 1,500, and a structure consisting essentially of phenol residues linked by methylene groups in the *ortho*- and *para*-positions; they are called Novolaks. No further condensation occurs on heating this product for longer periods; but it can be crosslinked by reaction with suitable polyfunctional components, e.g., with additional formaldehyde. On the other hand, in basic medium one obtains soluble, fusible hydroxymethylphenols, with a molecular weight between 300 and 700, containing one or more benzene nuclei; they are called Resols. In contrast to the Novolaks, the Resols undergo crosslinking through their reactive groups on heating, giving insoluble, infusible products (Resites).

By far the most important phenolic resins are those made from phenol and formaldehyde. They exhibit high hardness, good electrical and mechanical properties, and chemical stability. Very often they are used in combination with (reactive) fillers like sawdust, chalk, pigments etc.

4.1.3.1 Acid-Catalyzed Phenol-Formaldehyde Condensation (Novolaks)

The condensation of phenol with formaldehyde in acid medium is an exothermic reaction (97 kJ/mol) yielding soluble phenolic resins. The melting point(between 100°C and 140°C) is depending on the molecular weight of resulting resin (600–1,500 g/mol). The structure of the Novolaks can only be indicated schematically, since the coupling of the phenolic nuclei through methylene groups can occur in both *ortho*- and *para*-positions; coupling also occurs through oxymethylene groups so that there are a multitude of possibilities:

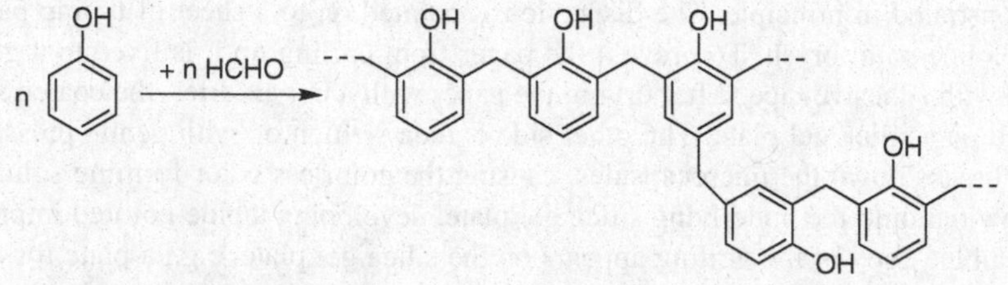

This polycondensation is practically always carried out in aqueous solution, either by dropwise adding an approximately 30% solution of formaldehyde at 80–100°C slowly into an acidified phenol solution, or by mixing the components at room temperature and then heating. In order to avoid premature crosslinking, the molar ratio of phenol/formaldehyde should not be higher than 1:0.8. After all the formaldehyde has been added, the mixture is allowed to react until the formaldehyde has disappeared. The aqueous layer is separated off and the product washed with hot water to remove the acid as completely as possible. The residual water and the unconverted phenol are then removed under vacuum at higher temperature. The resins so obtained are generally soluble in alcohols, lower esters, ketones, and dilute alkali. As catalyst one should choose an acid that is easy to remove from the final product by washing or distillation. Hydrochloric acid, oxalic acid, or mixtures of the two are very suitable. Oxalic acid can be removed both by washing and by heating to 180°C.

Since Novolaks have no reactive groups that can lead to self-crosslinking, they must be subsequently crosslinked (cured) by addition of suitable di- or polyfunctional compounds that react with phenols. Amongst these are formaldehyde, hydroxymethyl compounds, aminobenzyl alcohols, hydroxy- and aminobenzylamines, and bis(hydroxybenzyl) ether. The most commonly used crosslinking agent is hexamethylenetetramine (urotropine). Crosslinking is effected by mixing the components and heating for a short time (a few minutes) to 150–220°C. The structure of the crosslinked phenolformaldehyde resin is very complex. If crosslinking is carried out with hexamethylenetetramine the crosslinks consist mainly of dibenzylamine and tertiary amine bridges:

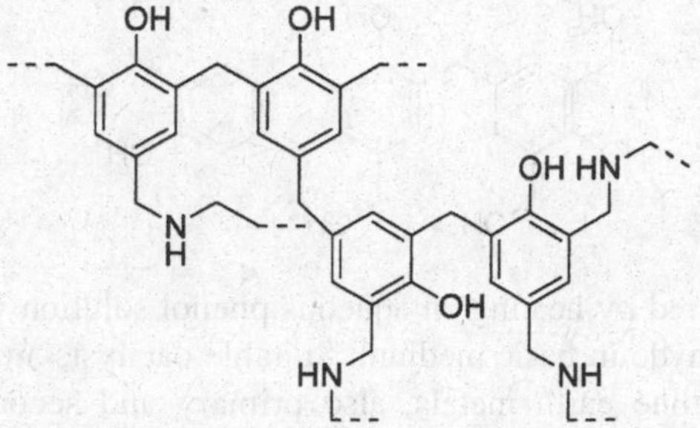

Example 4.14 Acid-Catalyzed Phenol-Formaldehyde Condensation

Safety precautions: Before this experiment is carried out, Sect. 2.2.5 must be read as well as the material safety data sheets (MSDS) for all chemicals and products used.

Caution: Because of the formation of gaseous formaldehyde all reactions must be carried out in a closed hood.

32.5 g (0.345 mol) of phenol, 23 g of a 37% aqueous formaldehyde solution (0.285 mol), 3.75 ml of water, and 0.5 g of oxalic acid dihydrate are placed in a 250 ml three-necked flask, fitted with stirrer and reflux condenser, and heated under reflux on an oil bath for 1.5 h. 75 ml of water are then added, stirred briefly, and allowed to cool, whereby the condensation polymer settles out. The aqueous layer is separated and the residual water distilled off at 50–100 Torr while slowly raising the temperature to 150°C. This temperature is maintained (at most for 1 h) until test samples solidify on cooling. The resin is poured out while still warm and solidifies to a colorless, brittle mass, soluble in alcohol.

The Novolak obtained is used to fabricate a molding by mixing with sawdust and hexamethylenetetramine as follows: 12.5 g of finely ground Novolak, 12.5 g of dry sawdust, 1.75 g of hexamethylenetetramine, 0.5 g of magnesium oxide (to trap residual acid from the condensation), and 0.25 g of calcium stearate (as lubricant) are thoroughly mixed (best in a ball mill or analytical mill) and then heated in a mold at 140 bar for 5 min at 160°C. The resulting molding is infusible and insoluble.

4.1.3.2 Base-Catalyzed Phenol-Formaldehyde Condensation (Resols)

Condensation of phenols with an excess of formaldehyde in basic medium yields phenolic resins (Resols), with an average molecular weight of 300–700, which are generally soluble in water or alcohol. Like Novolaks they consist essentially of phenol nuclei linked to one another through methylene groups; they differ, however, especially in their content of hydroxymethyl groups. The latter makes a number of reactions accessible (e.g., esterification and ether formation). The coupling of the phenol residues and the incorporation of hydroxymethyl groups again occurs in the *ortho-* and *para*-positions. The structure of Resols may be indicated schematically as follows:

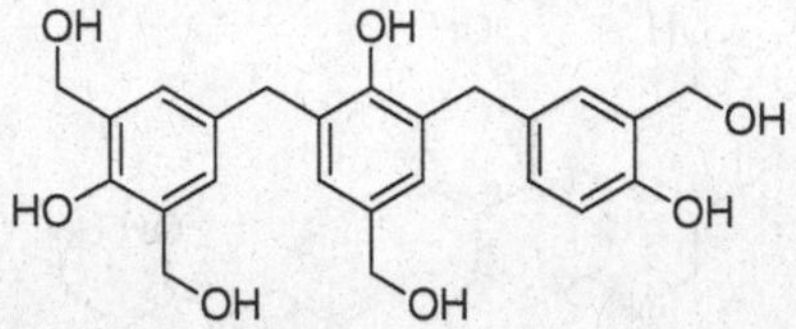

Resols are prepared by heating an aqueous phenol solution with a 1.1–1.5-fold excess of formaldehyde in basic medium. Suitable catalysts are the hydroxides of the alkali and alkaline earth metals, also primary and secondary amines, and especially ammonia; the use of amines can lead to their incorporation in the phenolic resin. The reaction temperature should be kept below 70°C if possible; otherwise the water solubility of the resulting Resol is reduced. Reaction times of 2–5 h are generally required. In contrast to the Novolaks, Resols can be transformed into insoluble, infusible products by self-crosslinking under mild conditions; water is eliminated from the hydroxymethyl groups, forming dimethylene ether bridges. Crosslinking, accompanied by structural transformations (formaldehyde elimination) and rearrangements, occurs on heating to 150–200°C; depending on the chemical composition and structure of the Resol, this requires a few minutes to several hours to achieve complete hardening.

4.1.4   Urea- and Melamine-Formaldehyde Condensation Products

4.1.4.1 Urea-Formaldehyde Resins

Condensation of urea with formaldehyde leads to products of various structures and properties according to the experimental conditions (pH, temperature, reaction time, molar ratio of components). The condensation is generally carried out in basic medium, resulting essentially in the formation of hydroxymethyl compounds; some oxymethylene groups are also formed, particularly on heating. Mono-, di-, and trihydroxymethylurea have been proved to be primary products of this condensation; whether tetrahydroxymethylurea is also formed is as yet uncertain:

$$H_2N \overset{O}{\underset{}{\|}} NH_2 + HCHO \xrightarrow{OH^\ominus} H_2N \overset{O}{\underset{}{\|}} \underset{H}{N} \diagdown OH + HO \diagup \underset{H}{N} \overset{O}{\underset{}{\|}} \underset{H}{N} \diagdown OH$$

4-Oxo-perhydro-1,3,5-oxadiazine (urone) is also formed by intramolecular condensation:

$$HO \diagup \underset{H}{N} \overset{O}{\underset{}{\|}} \underset{H}{N} \diagdown OH \xrightarrow{-H_2O} \text{[ring structure]}$$

These "pre-condensates" are soluble in water and alcohol; they are transformed by further condensation with elimination of water, first into high-molecular-weight, poorly soluble materials and finally into crosslinked insoluble products. The structure of the crosslinked (hardened) urea-formaldehyde resins is not yet entirely understood.

The soluble hydroxymethyl compounds can be chemically modified, before crosslinking, by reaction with monofunctional compounds (e.g., by esterification or ether formation). The properties of the starting materials as well as the crosslinked end products can thereby be substantially altered. For example, by partial etherification with butanol the hydroxymethyl compounds, originally soluble only in polar solvents, become soluble also in nonpolar solvents (toluene), without losing their ability to undergo self-crosslinking.

Urea-formaldehyde resins are generally prepared by condensation in aqueous basic medium. Depending on the intended application, a 50–100% excess of formaldehyde is used. All bases are suitable as catalysts provided they are partially soluble in water. The most commonly used catalysts are the alkali hydroxides. The pH value of the alkaline solution should not exceed 8–9, on account of the possible Cannizzaro reaction of formaldehyde. Since the alkalinity of the solution drops in the course of the reaction, it is necessary either to use a buffer solution or to keep the pH constant by repeated additions of aqueous alkali hydroxide. Under these conditions the reaction time is about 10–20 min at 50–60°C. The course of the condensation can be monitored by titration of the unused formaldehyde with sodium hydrogen sulfite or hydroxylamine hydrochloride. These determinations must, however, be carried out quickly and at as low temperature as possible (10–15°C), otherwise elimination of formaldehyde from the hydroxymethyl compounds already formed can falsify the analysis. The isolation of the soluble condensation products is not possible without special precautions, on account of the facile back-reaction; it can be done by pumping off the water in vacuum below 60°C under weakly alkaline conditions, or better by careful freeze-drying. However, the further condensation to crosslinked products is nearly always performed with the original aqueous solution. This can be done either by heating the neutral solution to 120–140°C (10–60 min) or catalytically in the presence of acids at low temperatures. The catalytic crosslinking (acid hardening) can be carried out not only with free acids (e.g., phosphoric acid), but also with compounds that become acidic on heating (latent hardeners). The latter include sodium salts of halo-genated carboxylic acids, esters of phosphoric acid, ammonium chloride, and pyridine hydrochloride. Addition of large amounts, for example, of phosphoric acid (pH ≈ 2), causes crosslinking even at room temperature (cold glue). All crosslinking reactions of urea-formaldehyde resins occur by further condensation of hydroxymethyl compounds with expulsion of water. This water can, during the curing of large moldings, lead to inhomogeneity and fissures; however, these difficulties can be overcome by addition of water-absorbing fillers such as cellulose or other polyhydric alcohols (Example 4.17).

Example 4.15 Urea-Formaldehyde Condensation

Safety precautions: Before this experiment is carried out, Sect. 2.2.5 must be read as well as the material safety data sheets (MSDS) for all chemicals and products used.

Caution: Because of the formation of gaseous formaldehyde all reactions must be carried out in a closed hood.

Preparation of a Urea-Formaldehyde Resin

30 g (0.5 mol) of urea are dissolved in 61 g of a 37% aqueous formaldehyde solution (0.75 mol) heated to 50°C (use a well ventilated hood) in a 250 ml three-necked flask fitted with thermometer, stirrer, and reflux condenser. 2.5 ml of concentrated ammonia solution are then added and the temperature is raised to 85°C. After about 20 min the solution becomes cloudy and the viscosity increases; at the same time the pH value falls to about 5. After a total period of 1 h the heating is removed and a small sample is tested to see if the condensation product is still soluble in water.

Cellulose powder is now stirred into the cooled solution until the mass can still just be stirred. The contents of the flask are then transferred to a large beaker and more cellulose powder is kneaded in by hand (wear rubber gloves!) until a total of 35 g has been added. The crumbled mixture is dried for 24 h at 50°C in the vacuum oven. Finally, the dried product is finely ground in a mortar, mixed with 1% of ammonium chloride and 1% of zinc stearate (as lubricant) and crosslinked (cured) by heating to about 160°C for 10 min in a simple press at 300–400 bar. A thin transparent plate is obtained that is no longer soluble in or attacked by water; this is more convincing if a small piece is allowed to stand overnight in a beaker of water.

If a heatable press is not available, one may proceed as follows: Two flat iron plates of about 2-cm thickness are heated in an oven to 240°C and then taken out. One of these plates is laid on the lower jaw of a horizontally mounted vice. It is allowed to cool until a crumb of the filled condensation product no longer decomposes (colors) when placed on the plate. A few grams of the mixture are then quickly put on the plate, the second plate placed on top and the whole tightened in the vice.

4.1.4.2 Melamine-Formaldehyde Resins

Formaldehyde resins with better water- and temperature-stabilities are obtained if the urea is partly or wholly replaced by melamine (aminoplasts). These condensations are likewise carried out mainly in alkaline medium, again yielding soluble "pre-condensates" consisting essentially of N-[tris- and hexakis-(hydroxymethyl)] compounds of melamine.

$$
\text{H}_2\text{N}\underset{\text{N}}{\overset{\text{N}}{\diagup}}\text{NH}_2 \quad \xrightarrow{\text{3 HCHO}} \quad \text{HN}\underset{\text{N}}{\overset{\text{N}}{\diagup}}\text{NH} \quad \xrightarrow{\text{3 HCHO}} \quad \text{HO}\diagdown\text{N}\underset{\text{N}}{\overset{\text{N}}{\diagup}}\text{N}\diagup\text{OH}
$$

These pre-condensates are most stable at pH 8–9; they are transformed by further condensation (essentially by elimination of water from hydroxymethyl groups and free NH groups) into poorly soluble and finally insoluble, crosslinked products. Chemical modification of the soluble pre-condensate, for example, by esterification or ether formation, is again possible.

The practical preparation of melamine-formaldehyde resins is done under the same conditions as for urea-formaldehyde resins. Melamine is at first insoluble in the aqueous reaction mixture but dissolves completely as the condensation proceeds. Because of the greater stability of the N-hydroxymethylmel-amines compared with the corresponding urea compounds the reaction can easily be followed by titration of the unconverted formaldehyde with sodium hydrogen sulfite (see Sect. 4.1.4.1).

Crosslinking (hardening) of these pre-condensates can be carried out exactly as for the urea-formaldehyde resins, best at a pH value of 3.5–5. Melamine-formaldehyde condensates crosslink most quickly if prepared using a 2.8–3-fold excess of formaldehyde.

Urea- and melamine-formaldehyde resins are used as moldings, lacquers, and adhesives (for wood), also as textile additives (increased crease resistance) and paper additives (improved wet strength).

Example 4.16 Melamine-Formaldehyde Condensation

Safety precautions: Before this experiment is carried out, Sect. 2.2.5 must be read as well as the material safety data sheets (MSDS) for all chemicals and products used.

Caution: Because of the formation of gaseous formaldehyde all reactions must be carried out in a closed hood.

Preparation of the Polymer

The experimental arrangement consists of a 250 ml three-necked flask equipped with stirrer, double-necked head with reflux condenser and thermometer, and a rubber stopper through which passes a glass rod. 63 g (0.5 mol) of melamine and 150 g of a 40% aqueous formaldehyde solution (2.0 mol) are placed in the flask; the aqueous suspension is stirred and adjusted to pH 8.5 by adding a few drops of 20% NaOH. The mixture is heated on a water bath to 80°C within 5–10 min, with continuous stirring. Complete solution is reached at 70°C to 80°C. During this warm-up period the decrease of pH of the solution must be continually compensated by dropwise addition of 20% NaOH. The stirred solution is now heated to 80°C at constant pH of 8.5 until the precipitation ratio (see below) reaches 2:2. The solution is cooled and filtered from small amounts of insoluble material.

Melamine dissolves in aqueous formaldehyde solution on warming, with the formation of N-hydroxymethyl compounds. The latter are crystalline substances that dissolve in hot water but are only slightly soluble in cold water. If a sample of the reaction mixture is cooled immediately after the melamine has been completely taken into solution, the poorly soluble N-hydroxymethyl compounds precipitate.

With further heating, condensation polymers are obtained that at first are completely miscible with water. However, they still contain a relatively large number of low-molecular-weight N-hydroxymethyl compounds so that the solutions quickly become cloudy on cooling. With longer heating times the aqueous solutions of the resins remain clear. At this stage of condensation, however, the solutions are clear only if a limited amount of water is present; they precipitate on dilution. With increasing condensation time the compatibility with water is diminished further until finally the melamine-formaldehyde condensation polymer separates out from the reaction solution.

Determination of the Precipitation Ratio

After about 50–60 min condensation time a small sample of the reaction mixture is dropped into iced water and the cloudiness observed. From this moment on, samples of exactly 2 ml are taken at regular intervals of 10 min and, after allowing cooling to 20°C, distilled water at 20°C is added dropwise with stirring. The condensation experiment is stopped when the sample becomes slightly cloudy after the addition of 2 ml of water.

Impregnation of Paper

The resin solution prepared above is transferred to a porcelain dish and in it are immersed about 10–20 circular filter papers (diameter 9 cm). After about 1–2 min the filter papers are lifted out with tweezers and the excess solution allowed to drip off. The impregnated filter papers are fastened with clips to a line and allowed to dry overnight.

Fabrication of a Laminated Molding

Ten of the resin-impregnated papers are stacked on top of each other. This stack is laid between two aluminum foils (15 × 15 cm) and pressed in a hydraulic press at 135°C and a pressure of 40–100 bar for 15 min. After releasing the pressure the sample is removed while it is still hot.

The fabrication of laminated plastics with good transparency requires pressures of the magnitude indicated. All heatable, hydraulic, laboratory or commercial presses are suitable for this work. It is also possible to use two nickel-plated iron plates, 2 cm thick, heated to about 140°C in an oven, and clamped in a horizontally mounted vice (see Sect. 2.5.2.1). The resistance of the hardened melamine-formaldehyde laminated plastic is tested against solvents and chemicals.

4.1.5   Poly(Alkylene Sulfide)s

The reaction of suitable aliphatic dihalogen compounds with alkali or alkaline earth polysulfides results in the formation of linear, rubbery or resinous, poly(alkylene sulfide)s:

$$n \; Cl-R-Cl \;\; + \;\; n \, Na_2S_x \;\; \longrightarrow \;\; \left[S_x-R \right]_n \;\; + \; 2n \; NaCl$$

The most widely used dihalide is 1,2-dichloroethane. The use of polyhalides (e.g., 2% 1,2,3-trichloropropane) results in the formation of branched or crosslinked products. Sodium tetrasulfide (Na_2S_4) is generally used as the polysulfide since it contains scarcely any of the monosulfide which reacts with dihalides to form cyclic by-products with unpleasant odors.

Sulfur can be removed from the poly(alkylene sulfide) with the aid of sulfur-binding agents; for example, by treatment of an aqueous dispersion with Na_2S, NaOH, or Na_2SO_3 at 30–100°C, the sulfur content can be reduced to two atoms per constitutional repeating unit of the macromolecule:

$$\left[S_4-R \right]_n \;\; \longrightarrow \;\; \left[S_2-R \right]_n \;\; + \; 2n \; S$$

Further desulfurization results in degradation of the poly(alkene sulfide)s to low-molecular-weight products. Vulcanization of the linear poly(alkene sulfide)s yields crosslinked elastic materials which are commercially important because of their solvent and oil resistance. They are also less sensitive to oxygen and light than most synthetic rubbers. The technical properties can be modified especially by changing the sulfur content, as well as by admixture of fillers. Poly(alkene sulfide)s are used as solvent-resistant rubbers, sealing compounds, and adhesives. Aromatic polysulfides [poly(arylene sulfide)s, PPS] possess substantially higher glass transition temperatures. Hence these polymers can be used as temperature-stable thermoplastics (see Sect. 4.1.6.2).

Poly(alkene sulfide)s are prepared by allowing the dihalide to drip slowly under vigorous stirring into a moderately concentrated aqueous polysulfide solution (generally in 10–20% excess). Temperature and reaction time depend mainly on the dihalide being used: for 1,2-dichloroethane, temperatures between 50°C and 80°C, and reaction times of about 5 h suffice; on the other hand, long-chain dihalides

Table 4.3 Different poly(arylene ether)s and their properties

X	Y	Chemical name	Properties
–O–	–O–	Poly(arylene ether)s	amorphous
–S–	–S–	Poly(arylene sulfide)s	crystalline
–O–	–SO$_2$–	Poly(arylene ether sulfone)s	amorphous
–O–	–CO–	Poly(arylene ether ketone)s	crystalline
–O–	Imide	Poly(arylene ether imide)s	amorphous

require 20–30 h at 100°C. Since the poly(alkene sulfide)s are insoluble in water and very easily agglomerate into lumps, thereby making further reaction and subsequent washing very difficult, it is expedient to carry out the polycondensation in the presence of a dispersing agent (for example, 2–5% magnesium hydroxide). Emulsifiers are not recommended since they make the work-up difficult. When the reaction has finished, the mixture is freed from sodium chloride and unreacted sodium polysulfide by slurrying several times with water. It is then acidified with concentrated hydrochloric acid in order to coagulate the poly(alkene sulfide). The resulting yellow-white spongy cake has an unpleasant smell. It is dried in a vacuum desiccator over P$_2$O$_5$. The dry product is partially soluble in carbon disulfide. Poly (alkene sulfide)s can also be prepared by ring-opening polymerization of episulfides.

4.1.6 Poly(Arylene Ether)s

According to investigations on oligomeric model compounds, macromolecules consisting of C–C linked, unsubstituted aromatic rings (polyphenylenes (I)) should be thermally and chemically very stable due to their low-energy crystal structures and their high melting points ($T_m > 500$°C). However, unsubstituted polyphenylenes are neither soluble nor meltable without decomposition and thus cannot be processed.

When groups are inserted into the main chain, its regularity is disturbed and chain mobility increases (reduction of the enthalpy of melting as well as increase of the entropy of melting). Thus, polymers are formed that can be further processed from solution or in bulk.

In cases where carbonyl or sulfonyl groups are inserted between the aromatic rings in addition to oxygen or sulfur, the resulting type of polymers is commonly classified as poly(arylene ether)s (II).

(I)   ⇒   (II)

X, Y = O, S, CO, SO$_2$

The five classes of poly(arylene ether)s of technical interest are shown in Table 4.3.

Poly(arylene ether)s generally exhibit high glass transition temperatures, resistance to hydrolysis, as well as an excellent thermo-oxidative stability.

4.1.6.1 Poly(Phenylene Ether)s

Dehydrogenation of aromatic compounds is a general method for the synthesis of poly(phenylene ether)s. This type of reaction, in which monomers are joined together to form macromolecules by loss of hydrogen, is formally to be classified as polycondensation reaction; here too, polymers are formed from monomers by continuous elimination of a low-molecular-weight compound (hydrogen), while the polymer retains its activity towards the growth reaction. The first step is the elimination of one electron (oxidation) from the highest-occupied molecular orbital (HOMO) of the phenolate molecule whereby a neutral radical is formed. This step of reaction is catalyzed by copper(I) ions (redox catalyst). Further steps of the oxidative polycondensation are abstraction of a hydrogen atom, combination of radicals with increase of the molecular weight and reoxidation of the electron-rich π-system.

To this type of reaction belongs the synthesis of poly(phenylene ether)s from substituted phenols, for example, poly(2,6-dimethylphenylene ether), PPE, from 2,6-dimethylphenol in the presence of pyridine and copper(I) chloride:

For the synthesis of polymers by dehydrogenation of phenols, the monomer structure and the reaction conditions must conform to certain requirements:

- The substitution of hydrogen atoms must occur in the 2-, 4-, or 6-position relative to the OH group.
- To obtain linear polymers, the phenolic nucleus must be substituted in the 2- and 6-positions.
- Halogen atoms can be present; they decrease the rate of the reaction and may be eliminated during the polymerization with the formation of branching points.
- The phenol must be relatively easily oxidizable; substituents that raise the oxidation potential lead to an inhibition of the dehydrogenation reaction (2,6-dichlorophenol gives only a low-molecular-weight polymer and 2,6-dinitrophenol does not react at all).
- The substituents in the 2- and 6-positions must not exceed a certain geometrical size. Otherwise, instead of regular –O–C– coupling leading to the poly (phenylene ether)s, there is simply a –C–C– coupling of the monomers to form diphenylquinones. This reaction is favored by higher temperatures. The pale-yellow coloration of poly(-2,6-dimethyl-1,4-phenylene ether) may be caused by the presence of quinones.

Suitable catalysts are copper(I) salts [e.g., Cu(I) chloride, bromide, and sulfate] in combination with amines to form oxidation sensitive phenolates. The amine/copper salt ratio must be made as large as possible, to minimize the formation of diphenylquinone and to give a high molecular weight.

Example 4.17 Preparation of Poly(2,6-Dimethylphenylene Ether)

Safety precautions: Before this experiment is carried out, Sect. 2.2.5 must be read as well as the material safety data sheets (MSDS) for all chemicals and products used.

Poly(2,6-dimethylphenylene ether) can be prepared by dehydrogenation of 2,6-dimethylphenol with oxygen in the presence of copper(I) chloride/pyridine as catalyst at room temperature. It is known that the mechanism involves a stepwise reaction, probably proceeding via a copper phenolate complex that is then dehydrogenated.

0.4 g of copper(I) chloride are slurried in a mixture of 100 ml of chloroform and 20 ml of pyridine contained in a 500 ml three-necked flask fitted with stirrer, gas inlet, and thermometer. Oxygen is passed through the flask under vigorous stirring. After 10–20 min a clear, dark-green solution is formed. 10 g (0.082 mol) of pure 2,6-dimethylphenol are added to this solution and more oxygen is passed through under vigorous stirring. The temperature of the solution rises slowly to about 40°C and the color of the solution becomes yellow-brown. After the temperature has reached its maximum (about 30 min), the viscous solution is poured into 1 l of methanol containing 10 ml of concentrated hydrochloric acid to destroy the copper complex. The polymer is washed with methanol and precipitated again from chloroform (100 ml) into 1 l of methanol containing 10 ml of HCl. Yield: quantitative.

When working with larger quantities, the temperature can be kept constant by external cooling. A temperature of 30°C is advantageous for the formation of a high-molecular-weight product. The higher the 2,6-dimethylphenol concentration, the higher is the molecular weight of the polymer formed.

Poly(2,6-dimethylphenylene ether) is amorphous and has a glass transition temperature of about 170°C. It is soluble in chlorinated hydrocarbons such as chloroform, carbon tetrachloride, as well as tetrachloroethane, and also in nitrobenzene and toluene. It is mainly used as a homogeneous blend with polystyrene (see Sect. 5.5).

To make a film, 2 g of the polymer is pressed for 5 min between two metal plates heated to 320°C (see Sect. 2.5.2.1). After chilling the metal plates with water, the film can be peeled off.

4.1.6.2 Aromatic Polysulfides [Poly(Arylene Sulfide)s]

The technically most important polysulfide is polythiophenylene or poly(p-phenylene sulfide), PPS. It is obtained by reacting sodium sulfide and p-dichlorobenzene in a polar solvent, for example, 1-methyl-2-pyrrolidone at about 280°C under pressure. The mechanism of the reaction is very complex and cannot be described by a simple aromatic substitution. This synthesis requires special autoclaves and is therefore not suitable for a laboratory course (for an experimental procedure see Table 2.3).

Because of the limited solubility of PPS – even at high temperature – the attainable molecular weights are relatively low ($M_n = 1,000-20,000$).

Nevertheless, PPS has an interesting property profile even at these molecular weights. To a large extent it is linear and crystalline with a glass transition temperature of 85°C and a crystallite melting point of 280°C. Due to its high chemical resistance – PPS is practically insoluble in common organic solvents – and inherent non-flammability, it has found applications in harsh environments. In addition, PPS can be filled with minerals or glass fibers at concentrations of up to 70 wt%. Hence, it can replace metals in many areas (machines, apparatus constructions, and electronics).

4.1.6.3 Poly(Arylene Ether Sulfone)s

The incorporation of SO_2 groups into the main chain of poly(phenylene ether)s leads to poly(phenylene ether sulfone)s. The preferential synthetic routes are:
- Electrophilic aromatic substitution, using Friedel-Crafts catalysts and
- Nucleophilic aromatic substitution, according to an addition-elimination mechanism.

The nucleophilic polycondensation has the advantage that the chemical structure and thus the properties of the poly(ether sulfone)s can be varied in a relatively simple way through the selection of the bisphenol components.

The technically most important poly(arylene ether sulfone) is obtained from bisphenol A and 4,4'-dichlorodiphenyl sulfone by nucleophilic aromatic polysubstitution.

The reaction is carried out in dimethyl sulfoxide at 130–160°C under an inert atmosphere. High purity of the starting materials is of prime importance in order to obtain high molecular weights. The water, which results from the neutralization of the bisphenols, can be removed via an azeotropic distillation with toluene.

Because of the angled structure of poly(arylene ether sulfone)s, they generally do not crystallize. They are thus amorphous and optically transparent with glass transition temperatures between 150–200°C. They are soluble in some polar solvents, hydrolysis resistant, and inherently flame resistant. Fields of application for these materials are found particularly in the area of electronics and membrane technology.

The chemical modification of poly(arylene ether sulfone)s has already been described in numerous papers. They relate to sulfonation, fluorination, and halomethylation. These derivatives are particularly suitable for the preparation of hydrolysis- and temperature-resistant separation membranes. They are used already for sea water desalination, and also for the separation of gas mixtures.

Example 4.18 Synthesis of Poly(Arylene Ether Sulfone) from Bisphenol A and 4,4'-Dichlorodiphenyl Sulfone

Safety precautions: Before this experiment is carried out, Sect. 2.2.5 must be read as well as the material safety data sheets (MSDS) for all chemicals and products used.

In a 500 ml three-necked flask, fitted with stirrer, thermometer, and water trap, 5.71 g (25 mmol) of bisphenol A are dissolved in a mixture of 60 ml dry dimethyl sulfoxide and 30 ml dry toluene. After thorough flushing with nitrogen, 7.0 g (50 mmol) potassium carbonate are added. The mixture is then heated under stirring at 170°C for 4 h, in order to remove the water by azeotropic distillation with toluene. Then 7.18 g (25 mmol) of 4,4'-dichlorodiphenyl sulfone are added to this suspension. After further 10 h at 170°C, the reaction mixture is cooled, diluted with 100 ml of THF followed by careful addition of 80 ml of concentrated hydrochloric acid. The aqueous phase is extracted repeatedly with THF. Phase separation is induced by addition of small amounts of methyl *tert*-butyl ether. The organic phases are collected and concentrated. This viscous solution is slowly dropped into 3 l of distilled water with intensive stirring, whereupon the polymer precipitates. After further 20 h stirring the suspension is filtered, and the product washed with distilled water. For purification, the polysulfone is dissolved in 80 ml of THF, precipitated again in 2 l of ethanol, and dried for 3 days at 80°C in vacuum.

The synthesis of poly(arylene ether sulfone)s with other bisphenols, e.g., 4,4-bis (4-hydroxyphenyl)pentanoic acid, can be carried out in a similar way.

4.1.6.4 Poly(Arylene Ether Ketone)s

Poly(arylene ether ketone)s have the following general structure:

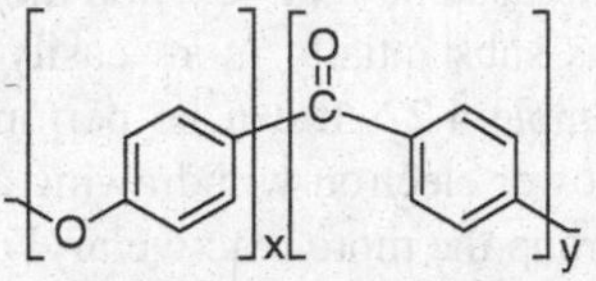

Depending on the number of ether and keto groups in the constitutional repeating units one distinguishes between poly(ether ketone)s (PEK, x = y = 1), poly (ether ether ketone)s (PEEK, x = 2, y = 1), poly(ether ketone ketone)s (PEKK, x = 1, y = 2), and poly(ether ether ketone ketone)s (PEEKK, x = y = 2).

In analogy to poly(arylene ether sulfone)s, there are two different polycondensation methods for the technical synthesis of poly(arylene ether ketone)s:
- Electrophilic aromatic substitution using Friedel-Crafts catalysts and
- Nucleophilic aromatic substitution according to an addition-elimination mechanism.

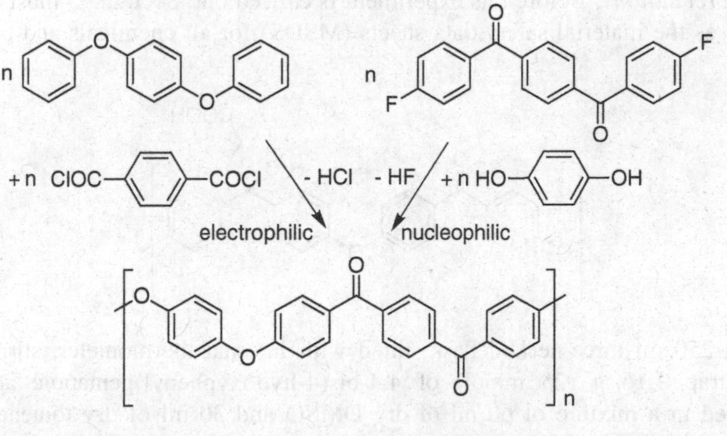

poly(ether ether ketone ketone) (PEEKK)

Depending on the type of the monomers, a large number of crystallizable poly (arylene ether ketone)s with different contents of keto and ether groups can be synthesized by employing one of the two procedures. The ratio of keto to ether groups determines essentially the thermal properties of the linear unsubstituted poly (arylene ether ketone)s. In addition to glass transition temperatures ranging from 120–190°C, crystallite melting points between 270°C and 400°C can be achieved. Unsubstituted poly(arylene ether ketone)s are generally poorly soluble. Often they can only be dissolved in concentrated sulfuric acid at room temperature, whereas in the proximity of their melting points only few organic solvents with high boiling points (e.g., diphenyl sulfone) can be used. Therefore, the synthesis of these polymers on an industrial scale is not an easy matter. The nucleophilic polycondensation is achieved in diphenyl sulfone as solvent and at temperatures above 320°C. It is experimentally not simple to separate the formed salt and the solidified solvent at room temperature from the rigid, stone-hard reaction mixture. Therefore, the synthesis of a substituted and thus substantially more easily soluble poly(arylene ether ketone) is described in Example 4.22. It can be performed in standard laboratory equipment. Because of the lower electron-withdrawing activity of the keto group in comparison to the sulfone group the more reactive aryl-difluoride must be used.

In addition to the excellent thermostability, which is due to their high crystallite melting points, poly(arylene ether ketone)s exhibit a very good chemical resistance because of their low solubility.

They display a high non-flammability, and for a partly crystalline material, an unusually high impact strength. The high stability of the melt permits one to process the material using conventional methods (injection molding, extrusion). Accordingly, poly(arylene ether ketone)s are used in the automobile, electrical, and electronic industry.

Example 4.19 Preparation of a Substituted Poly(Ether Ether Ketone) from 4,4-bis(4-hydoxyphenyl)Pentanoic Acid and 4,4'-Difluorobenzophenone

Safety precautions: Before this experiment is carried out, Sect. 2.2.5 must be read as well as the material safety data sheets (MSDS) for all chemicals and products used.

In a 250 ml three-necked flask, fitted with internal thermometer, stirrer, and water trap 7.16 g (25 mmol) of 4,4-bis(4-hydoxyphenyl)pentanoic acid are dissolved in a mixture of 60 ml of dry DMSO and 30 ml of dry toluene. After thoroughly flushing with inert gas, 7.0 g (50 mmol) of potassium carbonate are added. This mixture is heated under stirring to 170°C for 4 h to remove the water by azeotropic distillation with toluene. Then 5.45 g (25 mmol) of 4,4'-difluorobenzophenone are added to this suspension. After further 10 h at 170°C, the mixture is cooled, whereby a phase separation occurs. The upper phase is discarded, whereas the lower phase is diluted with a mixture of THF and concentrated hydrochloric acid (200 ml, 3:1 vol. ratio). The crude product is obtained by slowly dropping this viscous solution into 3 l of distilled water, followed by intensive stirring for further 20 h. The polymer is filtered and washed with distilled water. The resulting solid is dissolved in 80 ml of THF, precipitated again in 1.2 l of ethanol, and dried for 3 days at 80°C in vacuum.

Due to the side chains this polymer is amorphous with a glass transition temperature of 185°C. In contrast, unsubstituted poly(arylene ether ketone)s are crystalline and high melting ($T_m > 300°C$). The IR spectrum shows absorption bands at 3,300 cm^{-1} and 1,710–1,730 cm^{-1} for the acid group in the side chain and at 1,650 cm^{-1} for the keto group.

4.1.7 Polymers with Heterocyclic Rings in the Main Chain

In general, polymers with heterocyclic rings as structural element show a high chemical and thermal stability. On the other hand, they are insoluble and infusible, so that they could not initially be used for practical applications in this form.

It was not until the synthesis was accomplished in two steps that it became possible to process these polymers from solution. In the first reaction step, two tetrafunctional monomers form a linear and soluble macromolecule by a polyreaction of two of the functional groups of each molecule. Subsequently, cyclocondensations take place in a second step, in which the heterocyclic rings are formed. This side reaction occurs intramolecularly. Therefore, it does not lead to a further molecular enlargement. The resulting polymers have heterocyclic units in their main chain; they are mostly insoluble and infusible:

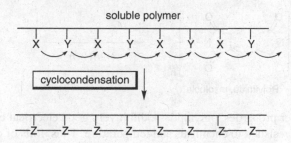

soluble polymer

cyclocondensation

insoluble polymers with heterocyclic units in the main chain

Stepwise addition polymerizations (polyaddition) or condensation polymerizations (polycondensation) are possible polyreactions for the first step. The two latter combinations attained interest in the technical synthesis of polyimides and polybenzimidazoles.

There are two possibilities for practical implementation. If the polymers with the heterocyclic groups are to be isolated in powder form, the polycyclocondensation can be achieved in the same reaction mixture, i.e., in solution. However, if the polymer is to be isolated in the form of fibers, foils, coatings, or adhesives, the following procedure is appropriate: The soluble, still reactive polymers of the first reaction step are poured out, e.g., as a film, or spun into a fiber. Then the solvent is evaporated and afterwards the polycyclocondensation is carried out by heating the finished article in the solid state.

4.1.7.1 Polyimides

Polyimides are obtained by a two-step synthesis, i.e., a combination of stepwise polyaddition followed by a polycyclocondensation. The first reaction consists of the addition of an aromatic diamine to a dianhydride of a tetracarboxylic acid and occurs rapidly at 20–40°C in a strongly polar solvent (N,N-dimethylformamide, N,N-dimethylacetamide, N-methylpyrrolidone). The resulting poly(amide acid) is soluble. The highly viscous solution of the poly(amide acid) so produced, is cast in a thin layer and heated to 150–300°C; the solvent evaporates and the second reaction occurs at the same time, resulting in ring closure with intramolecular elimination of water to give the polyimide. The commercial polyimides are obtained from pyromellitic- or 3,3′,4,4′-benzophenonetetracarboxylic dianhydride and phenylenediamine or dianiline derivatives (which are linked by methylene groups or a heteroatom).

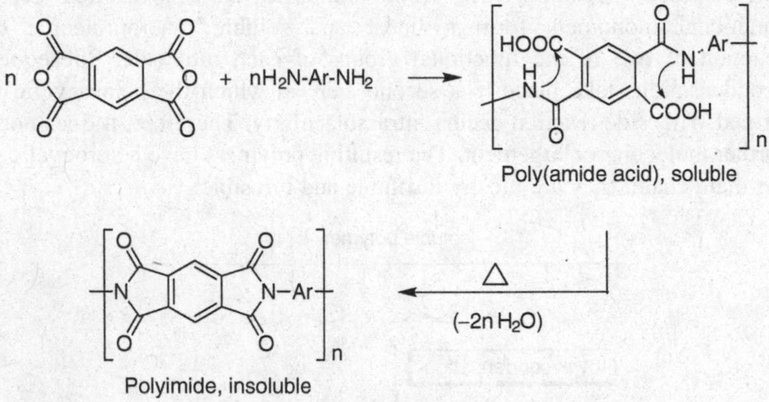

Poly(amide acid), soluble

Polyimide, insoluble

Among other properties, polyimides exhibit very good electrical characteristics. They are very resistant to chemicals and temperature. Thus, foils or coatings made of polyimides are applied in the electric and microelectronic industry (cable insulation, flexible printed circuit boards) or in membrane technology. By using a special sintering technology, plates and rods can be produced from polyimide powder, and from these, finished parts for mechanical engineering (valves, bearings, seals) can be made by metal cutting.

Example 4.20 Preparation of a Polyimide from Pyromellitic Dianhydride and 4,4'-Oxydianiline by Polycyclocondensation

Safety precautions: Before this experiment is carried out, Sect. 2.2.5 must be read as well as the material safety data sheets (MSDS) for all chemicals and products used.

First Reaction (Stepwise Addition Polymerization): Preparation of the Poly(Amide Acid)

4.0 g (0.02 mol) of 4,4'-oxydianiline (recrystallized from toluene and dried in vacuum for 8 h at 50°C) are dissolved in 30 ml of pure dimethylformamide in a dry 100 ml round-bottomed flask, fitted with a drying tube and magnetic stirrer. 4.36 g (0.02 mol) of pyromellitic dianhydride (sublimed in high vacuum) are added in portions over a period of a few minutes, with continuous stirring. The stepwise addition polymerization begins immediately and is completed by stirring the viscous solution for a further hour at 15°C.

About 20 ml of the poly(amide acid) solution are then dropped into 300 ml of water with vigorous stirring. The precipitated colorless polymer is filtered with suction, washed several times with water, and dried in vacuum at 100°C for 20 min. The limiting viscosity number is determined in dimethylformamide at 25°C (Ostwald viscometer, capillary diameter 0.4 mm).

Second Reaction (Polycyclocondensation): Preparation of the Polyimide

The polyimide is formed by the thermal polycyclocondensation of the poly(amide acid). For this purpose, 5 ml of poly(amide acid) solution are placed on a watch glass (diameter 10 cm) and kept in a vacuum oven at 50°C for 24 h. The solvent evaporates and at the same time cyclization to the polyimide takes place; the resulting film is insoluble in dimethylformamide. The formation of the polyimide can be followed by IR spectroscopy: the NH-band at 3,250 cm^{-1} disappears while imide bands appear at 1,775 and 720 cm^{-1}. Once the initial drying process has raised the solid content to 65–75%, the polyimide formation can be accelerated by heating the poly(amide acid) film to 300°C in a vacuum oven for about 45 min. The polyimide made from pyromellitic dianhydride and 4,4'-oxydianiline exhibits long-term stability in air above 200°C.

4.1.7.2 Poly(Benzimidazole)s

The technical production of poly(benzimidazole) (PBI) is also carried out in two steps. In the first step an aromatic tetramine is condensed with the diphenyl ester of an aromatic dicarboxylic acid at 220–260°C, yielding a poly(amino amid) with elimination of phenol. Ring closure with elimination of water occurs in the second step (solid-phase polycyclocondensation), conducted at 400°C and yielding the polybenzimidazole (experimental procedure, see Table 2.3).

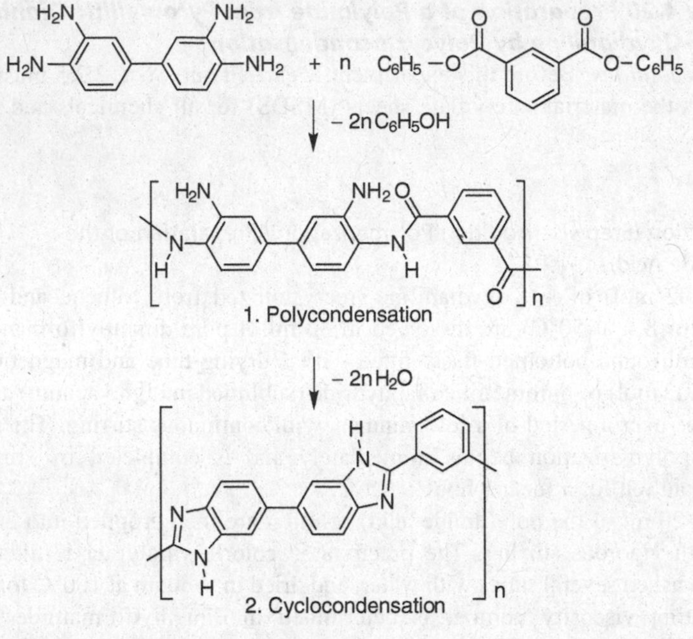

1. Polycondensation

2. Cyclocondensation

The resulting polybenzimidazole is amorphous and shows a very high glass transition temperature of 425°C. It is highly flame resistant. Therefore, PBI fibers that are spun from solution, are used for the production of fireproof protection clothes and also for fabrics of airplane seats.

By the use of a special sintering technology, moldings (semifinished materials) can be made from PBI powder. By mechanical processing, articles with exceptional properties can be produced. Due to the high pressure resistance PBI is suitable for bearings and also sealing rings for thermal and mechanical highly stressed parts.

4.1.8 Polysiloxanes

Polysiloxanes are macromolecules which have a backbone of alternating silicon and oxygen atoms, with the two remaining valencies of the silicon atoms linked to organic side groups, e.g., methyl groups. Polysiloxanes have the tendency to form helical conformations, in which the methyl groups are located at the outer side of the helix, thus causing a "shielding-effect" for the –Si–O–bonds. This explains the unusual combination of properties of these macromolecules, e.g., high temperature stability and resistance to weathering due to the strength of the –Si–O–bond, flexibility at low temperatures, low surface tension, and hydrophobicity, although their main chain is relatively polar. As a result, polysiloxanes either unfilled or filled with minerals, have found a multitude of quite different applications, like temperature-stable oils, greases, sealing materials, lacquers, and elastomers.

$$H_3C\underset{H_3C}{\overset{CH_3}{\underset{\Big[}{\overset{\Big[}{Si}}}}\underset{O}{\Big]_n}Si(CH_3)_3$$

In their properties, polysiloxanes are intermediate between purely organic polymers and inorganic silicates; the structure may be varied in numerous ways to shift the pattern of properties of silicones either in one direction or the other. Commercial polysiloxanes generally contain methyl substituents. Whereas in the scientific literature the name polysiloxanes is used, the name silicones (silicone oil, silicone grease, silicone rubber) is preferred in the technical literature.

Various functional silanes (e.g., R_2SiCl_2 or $RSiCl_3$) can be used as starting materials for the preparation of polysiloxanes. The silanes are first hydrolyzed to the corresponding silanols, which are very unstable and easily undergo polycondensation with the elimination of water and the formation of $-Si-O-Si-$ linkages:

$$n\ Cl\!\!-\!\!\overset{R}{\underset{R}{Si}}\!\!-\!\!Cl \xrightarrow[-HCl]{H_2O} n\ HO\!\!-\!\!\overset{R}{\underset{R}{Si}}\!\!-\!\!OH \xrightarrow{-(n-1)H_2O} H\!\!-\!\!\underset{O}{\Big[}\overset{R}{\underset{R}{Si}}\!\!-\!\!O\Big]_n\!\!H$$

Linear polysiloxanes obtained by the hydrolysis of dichlorosilanes are of relatively low molecular weight; they can, however, be condensed further through the terminal OH groups by thermal after-treatment.

The siloxanes have a great tendency towards ring formation so that in the hydrolysis products of dichlorosilanes one finds not only linear polysiloxanes, but also cyclic oligosiloxanes with 3 to 9 Si–O units in the ring; under the right reaction conditions these ring compounds can even be the main product (see Example 4.24). Cyclic siloxanes can be polymerized by ring-opening both cationically (e.g., with Lewis acids) and anionically (e.g., with alkali hydroxides); trimers and tetramers of dimethylsiloxane are especially suitable as monomers (see Example 4.21):

$$\text{(cyclic siloxane)} \xrightarrow{\text{(KOH)}} \text{(linear polysiloxane)}$$

The polysiloxanes obtained by anionic initiation have considerably higher molecular weights than those obtained by cationic initiation.

Ring-opening polymerization of cyclic siloxanes with cationic initiators allows the possibility of introducing stable end groups by the use of suitable chain transfer agents. Thus, polysiloxanes with trimethylsilyl end groups are formed when the cationic polymerization of octamethylcyclotetrasiloxane is carried out in the presence of hexamethyldisiloxane as transfer agent:

The acid-catalyzed degradation of a high-molecular-weight polysiloxane, containing OH end groups, when carried out in the presence of hexamethyldisiloxane, also leads to low-molecular-weight polysiloxanes with trimethylsilyl end groups (see Example 4.21). This reaction is also named "equilibration". Polysiloxanes that do not possess OH end groups cannot condense further on heating; the molecular weight and therefore the viscosity of this product remains constant on heating, which is a very important property of silicone oils.

Low-molecular-weight polysiloxanes are oily or waxy substances (silicone oils). High-molecular-weight polysiloxanes on the other hand are elastomeric; they can be converted to silicone rubbers by crosslinking. Depending on the nature of the organic side group this crosslinking can occur by condensation, by metal-catalyzed addition or with peroxides. Crosslinking by means of peroxides (see Example 4.21) probably occurs by abstraction of hydrogen atoms from methyl groups yielding carbon radicals on the polysiloxane molecules. Combination of carbon radicals of different macromolecules then causes crosslinking by the formation of C–C bonds.

Example 4.21 Ring-Opening Polymerization of a Cyclic Oligosiloxane to a Linear, High-Molecular-Weight Polysiloxane with Hydroxy End Groups; Curing of the Polymer

Safety precautions: Before this experiment is carried out, Sect. 2.2.5 must be read as well as the material safety data sheets (MSDS) for all chemicals and products used.

(a) *Preparation of Octamethylcyclotetrasiloxane*

200 g of dimethyldichlorosilane are dripped slowly from a dropping funnel (protected from moisture by a $CaCl_2$ tube) into 600 ml of water with vigorous stirring at room temperature. The organic phase is then taken up in about 200 ml of ether, separated from water, and washed twice with distilled water; the ethereal solution is dried over magnesium sulfate. The ether is taken off on a rotary evaporator or by distillation; the resulting oil consists essentially of cyclic oligosiloxanes that can be separated by distillation. A small amount (about 0.5%) of hexamethylcyclotrisiloxane comes over first (bp 134°C/760 Torr, mp 64°C), followed by the main product octamethylcyclotetrasiloxane (bp 175°C/760 Torr, bp 74°C/20 Torr, mp 17.5°C); yield: about 40%. Finally, a few percent of pentamer (bp 101°C/20 Torr) and hexamer (bp 128°C/20 Torr) distill over at temperatures around and above 200°C.

The residue from distillation is a viscous oil consisting of high-molecular-weight compounds. Further cyclic oligomers, especially trimer and tetramer, can be obtained by heating this oil to about 400°C. To do this, a new receiver is fitted to the distillation apparatus and the flask heated to 400–450°C. The more volatile products are carried over in a slow stream of nitrogen and the distillate is then fractionated as before.

(b) *Polymerization of an Oligosiloxane*

60 g of distilled octamethylcyclotetrasiloxane are mixed with 0.1 g of very finely powdered potassium hydroxide and 0.5 ml DMSO as solubilizer in a 250 ml conical flask, which is then placed in an oil bath at 140°C. The increase in viscosity of the mixture can easily be observed by occasional swirling. After 20–30 min, the liquid has reached the consistency of thin honey. Half the product is taken out and cooled [further work-up under (c) and (d)]. The residue is heated again until, after 2–3 h, a plastic, putty-like mass is produced. It is allowed to cool, whereby a rubbery polymer is obtained. Yield: 90–95%.

(c) *Hot Curing of the Polysiloxane*

The polysiloxane from experiment (b) is soluble in toluene. It can be converted by hot vulcanization into an insoluble silicone rubber. Using a small blender, 10 g of the polymer are kneaded with 10 g of quartz powder or 7.5 g of ground kieselguhr, and 0.6 g of dibenzoyl peroxide paste (50% in silicone oil). To work the additives into the silicone rubber without a mechanical blender is very tedious and difficult to achieve completely.

The mixture is then heated for 10 min in the oven at 110°C to bring about crosslinking. The solubilities of the mixture and the resulting silicone rubber are tested in toluene.

(d) *Cold Curing of the Polysiloxane at Room Temperature*

10 g of the honey-like polysiloxane from experiment (b) are mixed with 5 g of quartz powder or ground kieselguhr in a mortar. The viscous syrup is poured into a beaker and 0.3 g of tetraethyl silicate (crosslinking agent) are stirred in, together with 0.3 g of dibutyltin dilaurate as vulcanization accelerator. The mass solidifies in 1–2 h to an elastic silicone rubber. Such a silicone rubber retains its elasticity over an unusually large temperature range (-90–300°C); it is also very resistant towards harsh atmospheric conditions.

Example 4.22 Equilibration of a Silicone Elastomer to a Silicone Oil with Trimethylsilyl End Groups

Safety precautions: Before this experiment is carried out, Sect. 2.2.5 must be read as well as the material safety data sheets (MSDS) for all chemicals and products used.

10 g of the rubbery silicone with hydroxy end groups, made in Example 4.21 (b) are taken up in 20 g of toluene. 0.5 g of 96% sulfuric acid and 0.2 g of hexamethyldisiloxane are added to the very viscous solution, with stirring or shaking until the high-molecular-weight material has disappeared. 0.3 ml of water are then added and stirring is continued for 2 h. The solution is next washed with water in a separating funnel until the washings have a neutral reaction. The toluene is distilled off, leaving a clear, mobile silicone oil (viscosity 200–500 Pa·s). Yield: 70%.

4.2 Stepwise Addition Polymerization (Polyaddition)

The addition polymerizations described here involve a stepwise reaction of at least two bifunctional compounds, leading to the formation of macromolecules. In contrast to condensation polymerization, no low-molecular-weight compounds are eliminated; the coupling of the monomer units is, instead, a consequence of the migration of a hydrogen atom (cf. the equation for the formation of a polyurethane from a diol and a diisocyanate, Sect. 4.2.1). Like condensation polymerization, this kind of addition polymerization is also a stepwise reaction, consisting of a sequence of independent individual reactions, so that here, too, the average molecular weight of the resulting polymer steadily increases during the course of the reaction. The oligomeric and polymeric products formed in the individual steps possess the same functional end groups and the same reactivity as the starting materials; they can be isolated without losing their reactivity, in contrast to the products of addition polymerization. As stepwise reactions, they are governed by kinetic laws similar to those for condensation polymerization (see Eqs. 4.1–4.4 in Sect. 4.1). The experimental techniques are also similar.

4.2.1   Polyurethanes

Polyurethanes are macromolecules in which the constitutional repeating units (CRUs) are coupled with one another through urethane (oxycarbonylamino) groups. They are prepared almost exclusively by stepwise addition polymerization reactions of di- or polyfunctional hydroxy compounds with di- or polyfunctional isocyanates:

$$n \; HO\text{-}(CH_2)_x\text{-}OH \; + \; n \; O=C=N\text{-}(CH_2)_y\text{-}N=C=O$$

This addition reaction proceeds readily and quantitatively. Side reactions can give amide, urea, biuret, allophanate, and isocyanurate groupings, so that the structure of the product can deviate from that above; such side reactions are sometimes desired (see Sect. 4.2.1.2).

Linear polyurethanes made from short-chain diols and diisocyanates are high melting, crystalline, thermoplastic substances whose properties are comparable with those of the polyamides because of the similarity in chain structure. However, they generally melt at somewhat lower temperatures and have better solubility, for example, in chlorinated hydrocarbons. The thermal stability is lower than for polyamides: depending on the structure of the polymer the reverse reaction of the urethane groups begins at temperatures as low as 150–200°C with regeneration of functional groups; the cleavage of the allophanate groups begins at the still lower temperature of 100°C. Basically, polyurethanes are predominantly biphasic multiblock copolymers consisting of a sequence of more flexible elastomeric chain segments [e.g., $-(-CH_2-)_x-$ in the above formula] separated by corresponding hard domains formed by the diurethane groups with intermolecular hydrogen bonds (see Sects. 4.2.1.2 and 5.6).

A key factor in the preparation of polyurethanes is the reactivity of the isocyanates. Aromatic diisocyanates are more reactive than aliphatic diisocyanates, and primary isocyanates react faster than secondary or tertiary isocyanates. The most important and commercially most readily accessible diisocyanates are aliphatic and colorless hexamethylene-1,6-diisocyanate (HDI), isophorone diisocyanate (IPDI), and aromatic, brownish colored diphenylmethane-4,4'-diisocyanate (MDI), 1,5-naphthalenediisocyanate, and a 4:1 mixture of 2,4- and 2,6-toluenediisocyanates (TDI).

As already mentioned, polyurethanes decompose on heating into isocyanates and hydroxy compounds, the decomposition temperature depending on the structure of the urethane. Use is made of this fact when reacting polyhydroxy compounds with so-called "capped isocyanates" or "isocyanate splitters". One may use, for example, diurethanes of aromatic diisocyanates and phenols that revert to the original components at 150–180°C. Such diurethanes can be mixed with polyhydroxy compounds at room temperature and stored without change; on heating, the diisocyanate is split off and reacts with the free hydroxy groups forming urethane groups of greater stability, while the phenol distills away.

The addition of isocyanates to hydroxy compounds is inhibited by acid compounds (e.g., hydrogen chloride or p-toluenesulfonic acid); on the other hand, it can be accelerated by basic compounds (e.g., tertiary amines like triethylamine, N,N-dimethylbenzylamine, and especially 1,4-diazabicyclo[2.2.2]octane) and by certain metal salts or organometallic compounds (e.g., dibutyltin dilaurate, bismuth nitrate). These catalysts are often effective in amounts of much less than 1 wt%.

Polyurethanes are used for the fabrication of fibers, while crosslinked polyurethanes are employed as lacquers and adhesives, as coatings for textiles and paper, and as elastomers and foams.

4.2.1.1 Linear Polyurethanes

The addition polymerization reaction of dihydroxy compounds with diisocyanates sets in on mixing the two components and gentle warming. Under proper conditions, linear polyurethanes with molecular weights up to about 15,000 can be obtained. As in the case of polyamides and polyesters, the softening point of the aliphatic polyurethanes depends on the number of carbon atoms between the urethane groups.

The polymerization in bulk requires relatively high temperatures, and, in addition, the polyurethane formed is exposed to the action of the diisocyanate throughout the duration of the reaction, so that secondary reactions can easily take place (see Sect. 4.2.1). For the preparation of polyurethanes with a high molecular weight and with as linear a structure as possible, polymerization in solution is, therefore to be preferred. Suitable inert solvents are toluene, xylene, chlorobenzene, and 1,2-dichlorobenzene. The diisocyanate is normally dripped into the solution of the dihydroxy compound at the desired temperature, which may conveniently be the boiling point of the solvent. The resulting polyurethane often separates from the reaction mixture and is so much less vulnerable to secondary reactions than when the polymerization is carried out in bulk.

Example 4.23 Preparation of a Linear Polyurethane from 1,4-Butanediol and Hexamethylene Diisocyanate in Solution

Safety precautions: Before this experiment is carried out, Sect. 2.2.5 must be read as well as the material safety data sheets (MSDS) for all chemicals and products used.

22.5 g (0.25 mol) of pure 1,4-butanediol, 42 g (0.25 mol) of pure hexamethylene diisocyanate, and 125 ml of anhydrous chlorobenzene are placed in a dry 500 ml three-necked flask fitted with stirrer, thermometer, reflux condenser with attached drying tube, and nitrogen inlet; the air is removed by evacuation and the flask filled with nitrogen. The mixture is then heated carefully in an oil bath under a slow stream of nitrogen. At about 95°C the initially cloudy reaction mixture suddenly becomes clear; the temperature now climbs rather rapidly and under some circumstances may exceed the prevailing oil bath temperature. About 15 min after the solution has come to boiling (132°C), a faint cloudiness appears (recognizable as a blue rim at the vessel wall) which visibly strengthens. Finally, the high-molecular-weight polyurethane settles out in the form of sand. The reaction is completed by heating for a further 15 min. After cooling, the polyurethane powder is filtered off with suction. [If the mixture is filtered hot (100°C), about 1–3% of the low-molecular-weight portion remains in solution and can be isolated by precipitation with methanol]. The residual absorbed chlorobenzene is best removed by steam distillation. In this way, one obtains 62 g (96%) of a white powder that melts at 181–183°C and is soluble in *m*-cresol and formamide.

4.2.1.2 Branched and Crosslinked Polyurethanes

Essentially two methods can be used for the preparation of branched and crosslinked polyurethanes.

(a) The reactions of diisocyanates with compounds that possess more than two hydroxy groups per molecule.

The degree of crosslinking here depends essentially on the structure and functionality of the polyhydroxy compound so that the properties of the polyurethane can be altered by variation of this component. This procedure is applied mainly to the preparation of lacquers (reactions with diisocyanates at low temperature in anhydrous solvents such as butyl acetate)' or moldings (usually with "capped" diisocyanates at higher temperatures).

(b) The reaction of linear oligourethanes, which possess either hydroxy or isocyanate end groups, with suitable reactive compounds, followed by crosslinking through one of the mechanisms described below.

According to O. Bayer, the latter procedure, which is used especially for the preparation of elastomeric polyurethanes, is carried out in two separate stages. First, a carefully dried, relatively low-molecular-weight, aliphatic polyester or polyether with hydroxy end groups is reacted with an excess of diisocyanate. A "chain extension" reaction occurs in which two to three linear diol molecules are coupled with diisocyanate, so as to yield a linear polymer with some in-chain urethane groups and with isocyanate end groups.

Suitable starting compounds are polyesters from poly(ethylene oxide) and adipic acid, also poly(propylene oxide) or poly(oxytetramethylene) with molecular weights around 2,000, whose hydroxy end groups can be reacted with very reactive diisocyanates such as 1,5-naphthalene diisocyanate, 1,4-phenylene diisocyanate, and diphenylmethane-4,4'-diisocyanate.

The chain-extended, linear poly(ester urethanes) so obtained can now be crosslinked in a second stage, involving reaction with – water – or glycols – or diamines.

In crosslinking with water, pairs of isocyanate end groups in the chain-extended polymer OCN–X–NCO first react with a molecule of water; this results in a linear coupling through urea groupings, with simultaneous elimination of CO_2:

$$O=C=N\mathrm{\sim\!\sim\!\sim\!\sim}N=C=O \qquad O=C=N\mathrm{\sim\!\sim\!\sim\!\sim}N=C=O$$

$$\Big\downarrow \begin{array}{l} + H_2O \\ - CO_2 \end{array} \tag{1}$$

$$O=C=N\mathrm{\sim\!\sim\!\sim}NH{-}\underset{\underset{O}{\|}}{C}{-}NH\mathrm{\sim\!\sim\!\sim}N=C=O$$

The subsequent crosslinking probably occurs by reaction of the hydrogen atoms of the resulting urea groups with isocyanate groups still present in the starting polymer or the chain-extended polymer, with the formation of biuret groups:

$$\tag{2}$$

Crosslinking with glycols or diamines proceeds according to a similar scheme (but without elimination of CO_2). The first reaction is again a linear chain-extension reaction. With glycols, this occurs with the formation of urethane groups, which can then react with residual isocyanate end groups to give crosslinking with formation of allophanate groups. With diamines the linear coupling occurs through urea groups and the crosslinking reaction then proceeds as formulated in structure (2).

Crosslinking with glycols and diamines plays a major role in the preparation of polyurethane elastomers. The properties of the resulting products can be widely varied by choice of starting components and the number of crosslinks ("mesh width").

Crosslinking with water is mainly used in the preparation of polyurethane foams (see Sect. 5.6).

4.2.2 Epoxy Resins

Epoxy resins are usually understood to be products of reaction of polyfunctional hydroxy compounds with 1-chloro-2,3-epoxypropane (epichlorohydrin) in basic medium. In the simplest case two mol of epichlorohydrin react, for example, with one mol of bisphenol A, according to the following scheme:

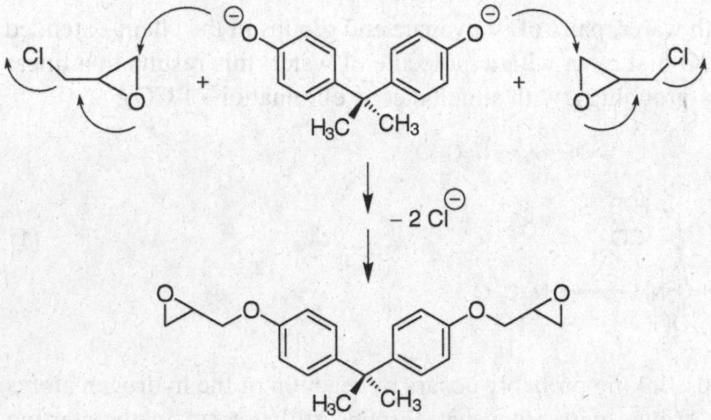

$$- 2 \, Cl^{\ominus}$$

(I)

Higher-molecular-weight products II result from coupling of epoxide I with further bisphenol:

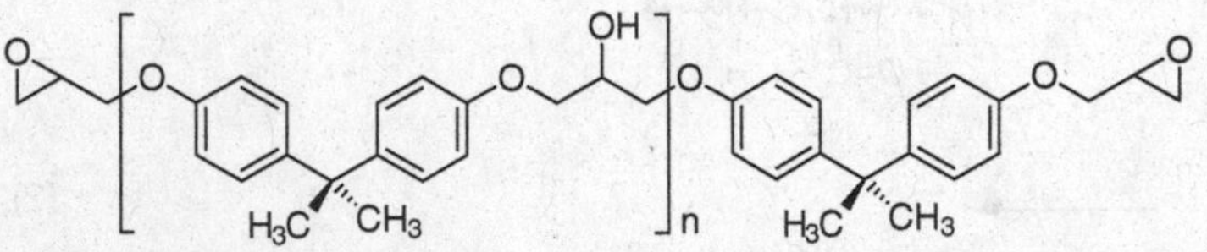

(II)

However, side reactions can also occur. Thus, bisphenol may add only one epoxy group and some of the very reactive epoxy groups may be hydrolyzed during the preparation so that the number of hydroxy groups will deviate from that shown in structure (II).

Depending on the conditions, reactions (I) and (II) can be carried out either concurrently or consecutively. If one works from the outset in alkaline medium, for example, by dropping the desired amount of epichlorohydrin into the mixture of hydroxy compound and the equivalent amount of aqueous alkali hydroxide at 50–100°C, then the addition reaction and the HCl elimination occur side by side (Example 4.24). On the other hand, if the hydroxy compound and epichlorohydrin are allowed to react in nonaqueous medium in the presence of acid catalysts, the corresponding chlorohydrin is first formed; this can then be transformed into the epoxy compound in a second step by reaction with an equivalent amount of alkali hydroxide.

The structure and molecular weight of the resulting epoxy resin are strongly influenced by the reaction conditions: A large excess of epichlorohydrin (about 5 mol per mol phenolic hydroxy groups) favors the formation of terminal epoxy groups; however, the molecular weight (and hence the softening point) of the product decreases with increasing amount of epichlorohydrin. The reaction temperature is also important: high temperatures promote secondary reactions such as hydrolytic cleavage of epoxy groups, leading to the formation of additional hydroxy groups.

The commercially most important epoxy resins are those prepared from 4,4′-isopropylidenediphenol (bisphenol A) and epichlorohydrin. They have molecular weights between 450 and 4,000 [n in formula (II) between 1 and 12] and softening points between 30°C and 155°C. Such epoxy resins are still soluble, but become insoluble and infusible through subsequent crosslinking reactions.

In principle, crosslinking (curing) can be brought about with any di- or polyfunctional compound that adds to epoxy groups (e.g., amines, aminoamides, thiols). Moreover, self-crosslinking can also be achieved by addition of catalytic amounts of a tertiary amine or acid compound, such as a sulfonic acid or a Friedel-Crafts catalyst (generally in the form of their adducts with ether or alcohols); this reaction often occurs even at low temperature, but does not proceed uniformly. Crosslinking of epoxy resins is an exothermic reaction, liberating 22–26 kcal per mol epoxy groups. In most cases, the amount of crosslinking agent used is equivalent to the analytically determined content of epoxy groups. Crosslinking is carried out always after application or molding. The crosslinking is generally carried out in bulk (casting resin), but sometimes in solution (lacquers, adhesives). A variant of crosslinking in solution is to use reactive thinners: These are substances which, on the one hand, lower the viscosity of the epoxy resin, but at the same time act as crosslinking agents (e.g., low-viscosity mono- and diepoxides or allyl glycidyl ether). Most crosslinking reactions only set in at higher temperatures so that the epoxy resin and crosslinker can be mixed and stored at room temperature. In practice, polybasic carboxylic acids, acid anhydrides, and amines are generally used as curing agents. In contrast to carboxylic acids, which only react sufficiently fast at high temperatures (above 180°C), carboxylic acid anhydrides (e.g., phthalic acid anhydride) allow crosslinking to be achieved at 100°C, especially in the presence of catalytic amounts of a tertiary amine. Optimum crosslinking is obtained with about 0.5 mol anhydride per mol epoxy groups.

Crosslinking with amines can be carried out either catalytically with tertiary amines (e.g., with N,N-dimethylbenzylamine), or especially by equimolar conversion with primary or secondary oligoamines at higher temperatures. This reaction is catalyzed by compounds that are capable of forming hydrogen bonds (water, alcohols, phenols, carboxylic acids, etc.). The most favorable ratio of amine/epoxide is not necessarily the stoichiometric ratio and therefore must be determined empirically in each case.

In the crosslinked state, epoxy resins are highly resistant to chemicals, temperature, and solvents and are also endowed with good electrical properties. They are therefore employed, for example, as casting resins in electro- and electronic industry as well as resistant lacquers and coatings. Moreover, they possess excellent adhesive power for many plastics, wood, and metals ("reaction adhesives"; "two-component adhesives").

Example 4.24 Preparation of Epoxy Resins from Bisphenol A and Epichlorohydrin

Safety precautions: Before this experiment is carried out, Sect. 2.2.5 must be read as well as the material safety data sheets (MSDS) for all chemicals and products used.

Epichlorohydrin, epoxy resins, and their curing agents are considered to be primary skin irritants. Some aromatic amines used as curing agents may be carcinogenic and should be handled with great care. Moreover, by prolonged action on the skin, epichlorohydrin is absorbed by the body and causes poisoning. All contact with these substances is, therefore, to be avoided; if necessary, wash off with plenty of water. Wear safety goggles and rubber gloves!

(a) *Preparation of an Epoxy Resin with a Molecular Weight of 900*

22.8 g (0.1 mol) of pure bisphenol A (recrystallized from dilute acetic acid) are mixed, under a hood, with a solution of 7.5 g (0.188 mol) of NaOH in 75 ml water contained in a 250 ml three-necked flask, fitted with thermometer, reflux condenser, and a powerful stirrer (good stirring is essential since the reaction mixture becomes heterogeneous). The contents of the flask is vigorously stirred and heated on an oil bath to 50°C within 10 min; 14.5 g (0.157 mol) of freshly distilled epichlorohydrin are then added in one batch. (The molar ratio epichlorohydrin/bisphenol A is thus 1.57). The temperature is now raised to 95°C within 20 min and held steady for 40 min. The temperature should not be allowed to exceed 95°C since high temperature favors side reactions. The reflux condenser is now removed and the stirrer switched off so that the resin formed can settle out. The clear aqueous upper layer is carefully siphoned off and the resin washed by vigorously stirring with hot distilled water at 80–95°C. It is allowed to settle again and the wash water removed. This washing procedure is repeated several times until 50 ml of the wash water is equivalent to less than 0.075 ml of 0.1 N HCl (indicator: methyl red). In order to remove the trapped water the washed resin is heated to 150°C with moderate stirring for 30 min, until it becomes clear. Finally, it is poured into a porcelain dish where it solidifies on cooling. The solid epoxy resin has a softening point of about 70°C and a molecular weight of approximately 900 [n = 2 in formula (II)]. The epoxy value (see below) is about 0.2, corresponding to 1.8 epoxy groups per molecule of resin (equivalent weight 500). It is soluble in aromatic hydrocarbons, tetrahydrofuran, and chloroform.

(b) *Preparation of an Epoxy Resin with a Molecular Weight of 1,400*

Under similar conditions to those used in (a), 22.8 g (0.1 mol) of bisphenol A, 5.55 g (0.14 mol) of NaOH (dissolved in 56 ml water), and 11.3 g (0.123 mol) of epichlorohydrin are allowed to react with one another. Because of the smaller molar ratio of epichlorohydrin to bisphenol A (1.22) the resulting epoxy resin is of higher molecular weight than that produced in (a). The epoxy value is approximately 0.1, corresponding to 1.44 epoxy groups per molecule of resin (equivalent weight approximately 970). The molecular weight of the resin, softening at 97–103°C, is found to be about 1,400 [n = 3.7 in formula (II)].

Determination of the Epoxy Value

This method of determination depends on the addition of hydrogen halide to epoxy groups (1 mol of hydrogen halide is equivalent to 1 mol of epoxy groups) and is carried out as follows: 0.5–1.0 g of epoxy resin are refluxed with an excess (50 ml) of pyridine hydrochloride solution (16 ml pure concentrated hydrochloric acid are made up to 1 l with pure pyridine) for 20 min and, after cooling, are back-titrated with 0.1 N NaOH, using phenolphthalein as indicator. The epoxy number represents the gram equivalents of epoxide-oxygen per 100 g resin:

$$\text{Epoxy number} = \frac{(B - A) \cdot N}{10E} \left[\frac{\text{Epoxide equivalent}}{100g} \right]$$

where A = titer for the sample (back titration), in ml; B = titer of the pyridine hydrochloride solution, in ml; N = concentration of sodium hydroxide used for titration, in mol/l, E = weight of resin, in g.

The equivalent weight is the amount of resin that contains one equivalent of epoxide; this is equal to 100/epoxy value.

(c) *Crosslinking (Curing) of Epoxy Resins*

With an amine: After determining the epoxy value and equivalent weight of the resin prepared according to (b), a small sample is melted in the oven at 150°C with the equivalent amount (0.25 mol per mol epoxy groups) of finely powdered 4,4'-methylenedianiline and the mixture well stirred for 30 s. After heating for 1 h at 150°C, the sample is taken out; it has now become insoluble and infusible.

With a carboxylic acid anhydride: 5 g of the resin prepared according to (a) are melted in a beaker at 120°C and 1.5 g of phthalic acid anhydride (0.6–0.8 equivalents per equivalent of epoxy groups) are stirred into the melt. The mixture is held at 120°C for 1 h (after this time the resin is still soluble in acetone or chloroform) and then cured at 170–180°C for 1–2 h.

Crosslinking of epoxy resins with carboxylic acid anhydrides is catalyzed by tertiary amines; thus, if 50 mg N,N-dimethyl aniline are added to the initial mixture in the above example, the curing process is already complete after 1 h at 120°C.

Bibliography

Allen G, Bevington J (eds) (1989) Comprehensive polymer science, vol 5. Pergamon, Oxford

Becker GW, Braun D, Oertel G (eds) (1993) Kunststoff-Handbuch, vol 7, Polyurethanes. Hanser, München

Brook MA (1999) Silicon in organic, organometallic, and polymer chemistry. Wiley, New York

Eisenbach CD, Baumgartner M, Guenter C (1986) In: Lal J, Mark JE (eds) Advances in elastomers and rubber elasticity (Polyurethanes). Plenum, New York

Elias HG (1984) Macromolecules. Plenum, New York

Flory PJ (1953) Principles of polymer chemistry. Cornell University Press, Ithaca

Houben-Weyl (1987) Methoden der organischen Chemie, vol E20, Makromolekulare Stoffe, Thieme, Stuttgart

Kricheldorf HR (ed) (1992) Handbook of polymer synthesis, part A and B. Marcel Dekker, New York

Mitsutoshi Jikei J, Kakimato M (2001) Hyperbranched polymers: a promising new class of materials. Prog Polym Sci 26:1233–1288

Rösch L, Weidner R Polymerization chemistry of silicones. In: Encyclopedia of materials: science and technology. Pergamon, Amsterdam, (Online Resource 2001)

Schlüter AD (2000) Synthesis of polymers. Wiley-VCH, Weinheim

Schmaljohan D, Voit BI, Jansen JFGA, Hendrdriks P, Loontjens JA (2000) New coating systems based on vinyl ether- and Oxetane-modified hyperbranched polyester. Macromol Mater Eng 275:31–41

Voit BI (2000) New developments in hyperbranched polymers. J Polymer Sci Part A Polym Chem 38:2505–2525

Modification of Macromolecular Substances

5

The term "modification of macromolecular substances" is used for chemical and physical processes that are carried out after the actual synthesis, i.e., on the finished macromolecule. *Chemical modifications* are, for example, the conversion of ester side groups to hydroxy groups, chemical degradation, and crosslinking reactions. *Physical modifications* are also of great importance in industrial practice. The utilization of additives to improve the processability (processing agents) or to increase the resistance to oxygen and light (oxidation inhibitors, photostabilizers) are among such modifications. Finally, there are some methods applied in order to modify the mechanical properties of polymers. These include the admixing of inorganic fillers ("filled polymers"), the introduction of inorganic or organic fibers ("reinforced polymers"), the admixing of other polymers ("polymer blends"), as well as stretching and foaming.

For almost all applications, polymers have to be optimized by one or more of the abovementioned modifications.

5.1   Chemical Conversion of Macromolecules

The preparation of macromolecular substances by addition and condensation polymerization relies on the reactivity of low-molecular-weight polyfunctional compounds. A second concept for synthesis is based on the conversion of ready-made macromolecular substances. Such reactions of polymers can occur at the functional groups of the constitutional repeating units (CRUs), with retention of the macromolecular skeleton and average degree of polymerization ("polymer-analogous" reactions), or they may involve degradation of the chains. In many cases both occur at the same time. There are also reactions involving enlargement of the macromolecules. They lead either to chain extension, chain branching, or crosslinking. Some reactions can, therefore, be used for the formation of *block* and *graft* copolymers. Conversion of macromolecules with bifunctional substances proceeds, depending on the concentration, by either intramolecular or intermolecular mechanisms. In the latter case crosslinking occurs causing the products to become insoluble and also no longer suitable for thermoplastic processing.

D. Braun et al., *Polymer Synthesis: Theory and Practice*,
DOI 10.1007/978-3-642-28980-4_5, © Springer-Verlag Berlin Heidelberg 2013

Numerous chemical conversions of macromolecular substances are also of technical interest, like reactions that attach or alter small parts of a polar group. Examples are the introduction of carboxyl or hydroxy groups to increase hydrophilicity, crosslinking (vulcanization) of polydienes with sulfur, or conversions that proceed only at the terminal groups under retention of the molecular backbone (chain-analogous conversion).

Conversions of these types are especially important for the capping of labile end groups. Thus, the poly(oxymethylene)s obtained from formaldehyde or 1,3,5-trioxane are thermally unstable because of the semiacetal end groups; but they can be stabilized by acetylation which, under suitable conditions, can be carried out without causing significant degradation (Example 5.9). Another reaction of this type is that of the amino end groups in the terminal amino acid units of polypeptides and proteins with, for example, 2,4-dinitrofluorobenzene. The end unit can then be removed by hydrolysis of the peptide link and identified as the dinitrophenyl-substituted amino acid. For the determination of end groups, see Sect. 2.3.3.3.

The various reactions of cellulose are amongst the important chemical conversions of macromolecular substances. The three hydroxy groups per CRU can be partially or completely esterified or etherified. The number of hydroxy groups acetylated per CRU are indicated by the names, i.e., cellulose triacetate, cellulose 2-acetate, etc. The conversion to dithiocarboxylic acid derivatives (xanthates) is another commercially important reaction of cellulose. Aqueous solutions of the sodium salt are known as "viscose"; they are spun into baths containing mineral acid, thereby regenerating the cellulose in the form of an insoluble fiber known as viscose rayon.

Sometimes these reactions allow the production of polymers whose monomers are unstable and consequently not existent [e.g., poly(vinyl alcohol), poly(vinylbutyral), poly(vinylamine)], or are difficult to prepare, or will either not polymerize, or do so only with great difficulty. An example of the last kind is vinylhydroquinone, which, like hydroquinone itself, is an inhibitor of radical polymerization. However, the "masking" of the phenolic OH groups by acetylation or silylation yields a polymerizable derivative; after its polymerization the protecting groups can be removed. The introduction of protective groups makes it possible to polymerize monomers ionically that are otherwise only radically polymerizable. For example, poly(methacrylic acid) with a narrow molecular-weight distribution and a defined tacticity can only be obtained in good yields by anionic polymerization of tert-butyl methacrylate and subsequent acid-catalyzed liberation of isobutylene. The anionic polymerization of methacrylic acid is not possible because of the presence of an acid proton.

Likewise, in the preparation of many ion-exchange resins, suitable functional groups are introduced by secondary reactions of macromolecular substances (that are generally crosslinked; see Sect. 5.2). In this context the utilization of crosslinked polystyrene resins or poly(acrylamide) gel in the solid-phase synthesis of polypeptides (Merrifield technique) or even oligonucleotides should be mentioned. After complete preparation of the desired products they are cleaved from the crosslinked substrate and can be isolated.

Crosslinked polymers with functional groups have recently been used even more frequently as reagents for the synthesis of low-molecular-weight organic compounds since they are easily separated after conversion and sometimes can easily be regenerated. The immobilization of enzymes by attaching them to crosslinked polymers should also be mentioned. This technique has already found industrial applications.

Under suitable conditions many of the known reactions in organic chemistry can, in principle, be applied to the corresponding functional groups of macromolecular substances, but there are differences in some respects between the conversions of macromolecular and low-molecular-weight substances. This is also true for the experimental implementation.

As long as all CRUs react in the same way during the reaction of a macromolecular compound, without any chain fission or significant side reactions, the molecular weight is changed, but not the degree of polymerization, i.e., the average number of CRUs per macromolecule remains constant. For example, in the quantitative hydrolysis of unbranched poly(vinyl acetate) to poly(vinyl alcohol), the molecular weight of the CRU and hence the average molecular weight of the polymer changes, but the average degree of polymerization remains the same. Those conversions in which the macromolecular skeleton remains intact are called "polymer-analogous" reactions because principally they are analogous to the corresponding reactions in low-molecular-weight chemistry. They play an important role in the elucidation of the structure of macromolecular compounds. In the development of macromolecular chemistry, polymer-analogous reactions on polystyrenes, for example, were also of fundamental significance in leading to the recognition that polymers are composed of macromolecules and do not consist of some form of associates of low-molecular-weight compounds.

In chemical reactions between low-molecular-weight compounds, new substances are formed that can, in principle, be separated from the unconverted reactants and by-products, e.g., by means of chromatography. With chemical reactions of macromolecular substances the situation is more complicated in that the main reaction and side reactions take place on the same molecular framework. If, for example, only 80 out of 100 CRUs in a polymer chain react in the desired sense while the rest either does not react at all or reacts in some other way, the remaining 20 units cannot be separated from the others since they all belong to the same macromolecule. Consequently, one cannot obtain a chemically uniform reaction product.

Due to the generally different molecular weights of the CRUs of reactant, main product, and by-product in the reaction of a macromolecule, it is necessary to define the yield or conversion in a special way (see Eqs. 5.3 and 5.4).

Changes in reactivity and reaction kinetics in comparison to conversions of low-molecular-weight substances can also be attributed to the fact that macromolecules in solution are in a more or less coiled state. During the course of a reaction this state changes through alteration of the solubility parameters, thus facilitating or aggravating conversion. Finally one observes influences on the type and grade of tacticity during the conversion of stereoregularly built macromolecular substances [e.g., saponification of poly(methacrylates)].

A limitation of the conversion for statistical reasons is always expected when two neighboring units of the macromolecule take part in the reaction with a low-molecular-weight compound, e.g., during the acetalization of poly(vinyl alcohol), in which hydroxy groups of neighboring CRUs undergo ring closure by reaction with the carbonyl group of an aldehyde (Example 5.2). According to calculations by Flory, the statistically maximum possible conversion of functional groups in such reactions is 86.5%. This becomes understandable by considering the fact that, for statistical reasons, an isolated functional group cannot react in the desired way since there is no neighboring group. These considerations are also true for polymers with several reactive groups in the fundamental building blocks or in branches.

The general considerations mentioned above can be summarized schematically as follows:

When two low-molecular-weight materials A and B react with one another, a by-product D is generally formed in addition to the desired product C:

$$A + B \rightarrow C + D \qquad (5.1)$$

Usually the reaction mixture also contains unreacted A and/or B.

To apply this general reaction scheme to the reaction of polymers, A must be regarded as the reactive group on the macromolecule M. Consequently, the reaction with reagent B yields group C (main product) and group D (by-product) that are likewise bound to the macromolecule by primary valencies; hence, in contrast to conversions of low-molecular-weight substances, C and D cannot be separated:

$$M[(A)_n] + B \rightarrow M[(A)_x - (C)_y - (D)_z] \qquad (5.2)$$

where $x + y + z = n$.

Depending on the structure and composition of the initial polymer (e.g., homopolymer or copolymer), n can have any value between unity and the degree of polymerization; in the latter case each CRU has a reactive group. The number of groups A, C, and D in the reacted polymer are denoted x, y, and z, respectively. It must be particularly emphasized that not all macromolecules in the product will contain the same number of these groups: x, y, and z are thus average values. The situation $x = z = 0$ and $y = n$ means that the desired reaction has gone to completion, but this is rarely attained.

In certain cases also the chemical conversion of functional end groups E of polymers or oligomers is of importance.

Reactions with monofunctional reagents are for example carried out in order to increase the thermal and/or chemical stability of the end groups (Polyoxymethylenes, Example 5.7). Reactions with bifunctional reagents can be used to enlarge the degree of polymerization or to synthesize block copolymers (see Sect. 4.2.1).

Continuous reactions at the side chains of branched or comblike structured polymers play an important role in the solid-phase synthesis of peptides or oligo-nucleotides.

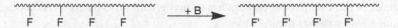

Here, besides the neighboring group participation along the chain, very often also the length of the spacer plays an important role for the reaction kinetics: if, for example, the reactive group F resides very far from the main chain due to the insertion of a flexible alkylene spacer $-(-CH_2-)_n-$, its pseudo-low-molecular-weight character increases intensely because of the higher mobility of the side chain in comparison to the less-mobile main chain. In these cases deviations from the kinetic behavior of corresponding low-molecular-weight model compounds are hardly found.

In the above-mentioned example of the polymer-analogous saponification of poly (vinyl acetate) the reactant and the product differ in their properties, for example, in their solubility; however, both compounds have the same average degree of polymeri-zation. The poly(vinyl alcohol) obtained by saponification can, in principle, be esterified back to poly(vinyl acetate) with the original molecular weight; the reacetylated polymer then has the same properties as the original material. The viscosity number may be used to check whether in fact any chain scission has occurred during the reaction sequence of saponification and reacetylation (see Example 5.1).

Before carrying out a specific conversion on a macromolecular substance, it is by all means expedient to first study the reaction of an appropriate low molecular model compound. The substance chosen as a model should be similar to the CRU in structure as well as in its reactive groups. In the case of addition polymers, the corresponding monomer is not generally useable, since it contains a double bond that is not present in the polymer. Thus, rather than using monomeric styrene as a model compound for polystyrene one should take cumene, or esters of isopropyl alcohol for poly(vinyl ester)s and the corresponding derivative of trimethylacetic acid for polymethacrylates. Due to the fact that neighboring reactive groups of a macromolecular substance exert influence on each other, it is sometimes useful to chose dimers or trimers as model compounds, e.g., 2,4-pentanediol as a model for poly(vinyl alcohol), or the derivatives of pentanedioic acid, namely α-methylpen-tanedioic acid or 1,3,5-pentanetricarboxylic acid as models for derivatives of poly (acrylic acid). These model compounds are then used for preliminary experiments to determine optimal reaction conditions as well as the type of by-products to be expected. Thereby one simultaneously obtains model compounds for macromolec-ular reaction products. These can then be used for dissolving tests and different analyses (e.g., determination of functional groups, comparison of NMR, IR, and

UV spectra, phase transitions, pyrolysis gas chromatography, etc.). The information thus obtained, however, is not necessarily valid for conversions of polymers; this is especially true for the choice of solvent and reaction temperature as well as work-up and purification processes.

While reactions of low-molecular-weight compounds can sometimes be carried out in the gas phase, this technique is not applicable to macromolecular substances since they are not volatile. However, it is indeed possible to let low-molecular reagents act upon solid or dissolved polymers in gaseous form. This is done, for example, in the commercial preparation of methylcellulose by conversion of alkali cellulose with gaseous methyl chloride.

If the polymer can only be applied in a solid form because it is poorly soluble, or is insoluble due to crosslinking (as in the case of ion exchangers, see Examples 5.9–5.10), or if the reaction is to take place only at the surface, it is convenient to work in dispersion of the finely divided polymer (for size reduction of polymers see Sect. 2.5.1). Therefore, the polymeric substance is suspended in an inert medium. Addition of a swelling agent is advantageous, because it lets the polymer swell superficially or throughout and thus favors access by the reagent. In some cases the reagent itself can act as a swelling agent, such as, for example, in the acetylation of the semiacetal end groups of poly(oxymethylene)s (see Example 5.7).

If the reaction is carried out in a homogeneous phase one has to take into consideration the high viscosities, and hence the impairment of mass and heat exchange in polymer solutions. For practical applications, this means that one must use strong and effective stirring units in order to obtain a good intermixing and to prevent local overheating.

When the low-molecular-weight reagent is liquid at the desired reaction temperature, and the polymer or the reaction product is soluble, sometimes no additional solvent must be used. In this case the reagent is used in large excess and the polymeric reaction product is isolated in the usual way from the finally formed solution (see Sect. 2.2.5.6). Examples for this procedure are the conversions of cellulose (see Examples 5.5 and 5.6) or polyacrolein. Whenever this procedure is not applicable, one has to use additional solvents. The right solvent is chosen by finding one that dissolves the polymer, the reagent, and the catalyst (when needed). However, it must be remembered that during the course of the reaction the solubility of the polymer may change quite sharply, even though only a small proportion of the reactive groups has been converted. The polymer will then precipitate prematurely, and thus hinder further conversion. In this case, it is advisable to use a solvent in which the reaction product is soluble or to add this solvent gradually with progressing conversion. Should it not be possible to find a common solvent or mixture of solvents for both starting polymer and end product, it is preferable to perform the reaction in a medium which is a solvent for the reaction product. Starting from a dispersion of very finely divided polymer, the partially converted polymers will go into solution with advancing conversion and finally can react to completion in a homogeneous system.

Also the reaction temperature has to be chosen with special care because, among others, this influences the ratio of main product to by-product decisively. Since with macromolecular substances it is principally not possible to separate the different reaction products within the polymer chain, the temperature should be chosen in such a way that primarily the main product will be formed, even though the reaction time may be prolonged by this choice. Lower temperatures are also advisable if crosslinking, thermal degradation, or chain scission by autoxidation (e.g., with polydienes) and hydrolysis (e.g., with cellulose) are to be anticipated. Such processes result in a substantial change of molecular weight and thus of the physical properties. These possibilities have particularly to be considered if the conversion should be kept polymer-analogous: it might then be advantageous to work under a nitrogen atmosphere and/or to add an antioxidant.

In most cases, the recovery and purification procedures for the macromolecular product differ from those of the model compound, except for its separation from low-molecular-weight reagents and by-products; this is further discussed in Sect. 2.2.5.7.

In order to assess the extent of reaction of a polymer, one cannot always apply the methods normally used to follow the reactions of low-molecular-weight compounds. Suitable methods for the characterization of polymers are therefore discussed in detail in Sect. 2.3.

One may first check qualitatively whether the desired reaction has taken place, especially by means of solubility tests and NMR or IR spectroscopy. It is also necessary to examine whether unreacted groups, or groups other than those desired, are present. Quantitative analysis is aimed at evaluating the proportions of A, C, and D in Eq. 5.2, for which one may employ not only the usual methods of determination, but also special procedures, such as spectroscopy and pyrolysis gas chromatography. It is also sometimes expedient to choose the low-molecular-weight reactant so that an easily determinable element is introduced (e.g., by using chloroacetic acid for esterifications).

The results of quantitative analysis can be presented in different ways. The experimental value may be expressed as a weight percentage conversion, i.e., as grams of analytically determined CRUs per 100 g of polymer. However, this method of expressing the conversion is not very informative since it does not indicate the extent to which the CRUs have been converted. It is, therefore, better to state the conversion in mol%, i.e., to indicate how many CRUs per hundred have reacted in the appropriate manner. It is then necessary to know the structure of any groups formed in side reactions, in order to calculate the molecular weight of all types of CRU present in the polymer. Making the simplifying assumption that no side chain reactions have occurred, the reaction product can be regarded as a two-component system made up of unconverted groups A and product groups C. The following equation can then be used to convert wt% into mol%:

$$\alpha = \frac{100}{1 + \frac{100-\alpha}{\alpha} \cdot \frac{E_{A^*}}{E_{C^*}}} \qquad (5.3)$$

$$c = 100 - a \qquad (5.4)$$

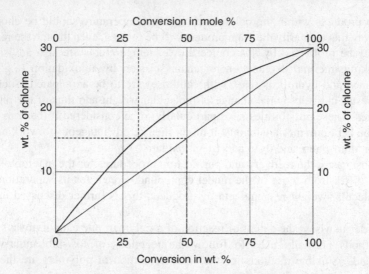

Fig. 5.1 Example of graphical determination of wt% conversion (*lower line*) and mol% conversion (*upper line*) from chlorine content in the esterification of poly(vinyl alcohol) with chloroacetic acid to give the CRU composition of $(C_4H_5ClO_2)_n$ at 100% conversion (29.5 wt% of chlorine). *Dashed lines* indicate values corresponding to an observed composition of 14.75 wt% of chlorine

A^* and C^* denote the molecular weights of the CRUs containing groups A and C, respectively, α is the fraction of CRUs containing group A, expressed in wt%, and a and c are the proportions of groups A and C, respectively, in mol%.

If many analyses of the same type have to be evaluated, it is convenient to prepare a diagram in which the analytically determined content of the appropriate group or element is plotted against the composition both in wt% and in mol% (see Fig. 5.1). It should again be pointed out that the observed values are generally less than the theoretical values for the reasons already mentioned.

Instead of giving the conversion in mol%, an average conversion factor f can be used:

$$f = [N \cdot \text{total conversion(in mol\%)}]/100,$$

with N = amount (in mol) of groups A per CRU.

This mode of expression is particularly recommended whenever the macromolecule contains more than one reactive group per CRU, e.g., 3 OH groups per CRU in cellulose (see Examples 5.5 and 5.6).

The calculations illustrated above can, of course, be extended to cover reactions involving the formation of three or more different groups, for example, when performing successive reactions on a polymer. This often means that the number of groups that has to be taken into account becomes unmanageably large because of side reactions or because quantitative analysis becomes very difficult. Calculations of conversion are then possible only on the basis of simplifying assumptions that must be decided individually. In such cases, the analysis will be limited to the determination of the expected product groups and the yield expressed in wt% appropriate to the particular reaction step.

Example 5.1 Poly(Vinyl Alcohol) by Transesterification of Poly(Vinyl Acetate); Reacetylation of Poly(Vinyl Alcohol)

Safety precautions: Before this experiment is carried out, Sect. 2.2.5 must be read as well as the material safety data sheets (MSDS) for all chemicals and products used.

(a) Preparation of Poly(vinyl Alcohol)

50 ml of 1% methanolic NaOH solution are placed in a 500 ml three-necked flask, fitted with stirrer, reflux condenser and dropping funnel, and heated to 50°C on a water bath. A solution of 15 g of poly(vinyl acetate) (see Example 3.4) in 100 ml of methanol are added dropwise with vigorous stirring over a period of 30 min. The transesterification sets in immediately as indicated by the precipitation of poly(vinyl alcohol). After the addition is complete, stirring is continued for further 30 min. The powdery precipitate is then filtered off, washed with methanol, and finally dried in vacuum at 30–40°C. Poly(vinyl alcohol) is soluble or swellable in only a few organic solvents (e.g., in warm dimethylformamide), but dissolves easily in warm water.

(b) Reacetylation of Poly(vinyl Alcohol)

5 g of the poly(vinyl alcohol) obtained above and 75 ml of the acetylating reagent (consisting of a mixture of pyridine, acetic anhydride, and acetic acid in the volume ratio 1:10:10) are placed in a 100 ml round-bottomed flask fitted with a reflux condenser. The mixture is then heated to 100°C and kept at this temperature for 24 h. The excess reagent is then removed in a rotary evaporator under vacuum at room temperature and the polymer precipitated from methanol into water.

The acetyl group content of the polymer is compared with that of the original poly(vinyl acetate), and the solubility behavior is also examined.

The acetyl content can be determined qualitatively by IR spectroscopy and quantitatively by ^{1}H-NMR spectroscopy. The limiting viscosity number is determined, both for the original poly(vinyl acetate) and the reacetylated sample, in acetone at 30°C. It will be found that the saponification and reacetylation has caused a fall in solution viscosity, indicating partial degradation of the macromolecules. This may be due to the presence of small amounts of ester moieties in the polymer chain, caused by the transfer reaction of a methoxy proton to the radical chain end during polymerization. Moreover, this viscosity effect may be attributed to the cleavage of side chains formed as a result of the branching reactions which occur when the polymerization of vinyl acetate is taken to high conversion.

The change of solubility of poly(vinyl alcohol) with the degree of acetylation can easily be followed by taking samples at various times and testing their solubility in water, acetone, and methanol. When the degree of acetylation is more than 20 mol% the solubility in water is lost, but the solubility in organic solvents increases.

Example 5.2 Preparation of Poly(Vinylbutyral)

Safety precautions: Before this experiment is carried out, Sect. 2.2.5 must be read as well as the material safety data sheets (MSDS) for all chemicals and products used.

3.3 g of distilled butanal (butyraldehyde) are placed in a 250 ml three-necked flask fitted with stirrer, reflux condenser, and dropping funnel. Into this is dropped a solution of 5 g of poly(vinyl alcohol) in 50 ml of water that has been warmed to 65°C and to which 0.3 g of concentrated sulfuric acid have been added. The addition should take about 2 min and during this time the stirring rate should be adjusted to give good mixing, but at the same time avoiding splashing the walls of the flask. The polymeric butyraldehyde divinyl acetal precipitates immediately. 1 g of 50% sulfuric acid is then added and the mixture allowed to react for another hour at 50–55°C. After cooling to room temperature the polymer is filtered off and washed with water until neutral. The polymer is reprecipitated from methanol solution into water, filtered, and dried under vacuum at 40°C. Yield: about 6 g. The solubility of the polymer is determined and compared with that of poly(vinyl acetate); it varies markedly with the degree of acetal formation. The course of the reaction can be monitored by IR or NMR spectroscopy.

A small amount of a Poly(vinyl butyral) solution is spread on a glass plate and evaporated slowly to create a thin film. Then the breaking behavior of the laminated glass plate is determined. Poly(vinyl butyral) foils are used for the manufacture of safety glass, for example, windshields in cars.

Example 5.3 Hydrolysis of a Copolymer of Styrene and Maleic Anhydride

Safety precautions: Before this experiment is carried out, Sect. 2.2.5 must be read as well as the material safety data sheets (MSDS) for all chemicals and products used.

2 g of an alternating copolymer of styrene and maleic anhydride (see Example 3.38) are heated to boiling with 50 ml of 2 M sodium hydroxide solution in a 100 ml round-bottomed flask fitted with a reflux condenser. The polymer dissolves within a few minutes. After 1 h the solution is cooled and the polymeric acid precipitated

by running about 500 ml of 2 M hydrochloric acid into the alkaline solution; it is allowed to settle and then filtered off. If the polymer has not settled out after some time, the acid dispersion should be shaken in a separating funnel with about 100 ml of diethyl ether; the polymer then separates into sticky rubbery lumps at the interface of the two liquids. The polymer is filtered and washed with a small amount of water (it is water-soluble!). It is now pressed well and allowed to dry in air. The polymeric acid is purified by dissolving it in 50 ml of tetrahydrofuran or 1,4-dioxane and precipitating in 500 ml of toluene. The polymer settles after some hours and can then be filtered off and dried under vacuum at 50°C.

The resulting styrene/maleic acid copolymer is soluble in hot water, in contrast to the starting material. The aqueous solution of the product gives a distinctly acid reaction. The disappearance of the anhydride moiety can be verified by IR or ^{13}C-NMR spectroscopic methods. The IR spectra of polymers should be recorded from a film of the sample prepared on a KBr pellet (freshly made from KBr powder). For this, a drop of a solution of the polymer in a low-boiling solvent (e.g., THF, methylene chloride) is placed on the pellet. The residual solvent can often be removed directly in the IR beam. The resulting spectra are characterized by their sharp bands.

Example 5.4 Preparation of Linear Poly(Ethyleneimine) by Hydrolysis of Polyoxazoline

Safety precautions: Before this experiment is carried out, Sect. 2.2.5 must be read as well as the material safety data sheets (MSDS) for all chemicals and products used.

In a 150 ml three-necked flask fitted with a thermometer and a reflux condenser 2 g of poly(*N*-acetyl ethyleneimine) from example 3.27 are dissolved in 30 ml of distilled water. After the addition of 2 g of sodiumhydroxide the mixture is heated under stirring to 100°C for 4 h.

Then the water is removed in a rotational evaporator resulting in two solid phases. The lower, yellowish phase is very compact and adheres to the flask; it contains only sodium acetate. The upper phase is a light powder and contains only poly(ethyleneimine).

Thus, the separation of this polymer can easily be done by scratching off with a spatulum. After isolation, the solid polymer is purified by dissolving it in 50 ml water at 80°C. After cooling down to room temperature, the poly(ethyleneimine) is isolated by filtration and washed with water and dried under vacuum at 80–90°C for 3–5 h. Linear poly(ethyleneimine) is obtained quantitatively as a colorless solid, which has to be stored under dry conditions.

The product is soluble in methanol, pyridine, dimethyl sulfoxide, and insoluble in nonpolar solvents, e.g., heptane and toluene. The ^{1}H-NMR spectrum of the polymer in D_2O shows one signal at 1.95 ppm ($-CH_2-CH_2$).

Example 5.5 Acetylation of Cellulose

Safety precautions: Before this experiment is carried out, Sect. 2.2.5 must be read as well as the material safety data sheets (MSDS) for all chemicals and products used.

10 g of cotton or shredded filter paper are covered with a solution of 0.5 g of concentrated sulfuric acid in 50 ml of glacial acetic acid in a 250 ml wide-necked bottle with a ground glass stopper (the type of starting material has a marked effect on the rate of reaction: filter papers of different kinds react at different rates). Uniform wetting of the cellulose is ensured by stirring with a glass rod, the closed bottle is then allowed to stand for 1 h at room temperature. After this pretreatment, a mixture of 50 ml of 95% acetic anhydride and 20 ml of glacial acetic acid is added, and the bottle is again closed and placed in a water bath at 50°C. The cellulose dissolves after about 15 min, and the reaction being complete after another 15 min. This so-called "primary solution" is divided into two equal parts, which are used for the preparation of cellulose triacetate and cellulose 2,5-acetate, respectively.

(a) *Preparation of Cellulose Triacetate*

25 ml of 80% acetic acid at 60°C are carefully stirred into one half of the primary solution in order to destroy the excess acetic anhydride. Care must be taken that there is no precipitation of cellulose acetate during this addition. The solution is held at 60°C for another 15 min, and then poured into a 1 l beaker; 25 ml of water are carefully stirred in. After the addition of another 200 ml of water, the cellulose triacetate precipitates as a white, crumbly powder. The product is filtered from the dilute acetic acid, slurried with 300 ml of distilled water, and the supernatant liquid decanted after 15 min. This procedure is repeated until the washing gives a neutral reaction. The polymer is dried as far as possible by suction or centrifugation and then in the oven at 105°C. Yield: about 7 g of cellulose triacetate. The product is soluble in methylene chloride/methanol (volume ratio 9:1), but practically insoluble in acetone.

(b) *Preparation of Cellulose 2,5-Acetate*

Into the other half of the primary solution, 50 ml of 70% acetic acid and 0.14 ml of concentrated sulfuric acid are slowly added with stirring at 60°C, in order to bring about partial saponification of the cellulose triacetate. The stoppered bottle is held at 80°C for 3 h and then worked up as described for cellulose triacetate. Yield: 6–6.5 g. Cellulose 2,5-acetate (acetyl group content 40%) is soluble in acetone and methylene chloride/methanol (volume ratio 9:1).

The progress of the partial saponification can be checked by a simple solubility test. About 1 ml of the solution is withdrawn and the cellulose acetate precipitated with water. The small sample is quickly washed free of acid and dried as much as possible by pressing between two filter papers. Some fibers of the still damp material are placed in a test tube with 15–20 ml of toluene/ethanol (volume ratio

1:1) and heated to boiling in a water bath. If the fibers go into the solution, then after about 15 min the whole charge can be worked up as described above.

For characterization, IR and ^{1}H-NMR spectroscopy are particularly suitable.

Example 5.6 Preparation of Sodium Carboxymethylcellulose

Safety precautions: Before this experiment is carried out, Sect. 2.2.5 must be read as well as the material safety data sheets (MSDS) for all chemicals and products used.

Filter paper, soaked in water, is worked into an aqueous pulp by kneading and shredding; it is then filtered and dried. 15 g of this finely divided cellulose and 400 ml of 2-propanol are placed in a 1-l three-necked flask fitted with stirrer, reflux condenser, dropping funnel, and nitrogen inlet; the air is displaced with a stream of nitrogen. 50 g of a 30% solution of sodium hydroxide are added with vigorous stirring over a period of 15 min, the stirring then being continued for another half hour. A solution of 17.5 g of monochloroacetic acid in 50 ml of 2-propanol is added dropwise over a period of 30 min. Stirring is continued for 4 h on a water bath at 60°C. The solution is then neutralized by addition of a few drops of glacial acetic acid using phenolphthalein as indicator, and is filtered while still hot. The raw fibrous sodium carboxymethylcellulose is dispersed in 400 ml 80% aqueous methanol at 60°C, filtered, and washed with a little aqueous methanol. This washing process is repeated two or three times until the product is free of sodium chloride. Finally it is washed with pure methanol and dried at 80°C. Yield: 22–25 g. The solubility of the dried sodium carboxymethylcellulose is tested in water. The presence of acid groups is shown by means of IR spectroscopy.

Example 5.7 Acetylation of the Semiacetal End Groups of Polyoxymethylene with Acetic Anhydride

Safety precautions: Before this experiment is carried out, Sect. 2.2.5 must be read as well as the material safety data sheets (MSDS) for all chemicals and products used. Because of the formation of gaseous formaldehyde the reactions have to be carried out in a closed hood.

Polyoxymethylene is prepared by the polymerization of anhydrous formaldehyde (see Example 3.22) or of 1,3,5-trioxane (see Example 3.24). Paraformaldehyde, obtained by polycondensation of formaldehyde hydrate, cannot be acetylated in heterogeneous medium. Acetic anhydride and *N,N*-dimethylcyclohexylamine are carefully fractionated and sodium acetate is dehydrated by heating.

(a) *Acetylation in Heterogeneous Medium*

In a 100 ml flask fitted with air condenser and calcium chloride drying tube, 3 g of finely powdered polyoxymethylene are refluxed (139°C) with 30 ml of acetic anhydride and 30 mg of anhydrous sodium acetate for 2 h with continuous stirring. The polymer is filtered off with suction and thoroughly washed five times with warm (50°C) distilled water to which some methanol has been added. It is then boiled with acetone for 1 h while stirring, and filtered again. The polymer is stored in a desiccator over calcium chloride and sodium hydroxide pellets. Yield: 92 wt%

of the original polymer. Properties of the acetylated polymer: melting range 174–177°C; thermally stable portion: 88% (see Example 5.13).

(b) *Acetylation in the Melt*

3 g of polyoxymethylene together with 6 ml of acetic anhydride and 2 ml of *N,N*-dimethylcyclohexylamine (to prevent acidolysis) are sealed in a glass ampoule and heated to 200°C for 30 min. The polymer is filtered off in a sintered glass crucible and thoroughly washed twice with ethanol. It is boiled with acetone for 1 h while stirring, and is filtered off under suction until the polymer is odorless. It is stored in a desiccator over calcium chloride and sodium hydroxide pellets. Yield: 90 wt%. Properties of the acetylated polymer: melting range 170–174°C; thermally stable portion: 97% (see Example 5.13).

5.2   Crosslinking of Macromolecular Substances

The preparation of covalently and three-dimensionally crosslinked macromolecular substances can be achieved in principle by crosslinking during the synthesis of macromolecules, by crosslinking of ready-made polymers with low-molecular-weight reagents, and by intermolecular chemical crosslinking reactions of preformed polymers. Depending on the reaction conditions it is possible to obtain either compact networks or swellable and/or, respectively, porous, networks with a defined interstitial structure. The latter are of interest, e.g., as ion exchangers, as a stationary phase for size exclusion chromatography (SEC), for affinity chromatography, as a substrate for the synthesis of peptides or oligonucleotides, or for the immobilization of enzymes ("biocatalysts").

Covalently crosslinked polymers lose their solubility and meltability and thus their processability. Hence, covalent crosslinking is first carried out after forming of the work piece, for example, on films, coatings, or on molded rubber articles. In addition to this, there is also the possibility of *physical crosslinking*, e.g., via hydrogen bonds as in case of polyamides or polyurethanes, which can be formed exclusively or in addition to covalent crosslinking. Another possibility is the crystalline domains of phase-separated *block* copolymers, composed of amorphous segments and segments that tend to crystallization. When the polymers are stable enough, physical crosslinking is, in contrast to the covalent chemical crosslinking, thermally reversible and the materials can be processed thermoplastically.

Network structures can be obtained directly through polyreactions using higher functional monomers, such as, e.g., hexamethylenebis-(meth)acrylates or divinylbenzene for radical homo- or copolymerizations, or at least trifunctional substrates for polycondensations and polyadditions, e.g., glycerol or trimethylolpropane with a triisocyanate. These reactions can take place in films placed on suitable carriers (e.g., hardening of coatings). Likewise, network formation can take place in adequately formed castings, which can be removed after the reaction if necessary. In this context, photolithography and the photochemical hardening of plastic tooth fillings, using bis(meth)acrylates and camphorquinone as photoinitiator, have to be mentioned (see also Sect. 3.1.4 and Example 3.13, photopolymerization of a hexamethylene bisacrylate).

Likewise, network structures can be built up by polymer-analogous conversions of ready-made, functional groups carrying polymers with an at least bifunctional, low-molecular-weight crosslinking agent. An example for this is the conversion of an epoxide groups-containing polymer, obtained by copolymerization using glycidyl methacrylates, with a diamine or a dicarboxylic acid. Also rubber vulcanization is an example for the crosslinking of a pre-formed macromolecule with a low-molecular sulfur component. An industrially significant crosslinking system is composed of an unsaturated polyester resin (obtained through polycondensation of maleic anhydride and diols) and of monomeric styrene. If such a mixture is subjected to a radical polymerization, styrene copolymerizes with the double bonds of the maleic anhydride CRUs, which leads to crosslinking (see Example 4.8).

Finally, it is possible to produce networks by intermolecular reactions of high-molecular-weight substances in the condensed phase. To ensure a sufficient mobility of the functional groups it is advantageous to use polymers with a low glass transition temperature. Addition of small amounts of a solvent may also increase chain mobility and thus may accelerate the reaction. Crosslinking reactions of cinnamic acid-containing polymers initiated by UV irradiation that proceed via a [2 + 2]cycloaddition are also belonging to this group. Also (meth)acrylic-modified macromolecules, obtained by esterification of hydroxy group-containing polymers with (meth)acryl chloride, can be intermolecularly transformed to networks with radical initiators.

Of technical importance are radically crosslinking reactions on finished articles of polyolefins by means of electron beams in order to increase, e.g., the thermostability. The technical importance of networks consisting of polydienes and other rubbers, polyurethanes, formaldehyde resins, alkyd resins, and silicones has already been explained in Sects. 4.1 and 4.2.

Example 5.8 Vulcanization of a Butadiene-Styrene Copolymer (SBR)

Safety precautions: Before this experiment is carried out, Sect. 2.2.5 must be read as well as the material safety data sheets (MSDS) for all chemicals and products used.

Industrially the curing (vulcanization) of diene homopolymers and copolymers with elementary sulfur is carried out in a heated press at 100–140°C (hot curing). This cannot be done in a normal laboratory on account of the expensive apparatus required. However, the principle of curing can be illustrated by crosslinking a butadiene-styrene copolymer (SBR 1500) with disulfur dichloride (S_2Cl_2) at room temperature (cold curing):

(a) In a hood, a small piece of a butadiene-styrene copolymer (see also Example 3.44) is placed in a test tube and covered with disulfur dichloride. The stoppered sample is allowed to stand under nitrogen for 1 h. The S_2Cl_2 is then poured off and toluene added. A sample that has not been treated with S_2Cl_2 is likewise covered with toluene. The solubility and swellability of the two samples are compared.

(b) The progress of the crosslinking during cold curing can be observed very nicely by the following experiment. 2 g of an SBR are dissolved under nitrogen in 100 ml of toluene in a 250-ml conical flask. 1 ml of disulfur dichloride is added to half of this solution and vigorously shaken after closing the flask. The other half of the solution is likewise kept stoppered, but without the addition of S_2Cl_2. After 5 min the solution treated with S_2Cl_2 is already significantly more viscous than the reference solution; after 10 min gelation sets in and after 20 min a pudding-like mass is formed. After some hours the crosslinking is so far advanced that a phase separation occurs as a consequence of the high crosslink density and the solvent is partially exuded from the shrunk gel.

For this teaching experiment it is not absolutely necessary to work under nitrogen.

5.2.1 Polyelectrolytes from Crosslinked Macromolecules

5.2.1.1 Ion Exchanger

A versatile application for crosslinked polymers is in ion exchangers. Ion exchangers are polyelectrolytes that generally consist of solid, crosslinked, and hence insoluble macromolecular compounds carrying acidic or basic groups on the macromolecular framework. The long known inorganic, naturally occurring or synthetically prepared materials (e.g., zeolites and permutites, respectively) play today only a minor role as ion exchangers. Nowadays, most ion exchangers are made by subsequent introduction of ionic groups into crosslinked addition or condensation polymers (see Example 5.9 and 5.10). Thus, the usual ion exchangers are macromolecular, insoluble polyvalent acids or bases. Because of their insolubility they are well suited for the exchange of H^+ or OH^- ions. Thus, if an insoluble polyacid is suspended in water with a low-valent salt, the cations of the salt are exchanged with hydrogen ions (cation exchanger); correspondingly, the anions of a low-valent salt can be exchanged with hydroxy ions by using a basic anion exchanger. This principle is applied, for example, in the preparation of pure water from sea water (desalination process). The processes which occur in these exchangers can be represented schematically as follows:

$$\text{(P)}-SO_3H \; + \overset{\oplus}{Na} \; \rightleftharpoons \; \text{(P)}-\overset{\ominus}{SO_3}\overset{\oplus}{Na} \; + \overset{\oplus}{H}$$

$$\text{(P)}-\overset{\oplus}{NR_3}\overset{\ominus}{OH} + \overset{\ominus}{Cl} \; \rightleftharpoons \; \text{(P)}-\overset{\oplus}{NR_3}\overset{\ominus}{Cl} \; + \overset{\ominus}{OH}$$

Depending on the type of exchanger used, the acidic and basic groups may be contained in the same polymer, or the anionic and cationic exchangers can be mixed, or used in tandem. The encircled "P" denotes the insoluble crosslinked resin with various exchangeable ions bound to its functional groups.

Exchangers that have been loaded with cations or anions can be regenerated by treatment with acid or alkali, respectively, since one is always dealing with an equilibrium reaction. Commercially available ion exchangers are frequently delivered in the form of salts so that before use they must be converted into the free acids or bases. Metal cations can also be directly exchanged with one another.

The most suitable acidic groups for synthetic ion exchangers are sulfonic and carboxylic acid groups; phosphoric acid groups are less common. For anion exchangers, primary, secondary, and tertiary amino groups are often used, also polymeric quaternary ammonium bases. For the preparation of such ion exchangers it is common to start from a polymer. Copolymers of styrene and 1,4-divinylbenzene (see Example 3.41) are especially useful for this purpose, which can be conveniently used in the form of polymer beads with a particle diameter of about 0.1–2 mm. The content of 1,4-divinylbenzene in the monomer mixture used for polymerization determines the degree of crosslinking; it is generally indicated as wt% 1,4-divinylbenzene. The higher the degree of crosslinking, the lower is the swellability of the polymer, however, the degree of swelling also depends on the nature of the counterion and on some other factors. Under certain conditions, for example, in the presence of organic solvents in bead polymerization, so-called macroreticular or macroporous networks are formed. In this case the crosslinks are distributed irregularly over the whole volume of the material, unlike in normal networks. This results in a porous structure combining the properties of high permeability for solvent and comparatively low swellability. Sometimes, such polymers have advantageous properties as starting materials for making ion exchangers. The ionic functional groups are then introduced by chemical reactions. One can also make ion exchangers by polymerizing monomers that already contain functional groups. Thus, methacrylic acid can be polymerized in the presence of small amounts of a crosslinker to give a weak cation exchanger.

Another possibility which is used commercially is to prepare insoluble condensation polymers, e.g., from phenol and formaldehyde, into which ionic groups are subsequently introduced.

Ion exchangers can also be made from cellulose, especially for scientific applications. They are prepared from alkali cellulose by reaction, for example, with chloroacetic acid (for preparation of sodium carboxymethylcellulose, see Example 5.6). By conversion with 2-chloroethyldiethylamine one obtains so-called DEAE-cellulose, an anion exchanger carrying 2-diethylaminoethyl groups, $-C_2H_4N(C_2H_5)_2$.

Besides grain size, degree of crosslinking, and swellability, an important characteristic of an ion exchanger is its capacity. This denotes the number of equivalents of exchangeable counterions on a high polymer network, with respect either to the weight or volume of dry or swollen exchanger. For laboratory use the capacity is usually expressed in meq/g; for the softening of hard water the usable exchange capacity is often given in g of CaO/l of exchanger.

Various procedures can be applied to effect ion exchange. The simplest method is to work batchwise whereby the exchanger is left in contact with a solution of the ions to be exchanged until equilibrium is reached. This method is applicable to those exchange reactions where the equilibrium is in favor of the desired product; this can of course always be achieved by employing a sufficient excess of a cation exchanger in its acidic form to a metal salt solution.

However, ion exchange is mostly carried out using columns. As in column chromatography, the solution to be exchanged is allowed to run through the column of ion exchanger from top to bottom. The ion exchanger used to fill the column must already be in the swollen state before it is washed into the column, otherwise the pressure caused by swelling may lead to bursting of the glass tube.

Ion exchangers are not only used commercially (e.g., for water softening, recovery of metals from waste water, refining of raw sugar), but also to an increasing extent in the laboratory. Thus, ion exchangers provide a convenient and clean way of preparing free acids or bases (e.g., free thiocyanic acid by exchange of ammonium thiocyanate with an acid cation exchanger) or of purifying aqueous solutions (e.g., removal of formic acid from solutions of formaldehyde). Ion exchangers can also be used to catalyze chemical conversions, being readily removed by filtration after the required reaction has occurred. Examples of this type are esterifications or protein hydrolyses catalyzed by acid ion exchangers. By introducing complex-forming groups into crosslinked polymers, numerous exchangers have also been prepared which allow selective extraction of certain metals from solution or provide a means of enrichment of elements that are present in low concentration. The synthetic routes for this purpose are indeed very complicated, For example, the aromatic amino group of crosslinked poly(aminostyrene-co-styrene) can be diazotized and then coupled with suitable phenol derivatives, or, by using formaldehyde 8-quinolinol, can be bound covalently to acrylamide-containing copolymer networks via a methylene group. Numerous other applications for these polymeric networks are well-established in the field of carrier-bound synthesis of organic substances, especially of peptides and oligonucleotides.

Example 5.9 Preparation of a Cation Exchanger by Sulfonation of Crosslinked Polystyrene

Safety precautions: Before this experiment is carried out, Sect. 2.2.5 must be read as well as the material safety data sheets (MSDS) for all chemicals and products used.

(a) *Sulfonation of Crosslinked Polystyrene*

Insoluble polystyrene crosslinked with divinylbenzene can easily be converted by sulfonation to a usable ion exchanger. For this purpose a mixture of 0.2 g of silver sulfate and 150 ml of concentrated sulfuric acid are heated to 80–90°C in a 500 ml three-necked flask fitted with stirrer, reflux condenser, and thermometer. 20 g of a bead polymer of styrene and divinylbenzene (see Example 3.41) are then introduced with stirring; the temperature climbs spontaneously to 100–105°C. The mixture is maintained at 100°C for 3 h, then cooled to room temperature and allowed to stand for some hours. Next the contents of the flask are poured into a 1 l conical flask that contains about 500 ml of 50% sulfuric acid. After cooling, the mixture is diluted with distilled water, and the gold-brown colored beads are filtered off on a sintered glass filter and washed copiously with water.

(b) *Determination of the Ion-Exchange Capacity*

To determine the ion-exchange capacity, the sulfonated polymer is washed into a glass tube closed at one end with a stopcock (chromatographic column) above which is a plug of glass wool. 100 ml of 2 M NaCl solution are allowed to run through the column, followed by 100 ml of 2 M HCl. Finally, the column is washed with distilled water, the washings being collected in 10-ml portions and titrated with 0.1-M NaOH using phenolphthalein as indicator. When the concentration has fallen below 0.001 M the washing is stopped. The water remaining in the column is allowed to run off and the damp ion exchanger is poured into a beaker. Three samples, each of 2 g, are weighed into preweighed 100 ml conical flasks as quickly as possible. One is heated to constant weight in an oven at 110°C in order to determine the water content of the sample. 50 ml of 0.1 M NaOH are added to each of the other two flasks and vigorously shaken. After 30 min the mixtures are filtered, the resin is washed with a little water, and the filtrate back-titrated with 0.1 M HCl. The ion-exchange capacity is expressed in meq/g of dry exchanger.

Example 5.10 Preparation of an Anion Exchanger from Crosslinked Polystyrene by Chloromethylation and Amination

Safety precautions: Before this experiment is carried out, Sect. 2.2.5 must be read as well as the material safety data sheets (MSDS) for all chemicals and products used.

(a) *Chloromethylation of Crosslinked Polystyrene*

Chloromethyl methyl ether is an alkylating agent and very poisonous. Therefore this synthesis must be carried out with rubber gloves under a hood, preferentially in a glove box.

20 g of a crosslinked styrene bead polymer (e.g., the styrene/divinyl benzene copolymer from Example 3.41) and a solution of chloromethyl methyl ether in 40 ml of tetrachloroethylene are placed in a 250 ml three-necked flask fitted with stirrer and reflux condenser (with drying tube attached), the third neck being closed with a ground glass stopper. This mixture is stirred for 30 min at room temperature, causing the beads to swell somewhat. 10 g of anhydrous zinc chloride are added at 40–60°C over a period of 60 min with continuous stirring; stirring is continued at this temperature for another 2 h. The unconverted chloromethyl methyl ether is now destroyed by careful addition of water. The beads are washed several times with water and dried in vacuum at 50°C. The chlorine content of the beads should be around 15 wt%.

(b) *Amination of the Chloromethylated Polystyrene*

20 g of the chloromethylated polystyrene obtained in (a) are refluxed with 50 ml of toluene for 30 min in a 100 ml three-necked flask fitted with stirrer, reflux condenser, thermometer, and gas inlet. The mixture is cooled to 30–35°C and gaseous anhydrous trimethylamine is passed in with stirring, while the temperature is raised steadily to 50–55°C. The gaseous trimethylamine is prepared from an aqueous solution by dropping it into concentrated sodium hydroxide in a separate vessel and passing the gas through a drying tube filled with NaOH pellets into the three-necked flask. The flow of trimethyl-amine is stopped after 4 h and the mixture is allowed to stand for another 3 h at room temperature. The beads are filtered off, washed a few times with toluene, and dried in vacuum at 50°C. The dry beads are

then treated for 2 h with 100 ml of 5% hydrochloric acid and finally washed thoroughly with water until free of acid (test with methyl red).

(c) *Determination of the Ion-Exchange Capacity*

Three samples of the moist beads obtained in (b), each of about 2 g, are weighed into tarred 100 ml conical flasks. One flask is heated to constant weight in the oven at 110°C in order to determine the water content of the beads. 10 ml of 15% sodium hydroxide are added to each of the other two flasks and the mixture is stirred magnetically for 30 min to form the quaternary ammonium base. Next the excess alkali is removed by washing with water (test with phenolphthalein) and filtering. The moist beads are transferred quantitatively back to the conical flask, and shaken back and forth with 50 ml of 0.1 M hydrochloric acid. After 30 min the beads are filtered off, washed with a little water, and the filtrate will be back-titrated with 0.1 M sodium hydroxide. The ion-exchange capacity is given in meq/g of dry exchanger (in the form of the chloride).

5.2.1.2 Superabsorbents

Another very interesting class of crosslinked polyelectrolytes are the so-called superabsorbents. They predominantly consist of crosslinked and (partially) neutralized poly(acrylic acid) and, hence, represent a network of flexible polymer chains that carry dissociated, ionic groups. Due to this structure they can function as water-swellable gels. Although they are hard, sandy powders in a dry form, they are able to absorb rapidly large amounts of water, thereby turning into a soft, rubbery gel. They may absorb as much as 1,000 g of water per gram of polymer (in comparison fluffed cellulose pulp: 12 g/g) and up to 100 g of dilute salt solution per gram of polymer. Unlike the behavior of a soaked pad of cotton or a sponge, the superabsorbent gel will not release the water when squeezed with the fingers or cut with a sharp knife.

These superabsorbents are synthesized via free radical polymerization of acrylic acid or its salts in presence of a crosslinker (crosslinking copolymerization). Initiators are commonly used (water-soluble compounds e.g., peroxodisulfates or redox systems). As crosslinking comonomers bis-methacrylates or N,N'-methylenebis-(acrylamide) are mostly applied. The copolymerization can be carried out in aqueous solution (see Example 5.11) or as dispersion of aqueous drops in a hydrocarbon (inverse emulsion polymerization, see Sect. 2.2.4.2).

Main application of the poly(acrylic acid) super-absorbents is in disposable, sanitary and hygienic systems, for example, diapers.

Example 5.11 Superabsorbent Polyelectrolyte Based on a Crosslinked Acrylic Acid Copolymer

Safety precautions: Before this experiment is carried out, Sect. 2.2.5 must be read as well as the material safety data sheets (MSDS) for all chemicals and products used. The reaction should be carried out in a hood.

The following example describes the synthesis of a superabsorbent via crosslinking copolymerization of acrylic acid (sodium salt) with N,N'-methylenebis (acrylamide).

In a 600 ml glass beaker equipped with a magnetic stirrer and a thermometer 25 g acrylic acid, 0.1 g N,N'-methylene(bisacrylamide), and 0.12 g sodium carbonate are dissolved in 60 ml dist. water. 0.7 ml of an aqueous sodium peroxodisulfate solution ($Na_2S_2O_8$, 10 wt%) are added slowly at room temperature to the reaction mixture in the beaker. The mixture foams strongly and a temperature increase can be observed when the addition is too quick. One has to be careful that the temperature does not exceed 25°C. If necessary, cooling is applied briefly but removed afterwards. The mixture should be set to pH 4.9.

The mixture is slowly stirred and after 3–5 min. the reaction proceeds leading to a strong increase in viscosity and temperature. Stirring is stopped and after further 10–15 min. the reaction reaches completion having a final reaction temperature of about 70°C. Large bubbles can be observed in the resulting gel. After 2–3 h the gel is cooled to room temperature, a slightly pink color can appear. In order to ensure full conversion the gel is kept over night at room temperature. Then, the product is cut in smaller pieces and placed in the vacuum drying oven at 100°C under inert gas and 200 mbar for another day. Yield: about 33 g (please note: the product is obtained as sodium salt).

The final very brittle product can be further crushed in a mill or in a mortar and piston. The swelling degree of the superabsorber can be determined by comparing the final weight of the water-swollen powder with the starting weight. For example, 0.1 g of the powder swell quickly in 10 ml distilled water to give in 2 min a solid ("dry") material (the material binds 100 times its weight). In this case the swelling limit is not yet reached. Even 0.05 g polyacrylic acid gel bind 10 ml water, but it takes 30 min to reach a "dry" swollen gel.

5.3 Degradation of Macromolecular Substances

Macromolecules can be cleaved by physical as well as by chemical action. The most frequent initial result is the formation of macromolecules with the same chain structure but with a lower average degree of polymerization. The polymer homologous series can be gradually degraded further until finally low-molecular-weight fragments are produced. The analysis and characterization of the oligomeric and monomeric species formed by degradation can provide valuable evidence concerning the structure of macromolecules. However, with some polymers degradation proceeds by an "unzipping" mechanism. In this case monomer molecules are broken off continuously either from the chain end or from a cleavage point within the chain, while the molecular weight of the remaining, unaffected polymer chains stays the same.

Thermally and chemically initiated degradation of polymers is particularly important. Chain cleavage by mechanical forces, for example, in blenders and extruders, or by light or high-energy radiation is also of considerable practical significance. Ultrasonic degradation of macromolecules can also occur. Nowadays the degradation of macromolecular substances by microorganisms or enzymes is gaining increasing importance.

The generally undesired changes in the chemical and physical properties of polymers during use, under the action of air, light, and heat, are grouped under the term aging. Such processes can be retarded or entirely eliminated by the addition of protecting agents (stabilizers, antioxidants, UV absorbers etc.; see Sect. 5.4).

The *thermal degradation* of polymers is often a radical process. Chain cleavage can take place either at random within the chain or preferentially at weak links, for example, in the neighborhood of branches or structural irregularities, or from unstable chain ends. With some polymers thermal degradation gives either none or very little of the monomers used in the preparation of the polymer (e.g., polyethylene, polypropylene, poly(acrylic esters), polyacrylonitrile, polybutadiene); here one speaks only of degradation. On the other hand, with other polymers degradation results in relatively large amounts of monomer (e.g., polystyrene, poly (α-methylstyrene), polyisoprene, poly(methyl methacrylate), polyoxymethylene) in which case the degradation may also be termed depolymerization. If the thermal degradation proceeds from labile end groups, the stability of such polymers can be substantially improved by blocking these end groups [see Example 5.13, thermal depolymerization of polyoxymethylene].

The kind of fragments formed on degradation depends mainly on the structure of the polymer and on the decomposition temperature. Thus, under the same conditions, the thermal decomposition of polymers of acrylic esters yields practically no monomer while that of polymers of methacrylic esters gives the monomer almost exclusively. Degradation of polystyrene at 250°C gives mainly oligomers of styrene such as the dimer, trimer, and higher homologs, while at 350°C monomeric styrene is also to be found in the decomposition products. As the temperature is raised further, ethylbenzene, toluene, and benzene are formed in increasing amounts.

Finally, there are those polymers that undergo an elimination reaction on heating but without any initial breakdown of the molecular chains. Poly(vinyl chloride) belongs to this group, eliminating hydrogen chloride on heating. Such a degradation is very undesirable both on account of the corrosive action of hydrogen chloride vapor and of the concomitant darkening of the polymer. This decomposition process can be suppressed to a certain extent, and in practice for considerable lengths of time, by addition of stabilizers (see Sect. 5.4.1). Similarly, poly(vinyl acetate) eliminates acetic acid on heating. Poly(*tert*-butyl methacrylate) yields isobutylene, water, and poly(methacrylic anhydride) at about 250°C.

In the presence of oxygen the thermal degradation of polymers is complicated by oxidation reactions, making the course of the reaction rather obscure.

Chemical degradation reactions can be caused by oxidation (autoxidation or ozonolysis), hydrolysis, and also by the action of light. In the same context also macromolecular substances that are biologically degradable by enzymes or microorganisms have to be mentioned. In addition to the biologically degradable natural products like cellulose or poly(hydroxybutyrate)s also conventionally manufactured polymers with ester or amide functions in the aliphatic main chain have attained some interest for technical applications. Strictly aliphatic polyesters do not show adequate material properties due to their soft character. This disadvantage can be overcome by incorporation of hydrogen bond-forming amide groups.

Hydrolytic degradation is especially important in polymers with hydrolyzable links between the CRUs. Thus, polyesters can be saponified to yield the starting materials from which they were formed. Acetal links in synthetic polymers such as polyoxymethylene, or in natural polymers such as cellulose, can be hydrolyzed with acids. However, the resistance to hydrolysis depends very much on the structure of the polymer; for example, polyesters of terephthalic acid are very difficult to hydrolyze while aliphatic polyesters are generally easily hydrolyzed. Polyamides are normally much more resistant to hydrolysis than polyesters; they may be cleaved by the methods usually employed for polypeptides and proteins.

Under suitable conditions, especially at elevated temperatures, macromolecular substances are autoxidizable, like low-molecular-weight compounds. The hydroperoxides that are formed in the process are generally unstable at these temperatures and decompose into radicals, rendering the reaction autocatalytic. The secondary products formed can initiate further reactions. Both chain degradation and crosslinking are observed. In commercial practice such processes are suppressed by the addition of antioxidants such as certain phenols or amines. Some macromolecular substances are, therefore, "stabilized" immediately after their preparation. This problem is accentuated in polymeric dienes; the double bonds of which give rise to an easy autoxidation. Furthermore, those double bonds can be cleaved easily by ozone present in the air in an ozonolysis reaction. For example, the ozonolysis of natural rubber results in 4-oxovaleraldehyde and 4-oxovaleric acid. The oxidative degradation is of analytical interest in diene-blockcopolymers as a simple way for the determination of block-length (see Example 3.47).

In addition to oxidation of polymers by molecular oxygen, other oxidizing agents can also cause degradation. For example, the number of head-to-head linkages in poly(vinyl alcohol) can be determined by oxidation of the 1,2-diol groups with periodic acid; at each of these positions the carbon chain is broken during the oxidation process so that the degradation can easily be followed viscometrically (Example 5.14). Conclusions can then be drawn concerning the number of irregularly bound CRUs formed during the preparation of the precursor, poly(vinyl acetate), by the radical polymerization of vinyl acetate. Such irregularities can arise either by reverse addition of monomer in the propagation step or by combination of polymer radicals in the termination step. Photochemical degradation of polymers in the presence of atmospheric oxygen is generally accompanied by oxidation and is one of the most important causes of aging of synthetic polymers.

Grinding or milling causes degradation of many polymers. The process of mastication of natural rubber involves a mechanically initiated, autoxidative degradation which lowers the molecular weight to a level where the material is easier to process on a commercial scale.

The radical chain fragments resulting from mechanical handling can initiate the formation of *block* and *graft* copolymers in the presence of polymerizable monomers.

Example 5.12 Thermal Depolymerization of Poly(α-Methylstyrene) and of Poly(Methyl Methacrylate)

Safety precautions: Before this experiment is carried out, Sect. 2.2.5 must be read as well as the material safety data sheets (MSDS) for all chemicals and products used.

Exactly 5 g of polymer (from Example 3.18 or 3.19) are weighed into a 100 ml round-bottomed flask. The flask is then connected via a right-angled glass tube to two cold traps maintained at $-78°C$ in a methanol/dry ice bath. The apparatus is evacuated to about 0.1 Torr and the flask immersed in a metal bath whose temperature can be regulated to $\pm 3°C$. The metal is first liquefied by heating to 100°C, the flask inserted and then quickly heated to the appropriate depolymerization temperature (within 6–8 min). For poly(α-methylstyrene) a temperature of 280°C is suitable; for poly(methyl methacrylate), 330°C. Depending on the rate of decomposition, the temperature is maintained for 1–2 h while evacuating the apparatus with an oil-pump. The experiment is stopped by removing the heating bath, and also the cold baths of the traps. As soon as the solid monomer, collected in the traps, has melted, the vacuum is released. The yield is determined by weighing the residue and the trapped monomer. The monomer is identified by measuring the refractive index (α-methylstyrene: $n_{20} = 1.5386$; methyl methacrylate: $n_{20} = 1.4140$). The decrease of molecular weight of the polymer during depolymerization is determined from the limiting viscosity numbers of the starting polymer and the residue left in the flask (measured in toluene at 20°C). For comparison one can also decompose polystyrene which, under these conditions, yields very little monomer.

Example 5.13 Thermal Depolymerization of Polyoxymethylene

Safety precautions: Before this experiment is carried out, Sect. 2.2.5 must be read as well as the material safety data sheets (MSDS) for all chemicals and products used. Because of the formation of gaseous formaldehyde the experiments have to be carried out in a closed hood. The decomposition of polyoxymethylene can be conveniently performed in a thermobalance, respectively TGA-apparatus (see Sect. 2.3.5.7).

100 mg of each of the following samples are weighed into small test tubes with an as constant as possible internal diameter:

(a) Polyoxymethylene with OH end groups (from Example 3.22),

(b) Polyoxymethylene with acetyl end groups (from Example 5.7).

The tubes are placed in the decomposition vessel which is evacuated and filled with pure nitrogen three times. A slow stream of nitrogen is passed through the vessel which is heated in an oil bath or air thermostat to 190°C. At intervals of 1 h the decomposition vessel is taken out of the hot bath, the tubes are allowed to cool under nitrogen for 15 min. They are then individually weighed and heated again as described above. The wt% residue is plotted against time.

The thermal depolymerization of polyoxymethylene starts from the unstable hydroxy end groups, but the oxidative and acid-catalyzed hydrolytic degradation takes place within the main chain. Hence, if polyoxymethylene is heated in air or in the presence of strong acids samples with blocked end groups will also degrade.

Example 5.14 Oxidative Degradation of Poly(Vinyl Alcohol) with Periodic Acid

Safety precautions: Before this experiment is carried out, Sect. 2.2.5 must be read as well as the material safety data sheets (MSDS) for all chemicals and products used.

2.0 g of poly(vinyl alcohol) (see Example 5.1) are placed in a 250 ml beaker containing 70 ml of distilled water; dissolution is hastened by warming somewhat and stirring with a glass rod. Care must be taken not to splash the solution. As soon as a homogeneous solution is obtained it is cooled to room temperature and filtered through a sintered glass disc in order to remove dust particles, the filtrate being collected in a 100 ml graduated flask. The beaker is washed several times with a little water and the washings are likewise filtered into the graduated flask which is then immersed in a thermostat at 25°C and made up to the mark. A solution of 1.7 g of periodic acid ($HIO_4 \times 2\ H_2O = H_5IO_6$) in 45 ml of distilled water is also prepared; this is filtered into a 50 ml graduated flask and made up to the mark in a thermostat at 25°C.

The oxidative degradation of poly(vinyl alcohol) is followed at 25°C by viscosity measurements in an Ostwald viscometer (capillary diameter 0.4 mm). One proceeds as follows:

1. 5 ml of the periodic acid solution are diluted to 10 ml in a graduated flask at 25°C using filtered distilled water. The flow time t_0 of this solution is determined.

2. In the same way, 5 ml of the poly(vinyl alcohol) solution are diluted to 10 ml with distilled water and the flow time t of this solution is determined.

3. Finally, 5 ml of the poly(vinyl alcohol) solution are mixed with 5 ml of periodic acid solution and the flow time is determined immediately and then at short time intervals until a constant value t_a is reached after a few minutes.

If it is assumed that the contribution of the periodic acid to the viscosity is negligible then the specific viscosity of the original poly(vinyl alcohol) is given by:

$$\eta_{sp} = \frac{t - t_0}{t_0} \tag{5.5}$$

and the specific viscosity of the final degraded polymer by:

$$\eta_{spdeg} = \frac{t_a - t_0}{t_0} \tag{5.6}$$

The limiting viscosity numbers are calculated with the equation of Mark and Houwink ($K_\eta = 0.27$; see Sect. 2.3.3.3.1) and from them the average molecular weights. From these one can estimate the number of cleavages of the original chain and hence the frequency of 1,2-diol groups.

Example 5.15 Hydrolytic Degradation of an Aliphatic Polyester

Safety precautions: Before this experiment is carried out, Sect. 2.2.5 must be read as well as the material safety data sheets (MSDS) for all chemicals and products used.

1 g of a linear aliphatic polyester (e.g., a polyester of succinic acid and 1,6-hexanediol, see Example 4.2) are dissolved in dry tetrahydrofuran in a 100 ml graduated flask and made up to the mark at 30°C. 50 ml of this solution are mixed with 2 ml of 30% sulfuric acid and stored in a closed vessel, separate from the other 50 ml of solution, at 30°C (e.g., in the viscometer bath). The viscosities of the two solutions are determined immediately and then at hourly intervals in an Ostwald viscometer (capillary diameter 0.3 mm) at 30°C. For this purpose 3 ml of the appropriate solution are pipetted each time from the flask into the viscometer. The two flasks are allowed to stand overnight at 30°C and the hourly viscosity measurements continued on the next day until the values remain essentially constant over 1 h, which will be the case after a total of 20 h. The viscosity of the sample without the sulfuric acid remains unchanged.

The specific viscosity (or simply the flow time) is plotted against reaction time. For the calculation of the specific viscosity the flow time of the solvent, t_0, must be determined for a mixture of 50 ml of tetrahydrofuran and 2 ml of 30% sulfuric acid; the viscosity of pure tetrahydrofuran is considerably raised by the addition of the acid. Finally, 20 ml each of the hydrolyzed and unhydrolyzed solutions are dropped into 200 ml of methanol and the resulting precipitates are compared. If the hydrolysis of the first sample is complete no precipitate will appear.

Example 5.16 Hydrolytic Degradation of Cellulose and Separation of the Hydrolysis Products by Chromatography

Safety precautions: Before this experiment is carried out, Sect. 2.2.5 must be read as well as the material safety data sheets (MSDS) for all chemicals and products used.

100 mg of cellulose (e.g., filter paper that has been well shredded by hand) is mixed with 1 ml of 72% ice-cold sulfuric acid and, after pulping well with a glass rod, is kept overnight at 0°C. 1 ml of 25% sulfuric acid is added and the mixture held for 2 h at 50°C. After cooling, 40 ml of iced water is added and the mixture finally refluxed for 1 h.

To neutralize the hydrolysis products the solution is passed through a column (of about 15-cm length and 1-cm width) packed with an anion exchanger (e.g., anion exchanger from Example 5.10). Neutralization with alkaline earth metal carbonates is to be avoided at all costs, in order to prevent epimerization of glucose to mannose which is favored by complex formation between mannose and alkaline earth metal ions.

The column is subsequently washed with about 50 ml of distilled water. A drop of the final washings should give a negative test for sugar when applied to a piece of chromato graphic paper and sprayed with aniline phthalate reagent after drying the paper. When this is the case the solution is evaporated in vacuum down to a volume of about 5 ml. Thin layer chromatographic (TLC) separation of the resulting sugars is advantageous and also can yield quantitative results. About 1.5 mm from the lower edge of a silica gel plate, 5 µl of the cellulose hydrolysis product are applied with the aid of a micropipette; 0.7% solutions of glucose, mannose, and xylose in 70% ethanol are also applied for comparison. A mixture of butyl acetate, ethanol, pyridine, and water (volume ratio 8:2:2:1) is used as eluent. Elution is carried out four times in a rectangular glass chamber, the plate being dried after each elution (separation distance about 16 cm, time about 60 min). The chromatogram is finally dried with the aid of a hot-air blower, sprayed with aniline phthalate reagent and placed in the drying oven at 105°C for 10 min. Approximate R_f values after four elutions and a separation distance of 16 cm: glucose 0.51, mannose 0.68, xylose 0.88.

5.4 Modification of Polymers by Additives

Most of the macromolecular substances cannot be processed or used as received from the manufacturing process. This is especially valid for solid polymers.

Frequently, small amounts of additives have to be admixed as processing aids or to improve the properties of the polymers, before molding compounds or granulated compounds can be made out of them. But also polymers accruing as dispersions have to be converted into a ready-to-use form by additives before they can be applied.

The following *additives* are of practical relevance:
- Lubricants, that enhance the rheological properties during processing in the melt,
- Stabilizers against thermooxidative degradation during processing in the melt,
- Plasticizers to increase processability, flexibility, and impact strength,
- Antioxidants and light stabilizers as antiaging additives for protection against oxygen and light during usage,
- Fillers and reinforcing materials for selective modification of certain, predominant mechanical, properties.

Depending on the intended application of the polymers the following agents are also admixed:
- Antistatic agents for protection against electrostatic charging,

– Conducting additives, e.g., carbon based particles like carbon black or carbon nanotubes, to decrease electrical resistance,
– Flame-proofing agents to decrease inflammability and combustibility,
– Dyes for specific coloring, e.g., inorganic or organic pigments and
– Blowing agents for fabrication of foams.

5.4.1 Addition of Stabilizers

During the fabrication of moldings, but in some cases also during use, polymers are exposed to substantial temperature influences. For example, for many manufacturing processes they have to be heated 100–200 K above the glass temperature in order to obtain a sufficient melt viscosity. Sometimes they are used at elevated temperatures, for example, as plastic pipes for hot water, as plastic moldings in "under the hood" applications in motor vehicles, or during intensive exposure to sunlight. Thermal chain scission or even oxidation reactions initiated by atmospheric oxygen can occur. Oxidations, proceeding via radical mechanisms, can be promoted by the influence of light. Particularly sensitive towards atmospheric oxygen or ozone are polymers with olefinic double bonds, like rubberelastic polydienes. The results of such oxidation reactions, sometimes also called aging, are, for example, yellowing, loss of brightness and transparency, but also the formation of surface cracks and thus the degeneration of mechanical properties. In the end, this can lead to the complete loss of serviceability.

Unwanted degradation and oxidation processes can be avoided or at least suppressed for some time either by structural modification of the polymer or by special additives. In practice, the addition of so-called antioxidants is particularly effective. Chemical substances that slow down oxidations and the following aging phenomena serve for this purpose. Antioxidants are sufficiently effective even in concentrations below 1 wt% and are added as early as possible to the polymer to be protected, e.g., already during the drying of powdery polymeric materials or during the preparation of granulates. Some of the most important so-called primary antioxidants are sterically hindered phenols and secondary aromatic amines. Secondary antioxidants are thioethers as well as phosphites and phosphonites. Also fillers like carbon nanotubes can serve as antioxidants due to binding of free radicals.

Example 5.17 Suppression of the Thermo-Oxidative Crosslinking of Polyisoprene by Addition of an Antioxidant

Safety precautions: Before this experiment is carried out, Sect. 2.2.5 must be read as well as the material safety data sheets (MSDS) for all chemicals and products used.

Because of the content of double bonds and a tertiary C-atom, polyisoprene undergoes numerous chemical transformations by the action of oxygen, light, or heat, like chain scission and crosslinking. The addition of suitable stabilizers can suppress these reactions even over a period of years. In the case of the thermooxidative crosslinking of polyisoprene this effect can be shown with the following experiment.

A solution of 2 g (2 wt%) of polyisoprene (see Example 3.12) in 100 ml of dry toluene is prepared. To 50 ml of this solution 1 mg of IRGANOX 565 {2,6-di(*tert*-butyl)-4-[4,6-bis(octylthio)-1,3,5-triazin-2-ylamino]phenol,   Ciba   Specialty Chemicals} is added. From either solution a film is cast. To do this, 5 ml of the solution are evaporated in a Petri dish (d = 35 mm) at room temperature. The films are then placed in a drying oven at 100°C for 6 h. If one tries to dissolve the thermally treated films afterwards in toluene, only the film with stabilizer is completely soluble; the film without stabilizer shows only swelling and is therefore crosslinked.

Apart from the oxidation of polymers the decomposition under the influence of light plays a special role, since thereby the application possibilities are substantially limited. Accordingly, besides antioxidants also light stabilizers are often added. These agents are substances that interfere with the chemical and physical processes of the light-induced degradation. Carbon black and some other pigments provide a practical useful light protection for many polymers, but in practice organic and metallo-organic compounds are of particular importance because they do not color at all or only add little color to the polymer. Nowadays there is a great number of light stabilizers that differ from each other in their mechanism of action. The so-called UV-absorbers prevent UV absorption or reduce light absorption due to their chromophoric groups. These are especially hydroxybenzophenones and hydroxyphenyltriazines, but also many other substances with suitable chromophoric groups. Some light stabilizers act as quenchers. They take up the energy absorbed by polymer-bonded chromophores and dissipate it in the form of heat, fluorescence, or phosphorescence and thus prevent degradation reactions.

Finally, also the hydroperoxides formed during the oxidation of polymers play an important role in the photo-oxidative degradation. Therefore as a third group of light stabilizers, hydroperoxide scavengers are used, e.g., metal complexes of sulfur-containing compounds, like dialkyl dithiocarbamates. Radical trapping agents, too, can slow down the photo-oxidative degradation, e.g., sterically hindered amines, that interfere with the degradation process in the form of nitroxyl radicals. The testing of the efficiency of a light stabilizer is carried out either in a laboratory experiment by means of so-called weathering instruments or in outdoor weathering tests in open air. Compared to accelerated laboratory tests the latter meet practical conditions in a better way, but the test period is much longer (months to years).

Of great technical and economical importance is the stabilization of poly(vinyl chloride). This polymer undergoes a hydrogen chloride elimination (dehydrochlorination) during thermal stress as well as during thermoplastic treatment. The mechanistic details of this elimination have not yet been elucidated completely. The dehydrochlorination of poly(vinyl chloride) leads in an unzipping reaction to the formation of polyene sequences. This is accompanied by an increasing darkening and ultimately also by a change in the physical properties and an embrittlement of the polymer. For this reason during the processing of poly(vinyl chloride) thermostabilizers have to be added that prevent or slow down dehydrochlorination. Additionally such stabilizers should shorten polyene sequences already formed and prevent their oxidation as far as possible. In practice depending upon the quite different requirements in individual cases, numerous classes of stabilizers work satisfactorily. For a long time, stabilizer systems based on barium and cadmium carboxylate salts in different mixing ratios have been of special importance. Because of the toxicity of cadmium salts these stabilizers are hardly used today; they have been replaced partly by barium/zinc or calcium/zinc stabilizers. Also lead salts belong to the oldest group of PVC stabilizers, their proportion, however, being reduced. In the meantime great importance is attached to organotin stabilizers, to which especially organotin carboxylates belong, but also sulfur-containing organotin stabilizers, such as, dioctyltinbisisooctyl thioglycolate and numerous similar dialkyltin metal thiolates.

Example 5.18 Suppression of the Thermal Dehydrochlorination of Poly (Vinyl Chloride) by Addition of Stabilizers

Safety precautions: Before this experiment is carried out, Sect. 2.2.5 must be read as well as the material safety data sheets (MSDS) for all chemicals and products used.

Two test tubes are each charged with 1 g of poly(vinyl chloride). 100 mg of lead stearate having been previously added as stabilizer to one of the samples (well mixed in a mortar!). The tubes are then loosely stoppered with corks which carry on the lower side (pinched into a slit) a 4 cm-long strip of moistened universal indicator paper. Both tubes are now heated for 10 min in a beaker containing colorless silicone oil at 170–175°C. The indicator paper above the sample stabilized with lead stearate shows scarcely any change, but that over the unstabilized sample shows very clearly by its color that hydrogen chloride has been evolved. The coloration of the polymer during heating is also strongly indicative: the unstabilized polymer becomes red to brown, while the stabilized sample only darkens a little. The evolution of hydrogen chloride can also be proved by passing the vapor over a beaker containing ammonia, or by passing it into acidified silver nitrate solution.

5.4.2   Addition of Plasticizers

Plasticizers are substances added to polymers to improve their softness, stretchability, and processability. Plasticizers decrease the glass transition temperature and thus influence the melting and the softening temperatures, as well as some mechanical properties, like, for example, the elastic modulus and the hardness of polymers. Being composed of low-molecular-weight substances or oligomers, most of the plasticizers can change the degree of softness of polymers over a wide range, depending on the type and amount added. The softener molecules are taken up by the polymer particles; thereby they "gel" the polymer by being inserted between the chain molecules of the macromolecular substances. In this form of "external" plasticizers, they reduce the interaction between the polymer chains, which can be observed physically in the reduction of the glass transition temperature. The low-molecular-weight substances used as plasticizers should be high boiling, have a low vapor pressure, and should not migrate to the surface of the molded article, in order to avoid a later embrittlement of the softened products.

For specific purposes, the principle of "internal" plasticization is applied. In this case monomer units are built into the polymer chain, acting as irregularities in the otherwise regular chain structure and thus causing a decrease in the glass transition temperature compared to the unmodified homopolymer. For example, it is possible to reduce the glass transition temperature of polystyrene by copolymerization with methyl acrylate. Depending on the composition of the copolymer the softening temperature can be lowered from 100°C to room temperature or lower, if necessary (see Example 3.42). The advantage of internally plasticized polymers is that the plasticizing unit is chemically linked to the polymer backbone and, hence, cannot migrate. On the other hand, the number of "plasticizing" monomers is very limited. Hence, internal plasticizing is used much less than external plasticizing because this route offers more variations. The most important plasticizers for polymers are, for example paraffin oil, phthalic esters, phosphoric esters, some esters of aliphatic carboxylic acids and, in particular, relatively low-molecular polyesters with predominantly aliphatic structures, also called polymeric plasticizers.

Example 5.19 External Plasticization of Polystyrene Via Polymerization of Styrene in Presence of Paraffin Oil

Safety precautions: Before this experiment is carried out, Sect. 2.2.5 must be read as well as the material safety data sheets (MSDS) for all chemicals and products used.

0.1 g of 2,2′-azobisisobutyronitrile (AIBN) is weighed into two Schlenk tubes each, and subsequently 10 g destabilized styrene are pipetted in each tube. Into one Schlenk tube 1 g (10 wt%) paraffin oil (white petroleum oil) is added. Both tubes are filled with nitrogen after three freeze-evacuate-thaw cycles. Both batches are then polymerized for 24 h at 70°C in a thermostat. After cooling, the Schlenk tubes are smashed with a hammer and the polymers are dried in vacuum for 24 h at 40°C in order to remove residual monomer. The glass transition temperature of both samples is determined by means of DSC. The glass transition temperature of the polystyrene with added plasticizer is distinctly lower (approx. 40°C with 10 wt% plasticizer).

5.4.3   Addition of Fillers and Reinforcing Materials

Fillers are solid materials that are dispersed in plastics and elastomers. One distinguishes between "inactive fillers" that are used in the first place to make the plastics less expensive and "active fillers" (reinforcing fillers) that improve specific mechanical properties and thus effect a "reinforcement". With the aid of these fillers, the elastic modulus, hardness, and thermostability are enhanced predominantly, whereas the impact strength of thermoplastic materials is reduced. Often the substitution of metals with plastics is made possible only by these filled or reinforced polymers. Active fillers, like carbon black or silica play a decisive role in the manufacture of tires.

Small, irregularly formed pieces or ball-shaped materials made of organic (cellulose, wood flour, carbon black, chopped cotton clothes) or inorganic (silica, powdered stone, chalk, gypsum, talcum powder, glass beads etc.) substances are used as fillers for polymers. They are used especially for crosslinkable, so-called thermosetting plastics (duroplastics) as extending fillers to save polymer, but also to improve surface conditions, to diminish the often high brittleness and to enhance stiffness (e.g., Example 4.14, 4–.15, and 4.16). Fillers are also added to non-crosslinkable thermoplastics in order to change their mechanical properties and to improve the flow characteristics of their melts. Often, anisotropy of the mechanical properties is achieved as a result of the geometric shape of the fillers. Depending on the manufacturing process, nonuniform distribution of the fillers in the molded article is also observed frequently.

By the addition of glass fibers, textile fibers, or chopped fabrics to crosslinkable polymers molding materials are produced with increased tensile strength, stiffness, and thermal stability compared to the filler-free polymers. The so-called reinforcing fillers, like carbon black, have good adhesion to the matrix due to specific surface interactions and their characteristic geometry.

Of special importance are glass fibers, carbon fibers, or polyaramide fibers. Molding materials manufactured with these fillers are prepared not only in a dough-like or flowable form, but also as laminates, i.e., as resin-impregnated sheets. To reinforce thermoplastics, short fibers of glass, carbon, or organic polymers with a length of approx. 1 mm are incorporated into granulates. These are then processed in the usual molding machines, particularly by injection molding and extrusion; fiber contents of up to 40% can be processed. For particular purposes longer fibers, sometimes in the form of continuous fibers, are incorporated with special devices. The fiber reinforcement often causes an essential increase of elastic modulus (i.e., of the stiffness), a decrease of impact strength, and a decrease of shrinkage that occurs due to changes of density during the freezing of polymer melts. Of special interest are glass fiber mat reinforced thermoplastics and duroplastics that, e.g., can be processed into expanded semifinished materials (composite materials).

During the last decade, nanofillers of different geometry, but in at least one dimension in the nanometer range, became more and more important. Examples are nanoclay (layered silicates, like montmorillonite), carbon nanotubes (single- and multiwalled), expanded graphite, and even graphene sheets.

A speciality of such nanocomposites is the very high surface of well dispersed nanofillers, which results in a very high interfacial area. Here, the behavior of the polymer chains near the interface is influenced by the interaction with the filler, leading to an interphase with new properties which already at low amounts of nanofillers can determine the nanocomposite's properties. In addition, quite big effects are observed already at quite low filler loadings, especially if the filler has an anistropic shape. As an example, in thermoplastic polymers electrical conductivity can be reached with carbon nanotubes even below 1 wt% addition.

In case of other fillers, the nanofillers can introduce new functionality into the polymer, e.g. electrical conductivity in case of carbon based nanoparticles, barrier properties in case of platelet like nanofillers (nanoclay, expanded graphite), enhancement of mechanical properties, enhanced flame retardancy, and many others.

Example 5.20a Preparation of a Composite Material from an Unsaturated Polyester Resin and Glass Fibers

Safety precautions: Before this experiment is carried out, Sect. 2.2.5 must be read as well as the material safety data sheets (MSDS) for all chemicals and products used.

The increase in stiffness of an unsaturated polyester resin by incorporation of glass fibers can be demonstrated by the following small-scale experiment: two small dishes of equal size (10 × 10 × 3 cm) are formed from aluminum foil. Subsequently, 0.2 g of dibenzoyl peroxide are mixed into 20 g of an unsaturated polyester in styrene (see Example 4.8). 10 g of this solution are poured in one of the dishes. In the other dish, several layers of glass fiber mats that have been soaked in the polyester resin solution are placed successively until the layer thickness of the first dish is reached. Both samples are then heated to 80°C in a drying oven. The crosslinking copolymerization sets in immediately and is almost finished after 15–20 min. The clearly increased stiffness of the sample reinforced with glass fiber can be verified using a simple bending test. Precaution: Use protective towel and wear safety gloves.

Example 5.20b Preparation of Electrically Conductive Composites by Filling Polycarbonate (PC) with Carbon Black (CB)

Safety precautions: Before this experiment is carried out, Sect. 2.2.5 must be read as well as the material safety data sheets (MSDS) for all chemicals and products used.

In practice, composites are mostly prepared by mechanical mixing. Discontinuous and continuous melt mixing reactors (kneading chambers, extruders) are used. Principally, also laboratory equipment like metal stirrers can be used to mix fillers with a polymer melt but the typical high viscosity of polymer melts hampers a homogeneous dispersion of the filler particles in the melt.

In a laboratory kneading chamber polycarbonate (PC) of medium viscosity (e.g. *Makrolon®*2600, Bayer) is mixed with different amounts of highly structured carbon black (CB, e.g. Printex XE2 (Evonik), up to 10 wt% in steps of 1–2 wt%) at ca. 280°C with a suitable defined mixing rate till a homogeneous mixture is obtained (typically 5–15 min). The melt will be taken out of the mixing unit, cooled down, and flat plates of the composites are prepared by compression moulding at ca. 280°C between two planar steel plates by means of a heatable hydraulic press. To hinder sticking of the composite at the metal plates, the plates may be covered with a release film and a frame can be used as distance holder to control the sample thickness.

During mixing one should observe the torque necessary to keep the mixer running at the defined speed. With increasing filler load the viscosity and therefore the torque will raise and it may be necessary to change the mixing parameters at high filler contents (reduce mixing speed or increase temperature).

The influence of the CB content on the stiffness can be analysed by a simple bending test on specimens cut from the plate. At a certain concentration of CB (in this system typically in the range of 4–8 wt%, depending on the materials and processing conditions) a transition from the electrically insulating (conductivity $\sigma < 10^{-16}$ S/cm) to a conductive state ($\sigma > 10^{-2}$ S/cm) will appear. This concentration is called percolation threshold. Often it correlates with a ductile-brittle transition, which can be analysed by tensile test measurements.

The change in conductivity can be easily tested by a simple electrical circuitry and the degree of changes can be determined by an ohmmeter.

5.5   Mixtures of Polymers (Polymer Blends)

Similar to other raw materials, e.g., metals, it is principally possible to vary the properties of macromolecular substances by mixing two or more different polymers. Such systems are generally termed polymer blends. Though, one has to pay attention to the fact that in most cases phase-separated, i.e., heterogeneous products are obtained by mixing macromolecules. Mostly, these polymer blends consist of a continuous phase (matrix) in which a discontinuous (dispersed) phase in the form of more or less regularly shaped particles is included. However, also co-continuous and other complex morphologies are of practical importance.

The phenomenon of phase separation can be explained thermodynamically:

If two low-molecular-weight liquids are united they will either mix homogeneously or separate into two phases. Which of the two cases occurs depends on whether the free energy ΔG increases or decreases during the mixing process:

$$\Delta G_{\text{mix}} = \Delta H_{\text{mix}} - T\Delta S_{\text{mix}} \tag{5.7}$$

Single-phase systems, i.e., complete miscibility, are obtained for negative values of ΔG_{mix} and two-phase systems, i.e., immisibility, for positive values of ΔG_{mix}. When mixing two polymers, both in solution and in the melt, the condition $\Delta G_{\text{mix}} < 0$ (and therefore molecularly disperse distribution) is fulfilled by far less frequently than when mixing two low-molecular liquids. This could be attributed to the very low value of the entropy of mixing (ΔS_{mix}) for high-molecular-weight substances. Although the entropy of mixing is likewise positive for polymers because basically disorder is increased during mixing, the absolute values are substantially lower for polymers than for low-molecular liquids. The reason for this is the lower number of molecules per volume element due to the high molecular weight of polymers.

Therefore, it is only possible to obtain a homogeneous, single-phase mixture of two polymers when the generally slightly positive mixing enthalpy ΔH_{mix} is sufficiently small or its magnitude is smaller than $T\Delta S$. However, this is only to be expected for chemically very similar macromolecules.

Thus, most of the time one obtains phase-separated systems in which the macromolecules of component A are not at all or only to a limited extent miscible with the macromolecules of component B, i.e., polymer A is immiscible or only partially miscible with polymer B.

The miscibility or partial miscibility of polymers can be predicted quite well on the basis of Flory's and Huggins' lattice theory because the equation developed originally for polymer solutions can be applied to solvent-free systems as well. In case of blends based on copolymers this theory also explains the often observed miscibility, however, it fails for such polymer mixtures, where crystallization and/ or strong intermolecular actions (e.g., hydrogen bonds or ionic interactions) occur since the energy gains obtainable in this way are larger than the mixing enthalpies. Besides, there are also semiempirically determined "compatibility parameters" that are sufficient for practical use. Simple tests for a rough estimation of the miscibility can be used: Turbidity- and T_g-measurments (See Example 5.21 and 5.22).

The term "miscibility" is a thermodynymic parameter describing a homogeneous mixture on molecular level, "partial miscibility" means miscibility in a limited concentration range or miscibility only under special conditions (pressure, temperature) and "immicibility" describes strong phase separated systems. In literature sometimes the terms "miscibility" and "immiscibility" are mixed with the terms "compatibility" or "incompatibility", but these derive from the macroscopic character describing that the mixtures exhibit properties useful for technical applications or not. Also the terms "polymer blend", "polymer alloys" or "polymer mixtures" are often synonymously used and may denote miscible (homogeneous) or immiscible (heterogeneous) blends consisting of two or more phases.

5.5.1 Properties of Polymer Blends

By mixing polymers manifold possibilities are opened to prepare new macromolecular materials with special combinations of properties, which can be obtained only with great difficulties or not at all by copolymerization. The "classical" example of polymer blends is rubber blends, e.g., mixtures of natural rubber with synthetic rubbers. They have been used already for many years by the tire industry. Of increasing interest as engineering plastics are heterogeneous polymer blends with a hard but brittle thermoplastic as matrix and a soft but tenacious elastomer as dispersed phase. This morphological principle is the basis for many impact-resistant plastics. In recent years also blends of two hard thermoplastics gained interest due to their advantages in properties and processability compared to the single components. Examples of binary polymer blends of industrial importance are listed in Table 5.1.

The properties of polymer blends depend on the physical and chemical properties of the participating polymers and on the state of the phase (homogeneous or heterogeneous). Phase morphology (particle shape and size as well as size

Table 5.1 Binary polymer blends of industrial importance

Type of blend	Examples
Soft matrix + soft dispersed phase	Elastomer blends
	e.g., natural rubber + synthetic rubber
Hard matrix + soft dispersed phase	Impact resistant thermoplastics
	Polystyrene + polybutadiene
	ABS-polymers[a]
	PVC + polyacrylates
	Polypropylene + C_2/C_3-elastomers
	Polyethyleneterephthalate + polyacrylates
	Polybutyleneterephthalate + polyacrylates
	Polyamides + elastomers
	Polyoxymethylene + elastomers (polyacrylates; thermoplastic polyurethanes)
Hard matrix + hard dispersed phase	Polystyrene + polyphenyleneether (homogeneous)
	Polycarbonate + polybutyleneterephthalate
	Polycarbonate + ABS[a]
	Polypropylene + aliphatic polyamides

[a]ABS is a blend of styrene/acrylonitrile-copolymer with butadiene/styrene-copolymer

distribution) and interfacial bonding between the phases are additional parameters for heterogeneous blends. Glass transition temperatures are good indicators for the effect of the state of the phasemorphology on polymer properties. Heterogeneous blends consisting of two polymers with different glass transition temperatures also show two glass transition temperatures, that is the one of polymer *A* and the one of polymer *B*. Homogeneous (single-phase) blends always possess only one glass transition temperature that depends on the mass ratio of polymer *A* to polymer *B*. In many cases this temperature can be predicted with specific rules of mixture. With respect to the glass transition temperatures (and some of the mechanical properties) heterogeneous blends are thus comparable to *block* copolymers and homogeneous blends can rather be compared with *random* copolymers.

The phase morphology has a strong impact on the physical properties of the blends. Maximum impact strength, for example, demands a different phase morphology than a specific optical property like, for example, transparency or surface gloss. Phase morphology is set mainly during the manufacturing process. It is decisive that these phase morphologies are fixed in such a way, e.g., by crosslinking of the dispersed phase, that they are not changed substantially during subsequent manufacturing processes (extrusion, injection molding).

The interfacial bonding between the phases, thus a partial compatibility (partial miscibility) between matrix and dispersed phase, is of decisive significance for the solid-state properties of heterogeneous blends. This interfacial bonding causes, for example, that the two phases do not separate on macro scale during processing from the melt. This causes a splicing of the finished product and therefore leads to better strength properties. Due to the adhesion, energy-absorbing processes can run across the interphases. Therefore, the shear forces at the surface boundary matrix/dispersed phase arising during an impact (i.e., a rapidly occurring deformation) do not lead to a complete separation of the two phases. Thus a mechanical failure is avoided. There are only few pairs of polymers having sufficient interfacial bonding due to their favorable solubility parameters. In most cases the phase adhesion has to be increased either via hydrogen bonds or ionic forces or by improving the compatibility of the polymers. This can be achieved by:

- Incorporation of polar groups that enter a mutual chemical or physical bond, e.g., by ionic functions or hydrogen bonds,
- Graft polymerization of the monomer from which the matrix is built-up, onto the dispersed phase, e.g., graft polymerization of styrene onto polybutadiene during the preparation of impact-resistant polystyrene,
- Addition of polymeric phase mediators (compatibilizer) like *graft* or *block* copolymers, consisting of segments that are (partially) compatible with the polymers to be mixed.

Apart from binary polymer blends, also blends consisting of three different polymers have found technical applications. Belonging to this group are mixtures of polypropylene, polyethylene, and ethylene/propylene elastomers.

Even the outstanding mechanical properties of natural products are based on the principle of polymer blends, like, for example, wood, which is composed in a complicated way of cellulose and lignin.

5.5.2 Preparation of Polymer Blends

Due to the great importance of the phase morphology and the interfacial bonding for the material properties, the manufacturing process plays an important role for heterogeneous blends. First of all it should be ensured that a specific particle shape, particle size, and particle size distribution of the dispersed phase can be reproducibly adjusted. At the same time, phase morphology has to withstand processing steps from users unchanged, i.e., renewed melting and shearing in extruders or in injection-molding machines. This can be achieved by crosslinking the dispersed phase. Furthermore, measures have to be taken that increase the interfacial bonding between matrix and dispersed phase. In the laboratory as well as on a plant scale the following manufacturing processes are used for polymer blends: concerted precipitation from solution, coprecipitation of polymer latices, mixing of polymer melts, and polymerization of monomers containing other dissolved polymers.

5.5.2.1 Concerted Precipitation from Solution

If two different polymers can be dissolved successfully in a common solvent, a molecular intermixing of the dissolved macromolecules should occur due to the fast establishment of the thermodynamic equilibrium. The difficulty with this procedure is due to the fact that very many polymers become incompatible above a certain concentration when their solutions in a common solvent are combined. This means that the originally homogeneous solutions of polymers *A* and *B* separate into two phases when being combined, whereby each of the phases contain different

Table 5.2 Examples for 1:1 polymer blends from two amorphous polymers

Components	Thermodynamic miscibility	Solvent	Film	T_g (°C) (DSC)
PS/PVC	Immiscible system	THF	Turbid	105 and 85
PS/PC		CH_2Cl_2		105 and 150
PMMA/PS	Partially miscible system	Toluene	Opaque	120 and 105
PMMA/PC		THF		120 and 150
PS/PB		Toluene		105 and 85
PMMA/SAN	Fully miscible system	THF	Transparent	115
PVC/NBR		THF		8
PS/PPE		THF		155

PS (polystyrene), *PVC* [poly(vinyl chloride)], *PC* (bisphenol A polycarbonate), *PMMA* [poly (methyl methacrylate)], *PB* (polybutadiene), *SAN* (styrene-acrylonitrile copolymer), *NBR* (acrylonitrile-butadiene rubber), *PPE* (polyphenylene ether)

quantitative proportions *A:B* [e.g., polystyrene and poly(vinyl acetate) in toluene]. But even when two polymers have been dissolved successfully in a common solvent, composition and phase morphology of the solid polymer blends obtained by precipitation or evaporation of the solvent depend strongly on the work-up method. Evaporation, which means slow increase of polymer concentration up to the solid state, can lead to inhomogeneities because the macromolecules have time to separate according to their molecular size and chemical composition. During fast precipitating processes this is far less the case, for example, by addition of a precipitating agent during spray precipitation (see Sect. 2.2.5.6) or by rapid evaporation.

Example 5.21 Preparation of Polymer Blends from Solution

Safety precautions: Before this experiment is carried out, Sect. 2.2.5 must be read as well as the material safety data sheets (MSDS) for all chemicals and products used.

The investigation of the mixing and demixing state and the characterization of the phase morphology are of importance for blends. Most amorphous polymers are soluble in a number of solvents, so it is possible to examine their mixing behavior in polymer blends on the basis of solution cast films. For crystalline polymers this is only possible with few high temperature solvents at temperatures in the region of the melting points of the polymers.

(a) *Solution from Two Amorphous Polymers*

Two separate solutions are prepared (same solvents, 5 wt%) from two amorphous polymers (see Table 5.2).

The solutions are then poured together in varying ratios. A few ml of the initially clear solutions are concentrated by evaporating the solvent in small glass dishes with flat bottoms in a drying oven until a film of the polymer blend forms (solvent not to be removed by distillation!).

The transparency of the polymer films is giving first information about the miscibility and phase morphology of the system which can be observed with the naked eye.

Opaque films are always of immiscible systems, which form a two phase morphology. The films are mostly clear, if the components are compatible and therefore form a homogeneous mixture. However, optical transparency is no certain sign of full miscibility as it is dependent on the light scattering of the phases. Two phase systems are opaque only, if the phases have large refractive Indexes differences and the phase dimensions are bigger than the wavelength of the radiated light.

The influence of the solvent can be observed, for example, with a polystyrene/polycarbonate-blend: CH_2Cl_2 allows a faster evaporation and leads to the expected turbid film consisting of the amorphous components PS and PC. Instead, THF and $CHCl_3$ evaporate slower and, hence, allow the PC to partially crystallize, which results in an opaque film.

(b) *Solution Made of a Crystalline and an Amorphous Polymer*

Due to light scattering, crystalline polymers mostly yield turbid films. Their blends with other polymers are always demixed because polymers are not able to form mixed crystals. Consequently, crystallizable polymers only yield homogeneous blends above their melting point. As soon as crystallization sets in, the components will separate.

Isotactic polypropylene (iPP, Example 3.30) and atactic polystyrene (PS, Example 3.1) are separately dissolved above the melting point of iPP ($>170°C$) in xylene (5 wt%) in small flasks. The temperature must be kept constantly high during the course of the experiment. The clear solutions are poured together in varying ratios in flat discs. The solutions immediately become turbid on mixing due to the distinct incompatibility of the components. During concentration in a (vacuum) oven coarsely separated films are formed with a white and a transparent phase. If cooled down during concentration, uniform white films are formed with a fine structure.

Because of the double refraction of the crystallites, the crystalline iPP phase appears bright under the microscope with crossed polarizers, the amorphous phase, however, appears black. If the iPP/PS film is molten ($>170°C$) and then briefly isothermally tempered at $130–135°C$, i.e., below the melting point of iPP, the growth of typical spherulite crystal structures can be observed under the microscope.

Since iPP and PS virtually do not mix with one another, one can observe the glass transition temperature of PS ($T_g = 105°C$) and the melting point of iPP ($T_m = 155°C$) in DSC.

5.5.2.2 Coprecipitation of Polymer Latices

If the two polymers to be mixed are present in the form of an aqueous latex, the mixing of the two latices and a subsequent joint coagulation or spray drying leads to a solid polymer blend. This technique is, for example, applied industrially for the manufacturing of ABS plastics (ABS is a blend of styrene/acrylonitrile-copolymers with butadiene/styrene-copolymers). The advantage of this method lies in the fact that the particle size of the phases can be controlled well by emulsion polymerization. Furthermore, the rubber phase can be partially crosslinked during polymerization. Hence, the particle size and size distribution remain the same not only during mixing and coprecipitating of the latices but also during processing in the melt.

5.5.2.3 Mixing of Polymer Melts

The simplest process for the preparation of polymer blends from thermoplastics is mixing the polymers in the melt in suitable devices like rollers, extruders, or kneading machines. The rate-determining step for the intermixing is the diffusion or mobility of the macromolecules inside the highly viscous melt. It is three to four times lower than, e.g., the diffusion of low-molecular additives in polymer melts; diffusion constants are in the region 10^{-12} cm^2/s. At a given temperature, the diffusion constants are inversely proportional to the melt viscosity and thus to the molecular weight. The morphological architecture of a polymer blend prepared by melt mixing is furthermore dependent on the mixing temperature, the shear gradient, the mixing time, and the rheological properties of the single components. Under suitable reaction conditions, chemical reactions, like transesterifications, transamidations, and chain scissions, can take place in the polymer melt. This can lead to *block* or *graft* copolymers which considerably improve the adhesion at the interphases as "phase mediators" (compatibilizer). In some cases, grafting reactions can also be achieved by adding suitable monomers to polymer melts in extruders. These types of melt mixing are named "reactive blending" ("reactive extrusion"). In many cases they are already applied for the industrial production of polymer blends.

Example 5.22 Preparation of Polymer Blends from the Melt

Safety precautions: Before this experiment is carried out, Sect. 2.2.5 must be read as well as the material safety data sheets (MSDS) for all chemicals and products used.

In practice, polymer blends are often produced by mechanical mixing in the melt. Most of the blend combinations are immiscible and the product is demixed but often in very small dimensions so that often scanning or transmission electron microscopy (SEM, TEM) have to be used to analyse the morphology in detail.
(a) Miscibility Test by Melt-Pressing a "Powder Blend"

In the laboratory simple tests for the miscibility of polymer melts can be carried out, in which the turbidity of the system is an indicator for the immiscibility:

Two polymer solutions (5 wt%) are successively dripped into a common precipitation bath under vigorous stirring. The precipitate is a "powder blend" that consists of small grains of the components.

If such a powder blend is fused on a hot plate after suction and drying, a turbid two-phase melt is formed with incompatible components (e.g., PS/PMMA), whereas a clear mixed melt is formed from compatible components (PS/PPE or SAN/PMMA).

For blends consisting of components with sufficiently different glass transition temperatures, like PS/PPE (T_g(PS) = 105°C, T_g(PPE) = 220°C), two phases (two glass transition temperatures) can still be detected for the blended powder. However, the melt and the solid obtained from the melt are only composed of a single phase (with only one glass transition temperature, depending on the composition of the blend).

If the components of the blend consist of polymers with sufficiently different glass transition temperatures, Tg-measurments can also be used for characterization: in the case of binary blend from polystyrene (T_g = 105°C) and from polyethylene ether (T_g = 220°C) two phases (two glass transition temperatures) are detected for the blended powder. Instead, the melt and the solid obtained from the melt show only one T_g-value, indicating a single phase system. It's T_g lies between those of the components and depends on the composition of the blend.

(b) Preparation of Polymer Blends

In a laboratory kneading chamber or small scale mixing reactor polypropylene (PP) and polyamide-6 (PA6) In a weight ratio of 90:10 are mixed under high shear at ca. 240°C until a homogeneous mixture is reached (typically 5–10 min). The material will be taken off and compression moulded between two planar steel plates at 240°C to ca. 2 mm thick plates by means of a heatable hydraulic press. To hinder sticking of the composite at the metal plates, the plates may be covered with a release film and a frame can be used as distance holder to control the sample thickness.

In a second experiment nonreactive PP will be replaced by PP grafted with maleic anhydride (PP-MA), e.g. Admer® PP-type grades (Mitsui Europe, MA content ca. 0.5 wt%). While the first system is nonreactive and just the physical factors like viscosity of the components, interfacial tension, shear rate and time of mixing, temperature and cooling conditions determine the final morphology, the PP-MA/PA6 blend is a reactive system in which during mixing at the PP-MA/PA6 interface in-situ compatibilizing agents (PP-g-PA6) are formed. These graft-copolymers stabilize the interface and increase the interfacial adhesion; which finally improves the mechanical strength of the blend.

The effect of the reactive compatibilization can be observed already during mixing by comparing the viscosity (detected e.g. as torque necessary to keep the mixing speed constant) over time curves. Furthermore, the size of the dispersed PA6 particles (mostly the minor phase forms particles dispersed in the major phase) can be analysed in the blends before and after compression moulding. By optical microscopy on thin cuts (or in the reflected light mode on cryo-fractures) in the nonreactive PP/PA6 blend particles in the size of few μm can be detected while in the reactive PP-MA/PA6 blend the PA6 phase is so small (100 nm scale) that scanning (SEM) or transition electron microscopy (TEM) on cryo-fractures or cryo-cuts (SEM) or ultra-thin cuts (TEM) have to be used to determine their size. To increase the contrast between PP and PA6 in cryo-fractured or cut surfaces the PA6 particles can be etched by immersing the surface into formic acid for short time.

Finally, the effect of the compatibilization on the mechanical properties can be studied by performing tensile or impact tests on test specimens cut from the compression moulded plates. In the blend containing nonreactive PP and PA6 tensile test values like yield stress and elongation at yield are not much changed as compared to the base PP, whereas modulus is enhanced due to the inclusion of harder PA. On the other hand, the impact strength values are reduced as compared to pure PP. In PP-MA based blends impact strength is enhanced compared to PP-MA due to the reaction at the interface. In the example with 10 wt% PA, an increase of Izod impact strength in the range of 35% can be found.

5.5.2.4 Polymerization of Monomers Containing Other Dissolved Polymers

The polymerization of monomers containing other dissolved polymers is used on a large technical scale for the fabrication of polystyrene/polybutadiene-blends (high impact polystyrene, HIPS). Hereby, up to 10% polybutadiene are dissolved in monomeric styrene and are then polymerized radically at temperatures between 60°C and 200°C until complete conversion is reached. In the course of polymerization several complicated chemical and colloid-chemical reactions take place. The precise control of which is essential for the product properties. The *prepolymerization* period starts with a solution of polybutadiene in styrene, which has to be carried out in bulk under controlled stirring. Two polyreactions set in simultaneously: the homopolymerization of styrene and, secondly, the graft-copolymerization of styrene onto polybutadiene. At the same time, the polybutadiene begins to precipitate because it is incompatible with polystyrene in the common solvent styrene monomer. At about 30% conversion the polystyrene phase becomes larger than the polybutadiene phase. Hence, a phase inversion takes place, resulting in a composite with polystyrene as matrix and polybutadiene as dispersed phase. During this phase inversion the above-mentioned graft copolymers act as polymeric emulsifiers and determine, inter alia, the particle size and particle size distribution of the dispersed polybutadiene phase. This morphology is fixed through another chemical reaction, that is the crosslinking of the polybutadiene phase. Therefore, the reaction mixture at the end of the prepolymerization period (at 30% conversion) does scarcely alter its morphology when it is polymerized to complete conversion, which is done without stirring mostly in bulk in a separate vessel.

Grafting of styrene (ST) onto polybutadiene (PB) can occur in two ways: Via a chain-transfer reaction with an allylic hydrogen of the 1,4- and the 1,2-units (Case 1); via copolymerization with $C = C$-double bonds of polybutadiene, in particular with the vinyl groups of the 1,2-units (Case 2):

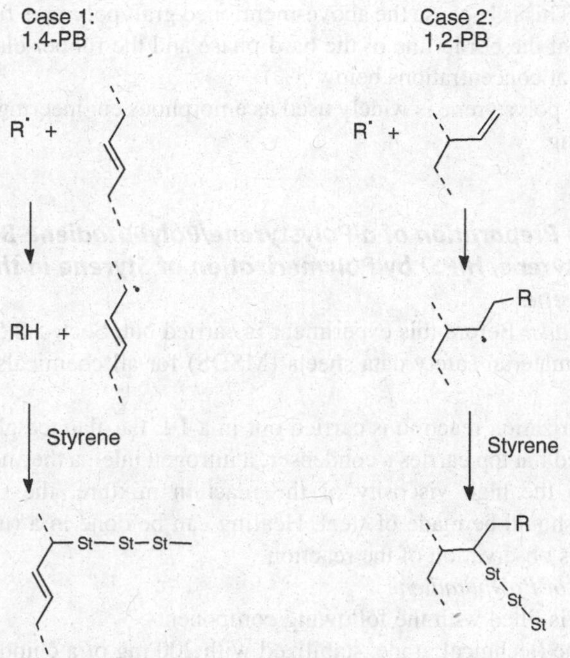

Case 1:
1,4-PB

Case 2:
1,2-PB

Under the conditions of Example 5.23 the rubber phase of the end product shows an interesting micro-morphology. It consists of particles of 1–3 μm diameter into which polystyrene spheres with much lower diameters are dispersed. These included polystyrene spheres act as hard "fillers" and raise the elastic modulus of polybutadiene. As a consequence, HIPS with this micro-morphology has a higher impact resistance without loosing too much in stiffness and hardness. This special morphology can be visualized with transmission electron microscopy. A relevant TEM-picture obtained from a thin cut after straining with osmium tetroxide is shown in Sect. 2.3.5.14.

The properties of such heterogeneous polystyrene/polybutadiene blends mainly depend on the following parameters:
- Molecular weight (MW) of the polystyrene matrix;
- Chemical structure and MW of polybutadiene;
- Volume fraction of the polybutadiene dispersed phase;
- Degree of grafting of the polybutadiene phase;
- Particle size and particle size distribution of the polybutadiene phase;
- Degree of crosslinking of the polybutadiene phase and
- Reaction conditions:

Variation of these parameters opens the possibility to influence the properties of HIPS over a wide range.

Unlike simple mixtures of polystyrene and polybutadiene such blends can be thermoplastically processed without macroscopic phase separation ("splicing"). Furthermore, they can to a certain extent withstand mechanical impact without disintegration. This is because the above-mentioned graft polymers function also as compatibilizer at the borderline of the hard phase and the rubber-elastic dispersed phase (already at concentrations below 3%).

High impact polystyrene is widely used as amorphous engineering plastic and in (food) packaging.

Example 5.23 Preparation of a Polystyrene/Polybutadiene-Blend (High Impact Polystyrene, HIPS) by Polymerization of Styrene in the Presence of Polybutadiene

Safety precautions: Before this experiment is carried out, Sect. 2.2.5 must be read as well as the material safety data sheets (MSDS) for all chemicals and products used.

The polymerization reaction is carried out in a 1 L flat-flanged glass autoclave. The four-necked flat top carries a condenser, a nitrogen inlet, a thermocouple, and a stirrer. Due to the high viscosity of the reaction mixture, the stirrer and the thermocouple should be made of steel. Heating can be done in a (transparent) oil bath that allows observation of the reaction.

(a) *Dissolving of Polybutadiene*

The reactor is filled with the following components:
- 538 g styrene (technical grade, stabilized with 200 mg of a chinon stabilizer),
- 48 g medium-*cis* polybutadiene with a solution viscosity of 170 mPas, cut into pieces of 1 cm^3,
- 12 g white oil with a viscosity of 70 mPas (e.g., WINOG 70) and
- 1.2 g IRGANOX 1076 as stabilizer.

This mixture is swept with nitrogen and kept over night at 50°C under stirring.

(b) *Prepolymerization*

After the polybutadiene is dissolved, 0.3 g *tert*-dodecylmercaptane (MW-regulator) are added and the reaction mixture heated under stirring (100 rpm) to 80°C. Now, 0.6 g bisbenzoylperoxide (BPO), dissolved in 1 ml of styrene, are added and the temperature kept at 80°C. After 2 h another 0.6 g BPO in 1 ml of styrene are added and the temperature further held at 80°C until a conversion of 35–40% is reached. This is the case after 2–3 h. In order to control the conversion, samples are taken every 30 min. They are weighed and then heated for 10 min at 220°C in vacuo in order to remove the styrene monomer. After a solid content of 35–40% is reached, the prepolymerization phase is finished.

(c) *Final Polymerization*

The viscosity of the reaction mixture is now too high for the final polymerization to be carried out in the same apparatus.[1] For laboratory purposes simple metal forms are sufficient. They consist of two metal plates covered with a Teflon film. A temperature- and chemicals-resistant hose (e.g., VITON A) serves as sealing and distance control. The whole set-up is pressed together with clamps until a distance

---

[1] The prepolymerization reactor can be cleaned with toluene.

of about 5 mm is reached. When filled with the hot prepolymerization mixture they are heated in a vacuum oven under nitrogen (!) according to the following program:
- 3 h at 120°C,
- 4 h at 140°C,
- 3 h at 175°C and
- 3 h at 220°C (under vacuum in order to remove traces of unreacted styrene).

(d) *Characterization of Process and Products*

During the prepolymerization phase several observations/tests can be made, for example:
- Turbidity of the reaction mixture already after a few percent conversion due to the incompatibility of polybutadiene and polystyrene in the monomer,
- Bending test with samples taken at different conversions: Increase of stiffness with conversion,
- Visualization of the micromorphology with transmission-electron microscopy (TEM) on ultrathin cuts after staining with osmium tetroxide.

With the monomer-free final products (c) one can determine, for example:
- Glass transition temperature (2 peaks, as is expected for a heterogeneous blend) and
- Mechanical data like stress/strain behavior, impact resistance in comparison to polystyrene.

Further tests are described in the literature.

5.6 Stretching and Foaming of Polymers

Stretching and foaming are some of the physical processes that can be used to subsequently modify the properties of polymers.

Stretching denotes a monoaxial or biaxial mechanical stress of a molded article close to the glass transition temperature. This leads to a controlled orientation of the molecular chains in the direction of stretching and thus to a substantial change in some physical properties. Fibers and foils made of synthetic polymers gain their optimal properties only by this mechanical post-treatment. Stability, stiffness, and dimensional stability of fibers, for example, increase nearly proportionally with the stretch ratio, whereas stretchability decreases. In practice, the stretch ratio is between 1:2 and 1:6, depending on the polymer material and the desired properties.

Natural fibers (wool, silk, jute, sisal, cotton) contain macromolecules that have already been aligned into fiber direction during the enzymatically catalyzed biosynthesis. Hence, stretching is not necessary and often not possible.

Foaming of polymers yields materials with a porous cell-like morphology. This porosity leads not only to a substantial decrease in overall density, but as a consequence some physical properties like stability, stiffness, and thermal conductivity are also altered. One distinguishes foamed plastics with opened pores from those with closed ones. Among the first are sponges and among the latter heat- and sound-absorbing insulating materials.

For the production of foamed plastics two methods can be essentially differentiated.

The first method starts from a finished polymer, which is foamed up in a separate step. All foamed plastics made of thermoplastics as well as the so-called foam rubber are manufactured in this way. For thermoplastics, the porous structure can be obtained by heating the polymer, containing a foaming agent, e.g., in an extruder above the softening temperature (T_g). Thus, it can be puffed up by the gas formed after emerging from the extrusion die; the product is then rapidly cooled down for solidification. Suitable foaming agents are low boiling, inert solvents (e.g., pentane) as well as compounds that decompose on heating under gas formation (hydrogen carbonate, aliphatic azo compounds). Solid foaming agents are applied to polymer beads by a subsequent impregnation step. Liquid foaming agents, however, can be fed into the polymer melt during the extrusion process. For the preparation of polystyrene foam, polymer beads containing a foaming agent, can be produced. These can be foamed with superheated steam and therefore processing aggregates simpler than extruders can be applied (see Example 5.24).

The inflation of the thermoplastic melt causes an orientation of the polymer chains by multi-axial stretching, whereby several mechanically characteristic values, especially the tenacity of the polymer foam, are positively affected.

During the production of rubber foam which starts from an aqueous emulsion of an elastomer, one utilizes the properties of an aqueous solution containing detergent. This means that the solution begins to foam very strongly during stirring, especially when air is blown in it at the same time. The finely pored cellular structure can be fixed by a subsequent crosslinking (vulcanization, see Example 5.8) of the emulsified polymer so that it is preserved even after work-up (e.g., washing with water).

In the second method for the production of foamed polymers, the foam-up process occurs simultaneously with the polyreaction, i.e., during the formation of the macromolecules. All foamed duroplastics are produced by this means, e.g., foams from urea-formaldehyde resins (see Example 5.25) or epoxide resins. The components necessary for the formation of the duroplasts are mixed together with a foaming agent (usually a low-boiling, inert liquid), poured into a mold, and are then allowed to react at the necessary temperature. The foaming agent evaporates due to the release of the reaction enthalpy, thus the reaction mixture foams up. A special case of this method is the preparation of polyurethane foams because there is no need for a foaming agent; under the special conditions of the polyaddition of diisocyanates to diols, CO_2 can be liberated from a part of the isocyanate moieties through the addition of small amounts of water, which causes the polymer to foam up (see Sect. 4.2.1.2, as well as Examples 5.26 and 5.27). This shows that in principle one could operate without foaming agents in this case, but this could be considered a waste of the expensive diisocyanate, among other factors.

In industrial practice the foaming processes are sometimes divided into chemical and physical processes. Chemical processes are those where the formation of gas takes place by decomposition of an unstable inorganic or organic compound or by a chemical reaction. Physical processes mean such techniques where the foaming gases are pumped into the polymer or are formed by the evaporation of liquids.

Example 5.24 Preparation of Foamable Polystyrene and of Polystyrene Foam

Safety precautions: Before this experiment is carried out, Sect. 2.2.5 must be read as well as the material safety data sheets (MSDS) for all chemicals and products used.

A process for the production of polystyrene foam very common in practice is the foaming of polystyrene beads, containing foaming agent. This process takes advantage of the fact that low-boiling aliphatic solvents (pentane, hexane) homogeneously dissolve in solid polystyrene in concentrations of 5–10 wt%. They evaporate during heating above the glass transition temperature and thus foam up the polystyrene. Since the dissolved aliphatic hydrocarbon lowers the glass transition temperature to below 100°C ("plasticizing effect"), the process can be carried out with superheated steam which is technically advantageous. The polystyrene beads, containing foaming agent (expandable polystyrene, EPS), can be prepared directly by a modified suspension polymerization. The difference to the normal bead polymerization (working technique see Example 3.7) is that now the pure monomer is not dispersed in water in the form of fine droplets. Instead, a homogeneous mixture of styrene with 5–10 wt% pentane is used and the polymerization is carried out under pressure in order to prevent the foaming agent from escaping at a reaction temperature of 80°C. When the polymerization is finished, the polystyrene beads, containing foaming agent, that have a diameter between 0.2 and 1 mm, are separated, washed, and dried at 50–60°C, whereby the pentane remains dissolved in the solid polystyrene. In closed tin cans, the EPS retains its ability to foam over a longer period. The polystyrene beads are heated to above 80°C with hot steam to induce foaming. During this process, the polystyrene softens and the evaporating pentane swells the beads that expand under the pressure to a multiple of their volume. At the same time, the beads stick together at the bonding surface and consequently form a mechanically quite stable agglomerate after cooling. This agglomerate is built up of polystyrene polyhedrons permeated with numerous fine and closed pores. The foaming process can be carried out in block molds with a volume of several hundred liters; from these polystyrene foam blocks, foamed plastic sheets of any size can be cut out using a hot wire. These sheets are then especially suited for insulation purposes. However, polystyrene foam for packing materials is produced by filling an exactly calculated amount of beads, containing foaming agent, into any complex mold that are then foamed directly in this mold.

For a large-scale production of polystyrene foam additional processing steps are needed with this process. First, additives (e.g., pore modifiers) have to be added to the styrene/pentane mixture before the polymerization. These additives must remain in the solid polymer particles and have to intervene during the foaming process taking place later, thus regulating pore size and pore size distribution. Furthermore, processing aids have to be applied to the surface of the polymer beads to optimize the flowability, to prevent electrostatic charging, or to improve the heat sealing properties of the foamed-up polymer particles. In practice the foaming process, during which a series of complex physical processes take place, is carried out in several steps.

Simple Demonstration of Foamability

For the above-mentioned reasons the performance of this procedure is not quite simple in the context of a practical course. Therefore the following instruction is intended only to experimentally demonstrate two effects: dissolving of pentane in solid polystyrene and foaming of the polystyrene, containing a blowing agent, in hot water.

2 g of polystyrene (e.g., prepared in Example 3.1 or 3.7 with 0.1 mol% of initiator) are thoroughly combined with a solution of 3 g of styrene, 0.3 g of pentane, and 0.04 g of dibenzoyl peroxide in a beaker until homogeneous. The resulting mass is transferred to a bomb tube using a wide-necked funnel until the bomb tube is about one-quarter filled. The tube is cooled under nitrogen atmosphere in a methylene chloride/dry ice bath to $-78°C$, sealed off, and kept at $30°C$ in a water bath for about 8 days (protection shield). Finally, the temperature is raised to $85°C$ for 6 h. The tube is then cooled and the upper end softened with the aid of a blow-torch so that the internal pressure can escape through the hole which is formed in that way. The foamable polystyrene, containing pentane, is obtained in the form of a transparent solid by breaking the tube. The polystyrene foam is obtained from this material as follows:

A piece of the polymer (about 1 g) is placed in a 500 ml beaker, containing boiling water. At this temperature, the polystyrene softens and is foamed by the vaporizing pentane. The test piece is held below the surface of the water for 5 min with the aid of a piece of bent wire and is then removed. The foamed material obtained in this manner is dipped into a graduated cylinder, containing methanol, in order to determine its approximate volume. It is also weighed after drying in a vacuum desiccator. The density is found to be below 0.1 g/cm^3 (cf., cork has a density of 0.2 g/cm^3).

Example 5.25 Preparation of a Urea/Formaldehyde Foam

Safety precautions: Before this experiment is carried out, Sect. 2.2.5 must be read as well as the material safety data sheets (MSDS) for all chemicals and products used.

Caution: Because of the formation of gaseous formaldehyde the experiments have to be carried out in a closed hood.

By a modification of the method described in Sect. 4.1.3 urea/formaldehyde foams can be prepared with the following method:

178 g (2.2 mol) of a 37% aqueous formaldehyde solution, neutralized with 1 M NaOH, are heated to $95°C$ in a 250 ml three-necked flask, equipped with thermometer, stirrer, and reflux condenser (the experiment must be carried out in a hood). In this aqueous formaldehyde solution 120 g (2 mol) of urea are dissolved under stirring and the pH value is adjusted to 8 with 1 N NaOH.

Within a period of 5 min, the pH value decreases again, after 15 min it is adjusted to pH 6 with NaOH. After another 1.5 h at 95°C the solution is acidified with 20% formic acid (pH 5) and the condensation finished after 15 min by cooling down and neutralization with NaOH.

For the preparation of the foam, a solution of 1 g technical sodium diisobutyl naphthalene sulfonate in 50 ml of 3% orthophosphoric acid is prepared. 20 ml of this solution are poured into a 1 l beaker and air is stirred in with a fast running mixer until the cream-like dispersion has reached a volume of 300–400 ml. Then, 20 ml of the prepared urea/formaldehyde resin are mixed in, whereby the resin must be evenly distributed. After 3–4 min the introduced resin gellifies into a molded article permeated with many water/air pores under the influence of the acidic catalyst. After 24 h, the crosslinking is completed. Drying for 12 h at 40°C in a circulating air dryer yields a brittle thermoset foam. The foamed plastic obtained is hydrophobic and has a large internal surface. It can take up about 30 times its own weight of petroleum ether.

This property can be used to separate highly volatile and low-viscous mineral oils from oil–water dispersions. To demonstrate this, a dispersion of 20 ml of ligroin or petroleum ether in 200 ml of water is prepared in a 400 ml beaker with a fast-running mixer. Then approx. 5 g of crushed urea/formaldehyde foam are added. After 5 min the solution is filtered through a folded filter. The aqueous filtrates are optically free from dispersed hydrocarbons. In the same way a crude oil/water dispersion can be separated.

5.6.1   Preparation of Polyurethane Foams

As described in Sect. 4.2.1 foamed polyurethanes can be obtained especially through the so-called water-crosslinking of suitable polyol/diisocyanate combinations. Carbon dioxide liberated during the reaction partially causes the expansion of the foam. For a number of reasons additional foaming agents (in the form of low-boiling liquids or CO_2) are added in practice, i.e., in order to vary the foam density and the pore size.

The properties of polyurethanes crosslinked in this way depend on the starting materials (polyol and diisocyanate) as well as on the additives and on the intensity of the mixing process.

The chain length, the chain length distribution, and the ratio of soft to hard multiblock sections determine the properties of the resulting foam. Flexible foams are obtained from long-chain polyol precursors (polyethylene oxide, polypropylene oxide, and copolymers, molecular weights >2,000 g/mol and OH numbers 40–60) forming the elastomeric domains whereas rigid foams are composed of shorter polyol segments (e.g., propylene oxides with molecular weights <1,000 g/mol). This results in a higher hard phase content originating from the initial diisocyanate leading i.e., to intermolecular interaction of urethane, urea and allophanate H-bridges (see Sects. 4.2.1 and 4.2.1.2). Additional crosslinks are obtained with multifunctional polyalkylene oxides.

Moreover, flexible foams are characterized by utilization of special emulsifiers in their synthesis yielding an open-cell architecture, whereas for rigid foams emulsifiers are chosen that create more closed-cell structures. As diisocyanate for both types, the commercially available mixture of 80% 2,4-toluene diisocyanate and 20% 2,6-toluene diisocyanate is especially suitable. If foam formation is to take place at room temperature, and especially when hydroxy compounds with secondary hydroxy groups are used [poly(propylene glycol)s], the presence of a catalyst is generally required (see Sect. 4.2.1).

As previously mentioned, the homogeneity of foamed materials and the proportion of open and closed cells can be influenced by additives such as emulsifiers and stabilizers. Emulsifiers (e.g., sodium, potassium, or zinc salts of long-chain fatty acids) cause a uniform distribution of water in the reaction mixture, ensuring homogeneous foaming, while stabilizers (certain silicone oils) prevent a breakdown of the cell structure at the beginning of the reaction and also act as pore regulators.

For laboratory experiments, it is sufficient to mix the industrial individual components with vigorous stirring. For the preparation of uniform foams with optimum properties, however, the proportions of polyhydroxy compound, diisocyanate, and additives must be very carefully balanced by using precision feeding and mixing techniques.

Example 5.26 Preparation of a Flexible Polyurethane Foam

Safety precautions: Before this experiment is carried out, Sect. 2.2.5 must be read as well as the material safety data sheets (MSDS) for all chemicals and products used.

10 g of a polyester with an OH number of 60 (see Example 4.1a) and 3.5 g of toluene diisocyanate are well mixed in a 250 ml plastic beaker for 1 min using a wooden stick. A mixture of 0.1 g of *N,N*-dimethylbenzylamine, 0.2 g of 50% aqueous solution of a non-ionic emulsifier, 0.1 g of an aqueous solution of sodium dodecyl sulfate, 0.025 g of a poly(dimethylsiloxane), and 1 g of water are then added with intensive stirring. After 1 min the expansion of the foam is practically finished; 20–30 min later (depending on the prevailing atmospheric conditions) the surface of the polyurethane foam is no longer tacky. The flexible foam (total volume about 150 cm^3) can now be removed from the beaker and mechanically stressed after a day.

Main applications of flexible polyurethane foams are in mattresses, upholstery, and packaging.

Example 5.27 Preparation of a Rigid Polyurethane Foam

Safety precautions: Before this experiment is carried out, Sect. 2.2.5 must be read as well as the material safety data sheets (MSDS) for all chemicals and products used.

In a 250 ml plastic beaker 10 g of a strongly branched polyester with a high OH number of about 350–370 (see Example 4.1b; or commercial sample, e.g., Desmophen 4070 X from Bayer AG) and 15 g of methylene di(phenylisocyanate) (technical product, liquid mixture of the diisocyanate with poly-MDI, e.g.,

Desmodur 44V20L from Bayer AG) are thoroughly mixed using a wooden stick. To this mixture successively 0.2 g silicon stabilizer (e.g., Tegostab B 8460 from Goldschmidt), 0.3 g water, and 0.1 g dimethylcyclohexylamine (catalyst) are added. The mixture is stirred vigorously until foaming starts. After standing for 1 h, the rigid, brittle foam can be removed from the beaker (total volume about 250 cm^3).

A small amount of nonreacted isocyanate groups in the hardened foam can be detected by IR spectroscopy.

Rigid polyurethane foams are mainly used as insulating and damping materials (construction industry, cooling equipment, ship construction).

Bibliography

Allen G, Bevington J (eds) (1989) Comprehensive polymer science, vol 6. Pergamon, Oxford

Eisenbach CD, Baumgartner M, Guenter C (1986) In: Lal J, Mark JE (eds) Advances in elastomers and rubber elasticity (Polyurethanes). Plenum, New York

Gächter R, Müller H (eds) (1993) Plastics additives handbook, 4th edn. Hanser, Munich

Houben-Weyl (1987) Methoden der organischen Chemie, vol E20, Makromolekulare Stoffe, Thieme, Stuttgart

Klemm D (1998) Comprehensive cellulose chemistry. Wiley-VCH, Weinheim

Manas-Zloczower I (ed) (2009) Mixing and compounding of polymers: theory and practice, 2nd edn. Carl Hanser Verlag, Munich

Paul DR, Bucknall CB (2000) Polymer blends. Wiley-VCH, Weinheim

Pionteck J, Wypech G (eds) (2007) Handboodk of antistatics. ChemTec Publishing, Toronto

Rabek JF (1996) Photodegradation of polymers. Springer, Berlin/Heidelberg/New York

Utracki LA (ed) (1991) Two-phase polymer systems. Hanser, Munich/Vienna/New York/Barcelona

Utracki LA (1998) Commercial polymer blends. Chapman & Hall, London/New York

Xanthos M (1992) Reactive extrusion, principles and practice. Hanser, Munich

Xanthos M (ed) (2005) Functional fillers for plastics. Wiley-VCH Verlag/GmbH&Co KGaA, Weinheim

Zweifel H (1998) Stabilization of polymeric materials. Springer, Berlin/Heidelberg/New York

Functional Polymers

6

The overwhelming number of known polymers can be basically sub-divided into those for structural applications and those, which come over with specific functions. The former ones as well as the later ones, called *Functional Polymers*, play important roles since the early days of polymer science and technology. The latter ones, however, usually are not so much characterized by their thermo-mechanical properties, but rather by their inherent functionalities instead, which they develop due to special constitutional features. For example, some of their atoms or groups of atoms may undergo specific interactions with solvents, ions, cells, surfaces, fillers or other polymers. Alternatively, specific optical or electronic properties may result from the molecular and supramolecular architecture of these macromolecules. Polyelectrolytes like poly(acrylic acid) or poly(diallyldimethylammonium chloride), and polymeric stabilizers like polyvinylalcohol or polyethyleneglycol [poly(ethylene oxide), PEO], all of them characteristic examples of functional polymers, are large-scale technical products nowadays. They have found widespread application in e.g. hygiene and cosmetics products. Moreover, polymeric additives and compatibilizers simplify polymer processing and morphology design, and allow the preparation of transparent nanocomposites with improved or even novel property profiles. Photoresins and photoresists are key compounds of photolithography. As such, they are revolutionizing the printing technology, and paved the way for semiconductor and microelectronics industry. In the year 2000, moreover, the Nobel prize was awarded to A.J. Heeger, A.G. MacDiaramid and H. Shirakawa for the development of electrically (semi)conducting polymers, which are key players in our today's organic (opto)electronic.

The first synthetic functional polymers were developed in the early days of the twentieth century already. They were prepared using standard polymerization techniques like free radical or step-growth polymerization reactions. The increased material needs in modern technology and the increased complexity of polymeric materials, however, led to a strong expansion of this field especially in recent years. Today's synthetic techniques allow a much better control of the polymers' structure,

D. Braun et al., *Polymer Synthesis: Theory and Practice*,
DOI 10.1007/978-3-642-28980-4_6, © Springer-Verlag Berlin Heidelberg 2013

architecture, functionality, molar mass and dispersity, and hence allow tailoring polymers according to the needs of a specific application. This accounts specially for the functional polymers, whose "function" is more in the focus than their mechanical properties – even though they have to fit the full range of application needs. For example, polymeric additives have to have a sufficient thermal stability to survive melt processing, and for the wide-spread thin-film applications the functional polymers require good mechanical film stability as well as sufficient thermal, chemical, and oxidative stability.

Representative classes of functional polymers include e.g.

- Polymers carrying ionically charged groups (polyelectrolytes), which undergo specific interactions and show specific solution and pH-responsive properties
- Amphiphilic block copolymers forming e.g. micelles and aggregates, and stabilizing additives and (nano)fillers
- Polymeric stabilizers (PVOH, PEG, various block and graft copolymers)
- Polymeric additives like rheology modifiers
- Polymeric carrier resins for catalytic or chromatographic applications (e.g. ion exchange resins)
- Intrinsically electron-conducting polymers
- Photoactive polymers
- Thermo-responsive polymers
- Bioactive polymers, bioconjugates and drug delivery systems

The scope of functional polymer systems has broadened tremendously in recent years. For example, various "hybrid" materials have been developed additionally, which combine synthetic and biological polymers as well as organic and inorganic/metallic components. Especially the field of biomaterials and polymers for biomedical applications cannot be addressed adequately in the context of this book, and only some selected examples of biodegradable polymers and biomaterials have been given earlier (see example 5.15: hydrolytic degradation of an aliphatic polyester and 5.16: hydrolytic degradation of cellulose). The synthesis of polylactide – as one of the most well-known and frequently used biodegradable polymers in medical application as e.g. wound dressing or degradable suture material – is covered by example 3.5. Many active components used in diagnostics, therapy as well as in biomedical studies are made biocompatible by "pegylation", the introduction of polyethylene glycol chains (PEG synthesis through ring-opening polymerization, see example 3.24 for poly(tetrahydrofuran)). Thus, many of the polymers described already in this book fit under the term "functional polymers", as prominent example polyelectrolytes are covered in Sects. 1.3.1.2 and 5.2.1, and some more references to already given examples will be referred to in the following chapters.

It cannot be the goal of this text book to address examples of all these and many further functional polymer structures and applications. Due to the very high – and continuously further increasing – importance of functional polymers in modern life and technology, this chapter will focus on some selected topics out of the broad

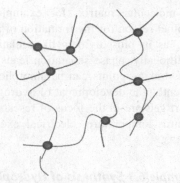

Fig. 6.1 Physically crosslinked gel

Fig. 6.2 Chemically crosslinked gel

scope of that polymer class instead, which have led to new applications in recent years. These are:

- Gels
- Responsive "smart" polymers
- Polymers with optoelectronic properties

6.1 Gels

Generally speaking, a gel consists of a crosslinked polymer and a liquid. The polymer or "gelling agent" absorbs large amounts of solvent without being dissolved. The term gel was established in 1861 by Graham and originated from gelatin. In food and coating industry, naturally occurring organic gelling agents like polysaccharides and proteins are frequently used. In food industry, hydrogels are used as additives in jelly, sauces, bakery products and in dairy products, in order to give them the desired texture and to prevent them from drying. Covalently cross-linked polyacrylates in superabsorbent polymers – used for hygiene articles such as baby diapers – have gained enormous economic importance.

Gels are classified by their source (natural or synthetic), chemical composition, the three-dimensional network structure and the liquid medium. Generally, one can differentiate between chemically (Fig. 6.1) and physically (Fig. 6.2) cross-linked gels. While chemical gels are characterized by covalent linkages between the structural units (see example 6.1), networks of physical gels are based on non-covalent interactions, such as dipole-dipole-interactions, hydrogen bonds, hydrophobic interactions, polyvalent metal ions, van der Waals forces, and host-guest interactions (see example 6.2). The outstanding feature of thermo-reversible organo- or hydrogels is their ability to form and cleave linkages depending on temperature (see example 6.3). These gels can be characterized by the type of their cross-linkage. Semi-crystalline polymers generate micro-crystallites in the low-molecular matrix, for example polyvinylchloride in dioctyl phthalate. Another reason for the formation of a thermo-reversible gel can be helix formation, as in bio-polymers like gelatin, agarose, carragennan and gellan gum. Additionally, phase-separation leads to formation of physical gels. In this case cross-linking points can be amorphous (e.g. atactic polystyrene) or crystalline. Recently, the development of hydrogels became increasingly important. Mainly smart gels are in the focus of research, due to their stimuli-responsive behavior towards temperature, electrical excitation, incident light, pH value and salt effects.

Example 6.1 Synthesis of Hydrophilic Physically Crosslinked Gel-Building Polymer

Safety precautions: Before this experiment is carried out, Sect. 2.2.5 must be read as well as the material safety data sheets (MSDS) for all chemicals and products used.

0.5 g of dried poly(acrylic acid) (1,800 g/mol) is mixed with 5 mol-% [with regard to the repeating units of poly(acrylic acid)] of dodecyl amine (64 mg). The mixture is grinded, filled into an 8 ml vial with a stir bar and flushed with argon. The vial is then closed with a septum and put into the MW cavity. The reaction takes place at 75 W for 5 min (preheating time = 2 min). The temperature is measured using a fiber optic temperature gage and controlled by permanent adjustment of the air cooling (T = 140°C). A gel-forming polymer is obtained after neutralization with NaOH. η_0 [PAA 1,800, 1 g/ml] $= 0.58$; η_0 [Gel, 1 g/ml] $= 1,168$

Example 6.2 Synthesis of Hydrophilic Chemically Crosslinked Gel-Building Polymer

Safety precautions: Before this experiment is carried out, Sect. 2.2.5 must be read as well as the material safety data sheets (MSDS) for all chemicals and products used.

7.2 g of acrylic acid and 20 ml of distilled water are placed in a 100 ml flask and stirred for 15 min under nitrogen. 200 mg of butanediol diacrylate (BDDA) is added to the mixture. 0.5 mol% of sodium disulfite and 0.5 mol% of potassium peroxodisulfate is dissolved in 1 ml water and added to the solution. The reaction mixture is stirred 1 h under nitrogen. Within 1 h the gel effect occurs and the polymerization ends. Yield: quantitative. The amount auf water sorption is quantified by the degree of swelling: the ratio of the swollen polymer mass to that of dry polymer. The dry polymer shows a degree of swelling in water of about a factor of 5 (after 1 h) and a factor of 130 (after 24 h), respectively.

Example 6.3 Synthesis of a Thermo-Sensitive Gel

Safety precautions: Before this experiment is carried out, Sect. 2.2.5 must be read as well as the material safety data sheets (MSDS) for all chemicals and products used.

5 g of n-isopropylacrylamide and 20 ml of distilled water are placed in a 100 ml flask and stirred for 15 min under nitrogen. 640 mg of butanediol diacrylate (BDDA) is added to the mixture. 0.5 mol% of sodium disulfite and 0.5 mol% of potassium peroxodisulfate are dissolved in 1 ml water and added to the solution. The reaction mixture is stirred 1 h under nitrogen. After 1 h a white solid precipitates. The solid is filtered out and dried at 70°C. Yield: quantitative. The polymer is soluble in water below 32°C and precipitates from aqueous solution above that temperature.

6.2   Responsive "Smart Polymers"

"Smart polymers" are macromolecular materials that change their properties and/or shape depending on their environment and their surroundings. Some examples of environmental changes that can affect polymers are temperature, the presence of water, chemicals or analytes, pH changes, the presence/intensity of light, electrical fields, mechanical forces etc.

Whereas in the beginning many standard polymers have been considered as "smart" due to their specific design needed in an application, the term is today used only for those polymers that respond in a *dramatic* way to very *slight* changes in their environment. In addition, the responses achieved today have been greatly enlarged due to the increasing preparation of various functional composites. Thus, the introduction of magnetic particles induces magnetic responses in polymeric materials and gels, and e.g. new sensoric properties can be introduced into polymers by incorporation of various quantum dots. Whereas specific responses have been known and studied for a long time in biological macromolecules, smart polymers are becoming increasingly prevalent as polymer chemists learn about the chemistry and triggers that induce conformational changes in polymer structures and devise ways to take advantage of, and control them. New polymeric materials are being chemically formulated that sense specific environmental changes e.g. in biological systems, and adjust in a *predictable* manner, making them useful tools for diagnostics, sensors, microsystems, drug delivery or other metabolic control mechanisms. But uses of smart polymers cover also well-known hydrogels, classical degradable polymers or photoresins.

Certainly, a highly important and well-known class of polymers are those which undergo major changes in solubility by a light-induced process, like photopolymerization, photodegradation (ablation) and photocrosslinking (see also Sect. 3.1.4 and example 3.13). These polymers, known as photoresins or photoresists, are the base of modern computer and communication technology: they allow the photolithographic creation of patterns which are then reproduced in mass production (printing technologies) or into the semiconductors (integrated circuits). A waste variety of materials have been developed adapted to the light used for patterning. The major challenge today is the increased miniaturization in the information technology which requires smaller and smaller features with high requirements on the responsiveness of the polymers used. Thus, polymeric resists are today available for UV-, deep-UV- or even electron-beam lithography. One distinguishes between positive and negative working photoresins. In the first case, solubility of the resin is enhanced upon irradiation (e.g. degradation process or activation of a dissolution enhancer). In the second case, the solubility is reduced by e.g. polymerization processes or crosslinking. Early examples for UV-resists are polyvinylalcohol/cinnamoyl acid (negative) and novolac/naphthochinone diazide (positive) resins. Today often chemically amplified resins are used, which are based on photocatalytic processes. The photopolymerization (see example 3.13) is mainly used in the printing technology e.g. for the preparation of large-scale printing plates according to letter press technique (so-called *flexoprint* process) used to print on packaging materials like plastic foils. Other photoactive polymers as well as polymers having specific electrical properties or those responding to electric fields (piezoelectric polymers) are covered in Sect. 6.3.

Another important class of responsive polymers is that which responds on chemical changes. The most important trigger is pH. It allows changes in the charged status of the polymers (see Sect. 1.3.1.2 and 5.2.1 on polyelectrolytes) leading to major changes in solubility upon pH changes. Especially, hydrogels – polar, slightly crosslinked materials (see Sect. 5.2.1.2, example 5.11) – are addressed here due to the possibility to adapt the degree of swelling significantly by the applied pH. Changes in the swelling allows for controlled uptake and release of water and included components like drugs or others. In addition, pH-responsive gels – mostly based on poly(acrylic acid) gels – are increasingly used in microsystems as actuators. They can open or close valves and channels e.g. in a lab-on-a-chip device upon changes in pH due to the changes in the degree of swelling. In addition to hydrogels, various polymers have been developed for sensors, diagnostic or separation techniques. They usually show a specific interaction with various analytes from water vapour and volatile organic compounds (VOC) to salts, metal ions, various analytes and chemicals to drugs and biomolecules. In the simplest way, thin polymer layers made up e.g. from polyimides are used in humidity sensors: they take up and release water vapour in a controlled fashion. The corresponding changes in the thin film (thickness, refractive index or conductivity) are measured in the sensor to quantify the uptake. In more complex systems, specific interacting groups are introduced (e.g. in ion exchange resins, see example 5.9 and 5.10 or in cyclodextrin-containing polymers, see example 3.14a) which allow a selective uptake of specific analytes. A further expansion of chemically responsive polymers is demonstrated by the concept of *Molecular Imprinting*. Here analytes are not only included in the functional polymer due to specific interactions but also by adaption the analytes' shape in three-dimensional cavities.

One of the more recently developed material classes is that of thermo-responsive polymers. This accounts mainly for those polymers showing a so-called lower critical solution temperature (LCST) or – to a much less extent – an upper critical solution temperature (UCST), meaning polymers that change their solution status significantly upon temperature changes. In general, this behaviour is the result of a delicate balance of hydrophilic and hydrophobic groups in the polymer. The major aspect in this response is thermodynamic, and is based on the dissociation of hydrogen-bonds upon heating. Below LCST, water-polymer hydrogen bonds dominate, whereas above LCST polymer-polymer interactions start to dominate leading to precipitation. One of the most well-known example in this regard is poly(N-isopropylacrylamide) (PNiPAAm) and others poly(N-alkylacrylamide)s that exhibit a lower critical solution temperature (LCST) and remarkable hydration-dehydration changes in aqueous solution in response to relatively small changes in temperature. Below LCST, PNiPAAm chains hydrate to form an expanded structure; above LCST, i.e. above the cloud point at a given concentration, T_c, PNiPAAm chains dehydrate and collapse. PNiPAAm-based hydrogels have attracted much attention since their LCST in water is in the physiologically interesting range. Thus, these polymers may be applied in the biomedical field as stimulus-sensitive materials. Other examples for polymers showing LCSTs are cellulose derivatives, poly(ethylene oxide) (PEO), polyethyloxazoline (see example 3.27), polyvinylmethylether, or poly(N-vinylcaprolactam). Similar as for pH-sensitive gels, thermoresponsive polymers are often prepared as gels which change the degree of swelling upon temperature changes, allowing for significant changes in volume as well as for changes in release and up-take properties. Applications again are seen in the biomedical field or in actuators. An example for a thermoresponsive hydrogel is given in Sect. 6.1 (example 6.3).

6.2.1 Polymer Networks with Shape Memory Effect

Further examples of thermo-responsive polymers are shape-memory polymers. Shape memory behavior is known for some time for metals and alloys with use in medical application, aerospace, and micro-machinery/actuators. However, it can also be observed in polymers of specific bulk structure. These smart materials have the ability to return from a deformed state (temporary shape) to their original (permanent) shape induced by an external stimulus (trigger). Most shape-memory polymers (SPM) can retain two shapes, and the transition between them is induced by temperature. In some recent SMPs, heating to certain transition temperatures allows to fix three different shapes. In addition to temperature change, the shape change of SMPs can also be triggered by an electric or magnetic field, light or solution. Shape memory polymers cover a wide property-range from stable to biodegradable, from soft to hard, and from elastic to rigid, depending on the structural units that constitute the SMP and include thermoplastic and thermoset (covalently cross-linked) polymeric materials. Representative shape-memory polymers are segmented polyurethanes with ionic or mesogenic components made by prepolymer methods. Other block copolymers also show the shape-memory effect, such as block copolymers of polyethylene terephthalate (PET) and poly(ethylene oxide) (PEO), block copolymers containing polystyrene and poly(1,4-butadiene), and ABA triblock copolymers made from poly (2-methyl-2-oxazoline) and poly(tetrahydrofuran), but also partially dehydrochlorinated poly(vinyl chloride) and nylon/polyethylene graft copolymers are reported to perform as SMPs.

Fig. 6.3 Photo series demonstrating the shape memory effect of a polymer network from a temporary shape to a permanent shape at 25°C after twisting and freezing at −20°C (from *left* to *right*)

Polymers exhibiting shape-memory effects have both a visible, current (temporary) form and a stored (permanent) form. Once the latter has been manufactured by conventional methods, the material is changed into another, temporary form by processing through heating, deformation, and finally, cooling. The polymer maintains this temporary shape until the shape change into the permanent form is activated by a predetermined external stimulus. The structural reasons of this effect lie in the molecular network structure of the polymers, which contains at least two separate phases e.g. as in segmented block copolymers. The phase showing the highest thermal transition, T_{perm}, is the temperature that must be exceeded to establish the physical cross-links responsible for the permanent shape. The switching segments, on the other hand, are the segments with the ability to soften above a certain transition temperature (T_{trans}) and are responsible for the temporary shape. Exceeding T_{trans} (while remaining below T_{perm}) activates the switching by softening these switching segments and thereby allowing the material to resume its original (permanent) form.

Industrial applications of shape memory polymers are e.g. as foams in the building industry and in sportswear. Further potential applications include self-repairing structural components. More recently, especially potential biomedical applications are discussed as, e.g., intravenous cannula, self-adjusting orthodontic wires and selectively pliable tools for small scale surgical procedures.

Shape memory polymers (SMPs) are generally applied as polymer networks. The existence of permanent cross-links is the reason for their performance. SMPs can be either chemically or physically cross-linked polymer networks. Molecular chains with a thermal transition at T_{trans} serve as molecular switch triggering the shape memory effect. Due to the flexibility of the network chains, the polymer can show extensive deformation at temperature above T_{trans}. In contrast, by freezing the SMP below T_{trans}, the network chains lose their mobility, and the polymer can be frozen in a temporary shape. The cross-links stabilize the permanent shape of the networks in the course of shape memorizing. The photo series demonstrates the shape memory effect of a polymer network (Fig. 6.3) showing the transition from a temporary shape to a permanent shape at room temperature after twisting and freezing at −20°C depending on time.

Example 6.4 Synthesis of a Bicomponent Polymer Network Through a Radical Polymerization of Methacrylic Acid 2-Ethoxyethyl Ester and bis(Acrylate)

Safety precautions: Before this experiment is carried out, Sect. 2.2.5 must be read as well as the material safety data sheets (MSDS) for all chemicals and products used.

20.0 g (126 mmol) of methacrylic acid 2-ethoxyethyl ester and 2.0 g (10 wt%) of 1,6-hexanediol diacrylate are added to a 25 mL two-neck round-bottom flask equipped with a septum. While stirring, the reaction mixture is purged with nitrogen. After 30 min. 15.4 mg (0.07 wt%) of a radical initiator 2,2′-azobis(2,4-dimethyl)valeronitrile are added to the reaction mixture. The liquid is injected into a mold and heated to 60°C for 24 h. The cross-linked polymer is removed from the mold. In a vacuum drying cabinet the rest monomer is extracted from the polymer network (125°C and 1.2 mbar, 2 days).

6.3 Polymers with Specific (Opto)Electronic Properties

Functional polymers are of increasing importance even in the field of (opto) electronics and related areas. Polymeric materials show plenty of desirable features when compared with traditional inorganic materials. For example, they are rather cheap, easily processable from the melt or solution (coating, printing, etc.), film-forming and flexible (unlike fragile inorganic materials). Also, well-established and convenient reactions of organic and organo-metallic chemistry can be used to tailor these materials to a specific purpose.

Optoelectronics in general deals with the conversion of electrical power into electromagnetic radiation (i.e., visible light), and vice versa with the conversion of (sun)light into electrical power (photovoltaics, PV). The devices commonly used for these two conversion processes are known as *light-emitting diodes* (LEDs) and *solar cells* (SCs), respectively. If organic polymers play the major role in these devices, they are named "organic" or "polymer-based". Abbreviations are used such as OLEDs, POLEDs, OSC, PSC, OPV, and so on. The principal architectures of OLEDs and OSCs are shown in Fig. 6.4, and additionally the most relevant elementary processes occurring in those devices are sketched.

In an operating OLED device, electrons are withdrawn from the functional polymer at its interface to the anode as soon as a sufficiently high voltage is applied. Chemically spoken, the polymer is oxidized, a so-called "hole" (electron-deficient site) is injected, which formally is characterized by a positive charge. Accordingly, this oxidation process is called "hole injection". From the cathode side, on the other hand, electrons are injected into the functional material, the polymer is reduced, and a negative charge carrier – called "electron" – appears at the interface.

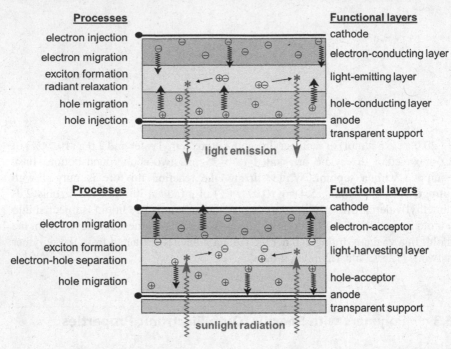

Fig. 6.4 Fundamental architectures and elementary processes occurring in operating OLEDs (*top*) and OSC (*bottom*) devices

Driven by the electric field, the injected charge carriers migrate into the polymer bulk, and hence move toward each other from the opposite interfaces. When inside the organic layer an electron meets a hole, they recombine: an electron–hole pair is formed, which represents an electronically excited state (exciton). The exciton finally can relax into the electronic ground state by emission of light (radiant relaxation, often termed as "electro-luminescence"). Such electro-luminescence is possible without restrictions if a singlet exciton is formed: in this case, radiant relaxation is spin-allowed fluorescence. Unfortunately, most of the excitons formed in OLEDs are triplet excitons. In this case, radiant relaxation is spin-forbidden. If conventional organic polymers are used as electro-luminescent materials, therefore, this (major) fraction of excitons gets lost, and heats up the device due to non-radiant relaxation processes. However, it is possible to offer alternative pathways to the radiant triplet relaxation: introduction of so-called triplet emitters, which open pathways for radiant relaxation with spin-inversion, help increasing power efficiency of such devices: they open the complementary electro-phosphorescence option for light generation. Therefore, development of tailored phosphorescence emitters (e.g., transition-metal complexes) is a hot-topic in current functional-polymer research.

In an operating solar cell, the fundamental processes are inversed to those in OLEDs: upon absorption of light, a chromophoric unit of the functional polymer, or of a dye embedded into the polymer matrix, changes from its electronic ground state into an excited state. In other words, one electron moves from the *highest occupied molecular orbital* (HOMO) of the absorbing entity into its *lowest unoccupied molecular orbital* (LUMO). The formed excited state (exciton) may relax back following radiant or non-radiant pathways – which either results in the emission of a fluorescence quantum (photo-luminescence) or in heating (excitation of vibration modes). Alternatively, the exciton will make use of its nature as an electron–hole pair, and separate into an "isolated" hole and an "isolated" electron located at different sites. After successful charge-carrier separation, electron and hole move in opposite directions to their respective electrodes. In order to minimize downstream recombination and deactivation processes, this latter migration process should proceed via different pathways for holes and electrons.

As one can see, functional polymers for (opto)electronic devices must feature many other abilities in addition to light absorption or light emission. First of all, they must be able to transport charge carriers, i.e. the electrons and holes, from the electrodes to the recombination and emission zone (in OLEDs), or vice versa from the absorbing dyes to the respective electrodes (in OSCs). Thus, they must be electrically conducting. More precisely, they should rank among the electrical *semi*-conductors: in the ground state, they act as insulators to prevent short circuits. After appropriate activation, however, which might be achieved due to injection or (light-induced) formation of charge carriers, they convert into a conductive state. Now, charge-carrier transport proceeds through the polymeric material predominantly via redox cascades enabled by hopping and tunnelling processes, and its principal direction is given by the applied electrical field. Depending on the molecular constitution of the polymer, either it can provide high mobility for electrons only, for holes only, or for both types of charge carriers simultaneously.

Most semi-conducting polymers that are used in optoelectronics have a backbone, which contains alternating carbon-carbon single and multiple bonds. Therefore, they are called "π-conjugated polymers". Nevertheless, one should keep in mind that in those macromolecules π-electron conjugation by far is not spread over the whole backbone, but is limited to rather short segments: the so-called "effective conjugation length" has been shown to include some few repeating units only, even if perturbations of chain geometry are minimized. As a consequence, the absorption and emission properties do not change with increasing chain length if a minimum contour length of the system is exceeded. On the other hand, changes of the local environment and the degree of ordering of the chain segments within the functional film may cause severe effects due to associated changes of the effective conjugation. Attachment of lateral substituents to the "conjugated" polymers therefore not only is a measure to improve solubility and to adjust electronic properties, but also to stabilize the material in the functional film in the morphology requested to develop the targeted properties. Important examples of semi-conducting organic polymers are shown in Fig. 6.5.

Moreover, semi-conducting polymers for OLEDs must ensure well-balanced hole- and electron-injection from the electrodes as well as comparable (but not too

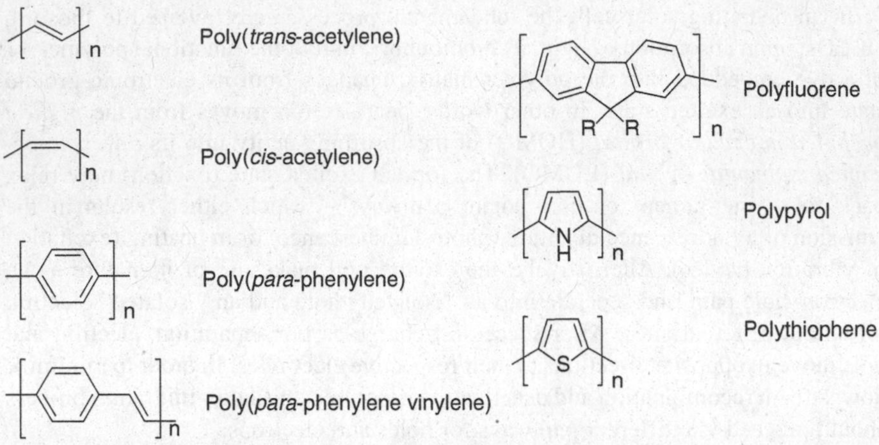

Poly(*trans*-acetylene)

Poly(*cis*-acetylene)

Poly(*para*-phenylene)

Poly(*para*-phenylene vinylene)

Polyfluorene

Polypyrol

Polythiophene

Fig. 6.5 Some important examples of the basic structures of intrinsically (semi)conducting polymers

high) charge-carrier mobility for electrons and for holes. In solar cells, on the other hand, the semi-conducting polymeric material should offer separated pathways for electrons and holes in order to minimize loss of energy by downstream electron-hole recombination processes.

Another hot topic in modern functional polymer research deals with the development of materials that can be used as electrode materials: these polymers must offer high charge-carrier mobility, well-adapted energy levels of their frontier orbitals, and light should be able to pass without scattering and absorption, i.e., the electrode material should be transparent for visible light even in the conducting state. Today, inorganic oxides such as indium tin oxide (ITO) are used as transparent anode materials. However, intense research has born first examples of organic polymers, mainly based on polythiophenes (such as PEDOT:PSS, also known as Baytron® P, from H.C. Starck Co.; see Fig. 6.6), that might be used as alternatives.

In addition to applications as functional materials in OLEDs and OSCs, semi-conducting polymers are needed for other (opto)electronic devices as well. With regard to displays, sensors, and radio-frequency identification tags (RFIDs) for example, it is a challenge to create polymer-based organic transistors (thin-film transistors, OTFT; field-effect transistors, OFETs). Figure 6.7 sketches an optional OFET design, and additionally shows schematically its principle of operation.

An OFET is composed of three electrodes, a source electrode, a drain electrode, and a gate electrode. The gate electrode is separated from the source and drain electrodes by an insulator layer (silicon dioxide, or insulating polymers like polyhydroxystyrene). Between source and drain electrode, moreover, an organic semi-conductor is placed. If no electric field is applied, the semi-conductor performs as an insulator: charge carriers cannot cross over from the source to the gate electrode. The transistor is in the "off state". If a sufficiently high electric field V_G is applied between source and gate electrode, on the other hand, charge carriers

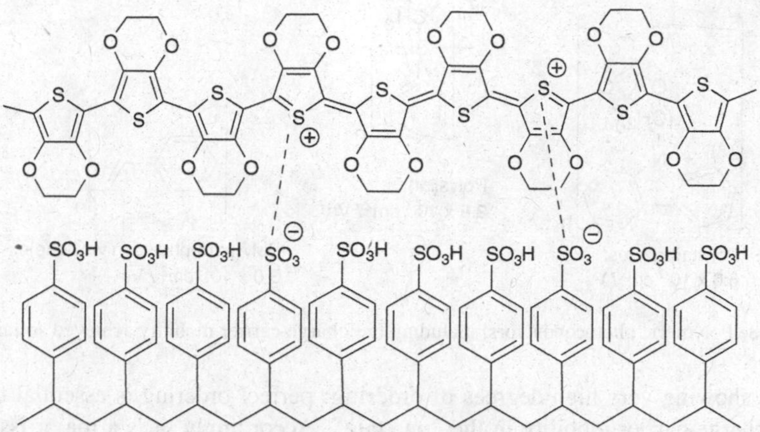

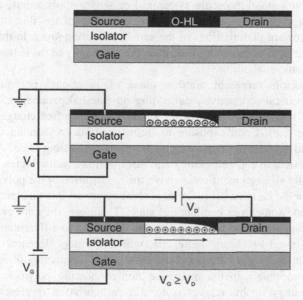

Fig. 6.6 Idealized chemical constitution of PEDOT:PSS

Fig. 6.7 Example architecture of a field-effect transistor and its basic function principle

appear at the interface between semi-conducting polymer and insulator. Due to the presence of these charge carriers, the organic material changes from its insulating state into the semi-conductor state. Consequently, charge transport is possible from the source to the drain electrode if there is an appropriate drain voltage V_D applied. Now, the transistor is in the "on state".

Such transistors can be used e.g. as switches in display applications, which manage the individual pixels. In this respect, it is an issue to develop functional polymers which can be processed from solution and nevertheless result in thin

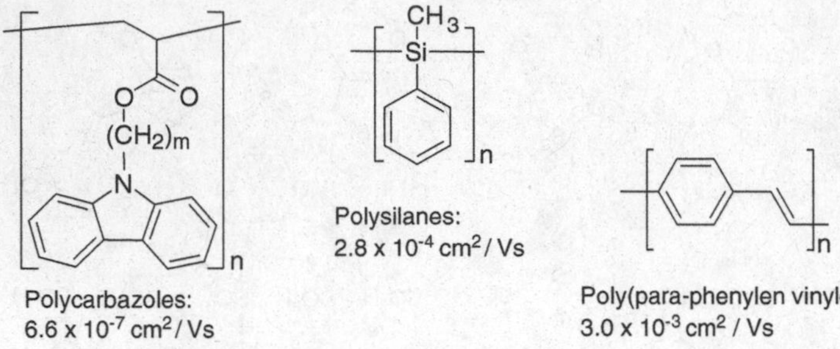

Fig. 6.8 Polymeric photoconductors, including the charge-carrier mobility achieved so far

layers showing very high degrees of ordering: perfect ordering is essential to get high charge-carrier mobility in the "on state". Accordingly, it is a major issue to design tailored polymers that self-organize in such a perfect fashion even when processed from solution. Otherwise, polymer-based transistors will not compete successfully with small-molecule systems. For sensor-applications, on the other hand, advantage can be taken of the high sensitivity of the thin-film transistor performance toward perturbations in the semi-conducting layer. In this context, it is an important field of research to increase the selectivity of those transistors with respect to specific analytes.

Photoconductors represent another class of functional polymers showing modulated electrical conductivity depending on external parameters: the material is electrically insulating in the absence of light since no free charge carriers are available (Fig. 6.8). Upon exposure to light, excitons are formed due to light-absorption and convert the material into its semi-conducting state. Advantage is taken of this effect in e.g. xerography and laser printing technologies.

In essentially all cases mentioned above the constitution of the polymers itself is responsible for the fact that the material is able to convert into an electrically semi-conducting (and sometimes conducting) state. Therefore, they are called *intrinsically conducting polymers*. This specification is applied to differentiate them from e.g. polymer-based *composite materials*, which develop electrical (and maybe thermal) conductivity just because of the presence of conducting fillers, which are dispersed appropriately in the otherwise non-conducting polymer matrix. One promising strategy in this respect is to add carbon-based (preferentially nano-scalic) fillers such as carbon black, carbon nano-tubes or graphenes into the insulating bulk polymer, and to process the composite subsequently in a fashion that allows the filler particles to develop a so-called *percolation network* through the whole material (Fig. 6.9). These percolation paths represent the highways for the charge carriers. If very high degrees of filling are realized and the processing is perfectly adapted, even heat transport can be observed through these composites. The latter option is expected to be of benefit for electro-mobility, heat management of batteries, bipolar plates of fuel cells, and cooler units of computer processors.

On the other hand, reliable prevention of nano-particle aggregation and agglomeration in the matrix of an insulating polymer as well is a key issue in the field of

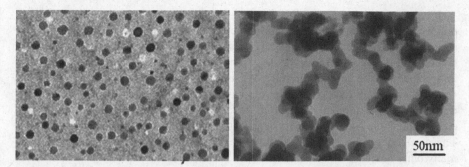

Fig. 6.9 Functional polymer nano-composites; *left*: well-dispersed TiO$_2$ nano-particles in a polystyrene matrix for increasing the composite's refractive index, *right*: percolating carbon-black nano-particles in a carbonate matrix to realize an electrically conducting material

functional polymers. For example, if conducting nano-particles are embedded in a polymer matrix without mutual contact, macroscopic conductivity is prevented, but the material may show very high dielectric constants, which again are desired for applications in e.g. miniaturized electronic devices. Based on closely related concepts, it is also possible to increase the refractive index of a transparent polymer: when mixed with nano-scale particles of high refractive indices such as titanium dioxide or zirconium dioxide, a composite material may result, which shows higher refractive indices than the pure polymer. Application of the materials as optical lenses is one option – if aggregation of the filler nano-particles can be prevented reliably (Fig. 6.9).

Finally, it should be mentioned that, in addition to the above examples, many further aspects may be encountered why and where functional polymers can be applied in combination with light and/or electrical power. For example, shape-anisotropy of polymers can cause the formation of lyotropic and/or thermotropic mesophases (liquid-crystalline polymers, LCPs). Alike, small-molecule mesogens, LCPs undergo specific interactions with light depending on the type and extent of mesoscalic order in the liquid-crystalline state. Since many semi-conducting polymers are rod like in shape (and thus anisotropic), not only modulated conductivity and emission can be observed depending on the respective structure, but even special effects such as emission of polarized light when applied in OLEDs.

Another topic related with π-electron conjugation in functional polymers is their nonlinear optical (NLO) behavior: such materials can be used to convert red light into light of shorter wavelength by frequency doubling or tripling. One advantage offered by shorter wavelengths is the higher capacity for processing and storage of information and optical computing. Moreover, these materials are of interest for photonic devices like optical switches, modulators and waveguides. High transparency and low light scattering are requested too. Polymeric materials showing excitingly high second-order and third-order nonlinear susceptibilities have been developed. One of the so far unsolved challenges, however, is ensuring sufficient long-term stabilities of material and effects.

Finally, it should be mentioned that conventional (electrically) isolating bulk polymers as well can develop highly attractive functional characteristics in combination with e.g. electrical voltage. This is if they are arranged in an appropriate layered or porous micro-morphology and charge carriers are introduced and immobilized at the internal interfaces or surfaces of these "piezoelectric" systems. Under these conditions, the materials expand or contract in an electrical field depending on the applied voltage. On the other hand, these materials can generate electrical voltage – and hence produce electrical power – if deformed mechanically. These effects can be used for microphones, loudspeakers, sensors, actuators, and many other devices.

Example 6.5 Synthesis of Poly{[2-(β-Ethylhexyloxy)-5-Methoxy]-Para-Phenylene Vinylene} ("MEH-PPV") Via Gilch Reaction

Safety precautions: Before this experiment is carried out, Sect. 2.2.5 must be read as well as the material safety data sheets (MSDS) for all chemicals and products used.

Preparation of 1-(β-Ethylhexoxy)-4-Methoxybenzene

Under an atmosphere of nitrogen, sodium methanolate (13.6 g, 251.0 mmol) is dissolved in methanol (50 ml). 4-Methoxyphenol (30.12 g, 251.0 mmol) dissolved in methanol (50 ml) is added dropwise over a period of 30 min. Stirring at room temperature is continued overnight to complete deprotonation. 2-Ethylhexylbromide (48.42 g, 251.0 mmol) is dissolved in methanol (100 ml) and added slowly over 3 h to the above solution. The reaction mixture is stirred at room temperature for 60 h and heated to reflux for an additional 30 h. The mixture is concentrated under reduced pressure. The residue is dissolved in a 1:1 (v/v) mixture of diethyl ether and water (400 ml). The organic phase is extracted with aqueous 0.2 M NaOH (3 × 100 ml) followed by pure water (3 × 100 ml), dried ($MgSO_4$) and concentrated in vacuo. The remaining solution is purified via flash chromatography using silica gel as the stationary phase and toluene as the eluent (Rf: 0.69). Yield: 42 g (72%).

Preparation of 2,5-*bis*(Chloromethyl)-1-(β-Ethylhexoxy)-4-Methoxybenzene

Under an atmosphere of nitrogen and vigorous stirring, aqueous HCl (37%, 87.9 g, 892.30 mmol) is added to a mixture of 1-(β-ethylhexoxy)-4-methoxybenzene (36.67 g, 155.43 mmol) and para-formaldehyde (12.80 g, 426.20 mmol). Acetic anhydride (162.31 g, 1589.88 mmol) is added dropwise. When addition is complete, the solution is heated to 70°C for 3.5 h. The reaction mixture is allowed to cool down to room temperature overnight. Concentrated aqueous sodium acetate (150 ml) is added within 2 h. Then, aqueous NaOH (110 ml, 25%) is added slowly within 4 h while the temperature is not allowed to rise above approx. 50°C. The white precipitate is molten at 60°C and precipitated under vigorous stirring and cooling using an ice bath. The precipitate is filtered off, washed with water (1,000 ml) and dissolved in *n*-hexane (500 ml). The organic phase is washed with water (3 × 200 ml) and dried (Na_2SO_4). The solvent is removed under reduced pressure. The residue is crystallized twice from *n*-hexane. Yield: 34 g (65%).

Polymerization Via Gilch Reaction

Under an atmosphere of nitrogen, the above monomer (0.5 g, 1.5 mmol) is dissolved in dry THF (100 ml). Potassium *tert.*-butanolate (1.0 M in THF, 6 ml, 6 mmol) is added quickly, and the mixture is stirred at room temperature for 20 h. Conc. acetic acid (2 ml) is added, the solution is concentrated down to 50 ml, and the polymer is precipitated by pouring the whole reaction mixture into methanol (200 ml). The formed solid is filtered off and dried in vacuo (40°C).

Example 6.6 Synthesis of Poly[9,9-bis(2-Ethylhexyl)Fluorene] Via Yamamoto Coupling

Safety precautions: Before this experiment is carried out, Sect. 2.2.5 must be read as well as the material safety data sheets (MSDS) for all chemicals and products used.

Preparation of 2,7-Dibromo-9H-Fluorene

In a 1 L round bottom flask wrapped in aluminum foil, a solution of fluorene (50.0 g, 300.8 mmol) in CHCl$_3$ (450 ml) is cooled to 0°C. Ferric chloride (0.716 g, 4.5 mmol) is added. At 0°C, bromine (32.6 ml, 632 mmol) is added slowly over 15 min. The ice bath is removed and the solution is allowed to warm slowly. The mixture is stirred overnight at room temperature, washed with aqueous Na$_2$S$_2$O$_3$ until the solution is clear and slightly yellowish only, and extracted with CHCl$_3$. The organic layer is washed with water, dried (MgSO$_4$), and the solvent is removed. The desired product is obtained in essentially quantitative yields (98 g) as colorless crystals.

Preparation of 2,7-Dibromo-9,9-bis(2-Ethylhexyl)Fluorene

2,7-Dibromofluorene 50 g (154 mmol) is dispersed in DMSO (200 ml). Benzyltri-methylammonium chloride (1.5 g, 8 mmol) is added as a phase-transfer catalyse. Nitrogen is passed through the suspension for 10 min, and subsequently aqueous NaOH (100 ml, 50%) is added at room temperature slowly. Stirring is continued at room temperature for a further 1 h. 2-Ethylhexyl bromide (65.4 g, 339 mmol) is added slowly, and the mixture is stirred overnight at room temperature. MTB ether (300 ml) and water (100 ml) is added, stirring is continued for further 15 min, the organic layer is separated, washed with saturated aqueous NaCl (3 × 100 ml) and water until the color of the solution changes from violet via green to reddish-orange. The organic layer is dried (Na$_2$SO$_4$), and the solvent is removed in vacuo. The oily residue is purified via column chromatography (eluent: pentane, $R_f = 0.85$). After removal of the solvent, slightly yellowish oil is obtained. In order to remove the monoalkylated product, the oil is dissolved in THF (250 ml) and a solution of KO^{t}Bu (20 g) in THF (200 ml) is added. The mixture is stirred for 30 min, filtered under an atmosphere of nitrogen over basic aluminum oxide pad, the solvent is removed, the residue dissolved in hexane and filtered. The hexane is removed in vacuo. This procedure is repeated until no discoloring is observed any more upon addition of KO^{t}Bu. The residue is dissolved in ethanol (200 ml, 60°C) and the solution stored in the fridge at − 21°C for 1 week. 2,7-Dibromo-9,9-bis-(2-ethylhexyl)-fluorene (63.5 g, 75%) is obtained as colorless waxy solid.

Polymerisation Via Yamamoto Coupling

Under an atmosphere of nitrogen, $Ni(COD)_2$ (1 g, 3,65 mmol, 2 äq.), 2,2'-bipyridine (569 mg, 2,2 mmol, 2,3 äq.) and cycloocta-1,5-diene (COD) (324 µl, 2.6 mmol, 1.45 äq.) are dissolved in a mixture of DMF (3 ml) and toluene (8 ml) and then heated to 80°C for 30 min. Subsequently, the above monomer (1.00 g, 1.82 mmol), dissolved in dry toluene (5 ml), is added under an atmosphere of nitrogen, and the reaction mixture is stirred at 80°C for 6 d. The reaction mixture is allowed to cool down to room temperature, aqueous HCl (50 ml, 2 M) is added, and the mixture is stirred at room temperature for 15 min. Methylene chloride (200 ml) is added, the organic layer is separated, washed with 2 M aqueous HCl (2×200 ml). To the resulting solution a saturated aqueous solution (400 ml) of Na_4EDTA is added vigorously and then washed with aqueous $NaHCO_3$ solution and dried (Na_2SO_4). The solvent is removed in vacuo; the residue is dissolved in a small volume of methylene chloride, filtered over a short pad of silica gel, concentrated down to 30 ml, and then poured into acetone (300 ml). The precipitate is extracted with acetone in a Soxhlet apparatus to remove lower oligomers, the remaining polymer is dissolved in methylene chloride and precipitated in methanol excess. The yield is approx. 70–80%.

Example 6.7 Synthesis of Poly(3-Dodecylthiophene) Via Grignard Coupling

Safety precautions: Before this experiment is carried out, Sect. 2.2.5 must be read as well as the material safety data sheets (MSDS) for all chemicals and products used.

Preparation of 2,5-Dibromo-3-Dodecylthiophene (Adv. Mater. 1999, 11 (3), 250)
3-Dodecylthiophene (19.41 g, 77.06 mmol) is dissolved in 100 ml of THF. N-Bromosuccinimide (27.43 g, 154 mmol) is added to the solution over a period of 5 min. The solution is stirred at room temperature for 2 h. The solvent is removed in vacuo and 250 ml of hexane is added. The mixture is filtered through a silica plug and the solvent is removed in vacuo. A Kugelrohr distillation (120°C, 0.02 Torr) affords the title compound (26.26 g, 83.3%) as the highest boiling fraction, a clear, colorless oil.

Preparation of Head-to-Tail Polydodecylthiophene (HT-PDDT)

2,5-Dibromo-3-dodecylthiophene (1.28 g, 3.12 mmol) is dissolved in 18 ml of dry THF. Methylmagnesium bromide (3.15 ml, 1.0 M solution in butyl ether) is added and the mixture is heated to reflux for 1 h. Ni(dppp)Cl$_2$ (16.9 mg) is added and the solution is stirred at reflux for 2 h. The mixture is poured into 150 ml of methanol and filtered into a soxhlet thimble. Soxhlet extractions are performed with methanol (to remove monomer and salts), hexanes (to remove catalyst and oligomers), and chloroform. The chloroform fraction is reduced and dried in vacuo to afford 0.510 g (65% yield) of the title polymer as a violet film.

Bibliography

Alvarez-Lorenzo C, Guney O, Oya T, Sakai Y, Kobayashi M, Enoki T, Takeoka Y, Ishibashi T, Kuroda K, Tanaka K, Wang GQ, Grosberg AY, Masamune S, Tanaka T (2000) Polymer gels that memorize elements of molecular conformation. Macromolecules 33:8693

Arias AC, MacKenzie JD, McCulloch I, Rivnay J, Salleo A (2010) Materials and applications for large area electronics: solution-based approaches. Chem Rev 110:3

Drobny JG (2012) Polymers for electricity and electronics- materials, properties, and applications. Wiley-VCH, Hoboken. ISBN ISBN-10: 0-470-45553-5

Kretschmann O, Schmitz S, Ritter H (2007) Microwave assisted synthesis of associative hydrogels. Macromol Rapid Commun 28:1265

Lendlein A, Jiang H, Jünger O, Langer R (2005) Light-induced shape-memory polymers. Nature 434:879

Liu C, Qin H, Mather PT (2007) Review of progress in shape-memory polymers. J Mater Chem 17:1543–1558

Marder SR, Lee KS (eds) (2008) Photoresponsive polymers I and II, vol 213/214, Advances in polymer science. Springer, Berlin

Matyjaszewski K, Gnanou Y, Leibler L (eds) (2007a) Macromolecular engineering: macromolecular engineering – precise synthesis, materials properties, applications, vol 1–4. Wiley-VCH, Weinheim

Matyjaszewski K, Gnanou Y, Leibler L (eds) (2007b) Functional polymers: macromolecular engineering – precise synthesis, materials properties, applications, vol 4. Wiley-VCH, Weinheim

Meller G, Grasser T (eds) (2010) Organic electronics, vol 223, Advances in polymer science. Springer, Heidelberg/New York

Nowak AP, Breedveld V, Pakstis L, Özbas B, Pine DJ, Pochan D, Deming TJ (2002) Rapidly recovering hydrogel scaffolds from self-assembling diblock copolymerpeptide amphiphiles. Nature 417:424

Osada Y, Rossi DE (eds) (2000) Polymer sensors and actuators. Springer, Berlin/New York

Schild HG (1992) LCST. Prog Polym Sci 17:163

Schmidt M (ed) (2004) Polyelectrolytes with defined molecular architecture I and II, vol 165/166, Advances in polymer science. Springer, New York

Wong W-Y, Tang BZ (2010) Essay, polymers for organic electronics. Macromol Chem Phys 211:2460

Yoshida M, Langer R, Lendlein A, Lahann J (2006) From advanced biomedical coatings to multi-functionalized biomaterials. Polym Rev (Phila) 46:347–375 (Smart and shape memory materials)

Zhou J, Schmidt AM, Ritter H (2010) Biocomponent transparent polyester networks with shape memory effect. Macromolecules 43:939

Index

D. Braun et al., *Polymer Synthesis: Theory and Practice*,
DOI 10.1007/978-3-642-28980-4, © Springer-Verlag Berlin Heidelberg 2013